Geometry From Africa

Mathematical and Educational Explorations

Paulus Gerdes

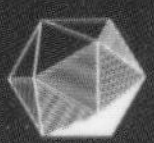

The Mathematical Association of America

Geometry from Africa
Mathematical and Educational Explorations

Cover photographs

Center photograph: Photograph of the basket weaver Luisa Ofice from Jangamo (Inhambane Province, southeastern Mozambique). Photograph by Gildo Bulafo, included as photograph no. 3 in the book: Paulus Gerdes and Gildo Bulafo, *Sipatsi: Technology, Art, and Geometry in Inhambane,* Ethnomathematics Research project, Maputo, 1994.

Outer photographs: Photograph of a collection of baskets from eastern Zimbabwe; photograph by Marcos Cherinda, Ethnomathematics Research Project, Maputo.

Library of Congress Catalog Card Number 99-62794
ISBN 0-88385-715-4

Printed in the United States of America

Current Printing (last digit)
10 9 8 7 6 5 4 3 2 1

Geometry from Africa

Mathematical and Educational Explorations

Paulus Gerdes

Universidade Pedagógica

Maputo, Mozambique

with a Foreword by

Arthur B. Powell

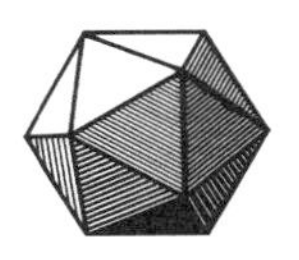

Published and Distributed by

The Mathematical Association of America

CLASSROOM RESOURCE MATERIALS

Classroom Resource Materials is intended to provide supplementary classroom material for students—laboratory exercises, projects, historical information, textbooks with unusual approaches for presenting mathematical ideas, career information, etc.

101 Careers in Mathematics, edited by Andrew Sterrett
Archimedes: What Did He Do Besides Cry Eureka?, Sherman Stein
Calculus Mysteries and Thrillers, R. Grant Woods
Combinatorics: A Problem Oriented Approach, Daniel A. Marcus
A Course in Mathematical Modeling, Douglas Mooney and Randall Swift
Elementary Mathematical Models, Dan Kalman
Interdisciplinary Lively Application Projects, edited by Chris Arney
Laboratory Experiences in Group Theory, Ellen Maycock Parker
Learn from the Masters, Frank Swetz, John Fauvel, Otto Bekken, Bengt Johansson, and Victor Katz
Mathematical Modeling in the Environment, Charles Hadlock
A Primer of Abstract Mathematics, Robert B. Ash
Proofs Without Words, Roger B. Nelsen
A Radical Approach to Real Analysis, David M. Bressoud
She Does Math!, edited by Marla Parker

MAA Service Center
P.O. Box 91112
Washington, DC 20090-1112
1-800-331-1622 fax: 1-301-206-9789

Foreword

I am honored to write the Foreword to this beautifully illustrated book, *Geometry from Africa: Mathematical and Educational Explorations*. In it, our Mozambican colleague, Paulus Gerdes elaborates and presents us a rare mathematical gift. Through him, we learn of the diversity, richness, and pleasure of mathematical ideas found in Sub-Saharan Africa. From a careful reading and working through this delightful book, one will find a fresh approach to mathematical inquiry as well as encounter a subtle challenge to Eurocentric discourses concerning the when, where, who, and why of mathematics.

Besides being a distinguished mathematician and mathematics educator, Paulus Gerdes is a productive researcher and prolific writer. He is a leading researcher in uncovering mathematical ideas embedded in African cultural practices and artifacts. He writes in English, French, German, and Portuguese. In just two decades, he has published well over one hundred journal articles and other books, several of which have won awards.

Gerdes has also given service to African educational and professional institutions. From 1986 to 1995, he served as a member of the Executive Committee of the African Mathematical Union and, from 1991 to 1995, as Secretary of the Southern African Mathematical Sciences Association. At present, he heads the Commission on the History of Mathematics in Africa of the African Mathematical Union and edits its newsletter (AMUCHMA). In an important position for furthering educational development in his country, for eight years (1988 to 1996), he was the Rector of the Universidade Pedagógica (formerly, Instituto Superior Pedagogico). Currently, continuing in the mathematics department, he devotes himself to teaching and leading a research team in ethnomathematics.

With this publication, Gerdes brings to our awareness geometrical ideas encoded in cultural products of Sub-Saharan Africa, ranging from woven and tiled designs to carved patterns, created by women and by men. For as he states in the first chapter, "women and men all over Africa south of the Sahara, in diverse historical and cultural contexts, traditionally have been geometrizing." He connects the encoded geometrical ideas of Africans to topics such as the Pythagorean Theorem, number theoretic notions, polyhedra, combinatorics, vector geometry, and trigonometry. Importantly, he goes even further to highlight geometrical ideas found in African artifacts. For instance, he draws fascinating and extraordinary connections between the geometry of hexagonal basket weaving and molecular chemistry as well as ideas of Smalley and Kroto for which they won the 1996 Nobel Prize in chemistry. Throughout the book, in the tradition of an engaging teacher, he stimulates our students and us with challenging reflections and questions.

This publication raises absorbing and challenging questions regarding the origin of particular geometrical ideas. It does so in ways reminiscent of two of his earlier books: *African Pythagoras* ([1992], 1994), for which one of the aims is to stimulate teachers on the African continent to africanize mathematics teaching; and *Sobre o despertar do pensamento geométrico* ([1985], 1992), where he discusses directly questions concerning the origin of geometrical thinking. Implicitly, this book raises the possibility of African origins of the theorem that asserts the equality between the sum of squares of the lengths of the two sides of any right triangle and the square of the length of its hypotenuse. Gerdes demonstrates how one can discover this and other important mathematical ideas by attending to quantitative, qualitative, and spatial features in a diversity of ancient and modern objects in the multicultural mix of African civilization. Is it not possible, then, to claim African originality of these mathematical ideas? Clearly, cultural groups in other geographical (perhaps even planetary) regions do and can make similar claims. It, therefore, is crucial to strive for a history of mathematics that is unfettered with nationalistic and ethnocentric bias, that acknowledges and valorizes multicultural manifestations of mathematical ideas, and that entertains no primacy claims.

Indeed, the theoretical framework for questioning the origins of mathematical ideas, calling for a new history of mathematics, and research projects from which this publication emerges are all part of a new disciplinary paradigm: ethnomathematics. Gerdes is a major theoretical and empirical contributor to this field (see, for example, Gerdes, 1995), whose theoretical origins are owed to the Brazilian mathematician, Ubiratan D'Ambrosio (see, for instance, D'Ambrosio, 1990). Theoretically, D'Ambrosio (1997) recognizes that

> [t]he complexity of every society, so different one from another, is responsible for the generation of codes, norms, rules and values in the direction of organizing, classifying, comparing, and ordering the action of its individuals. Instances of these codes, norms, rules and values are instruments of analyses, of explanations, and of actions, such as more or less, small and big, few or many, near and far, and in and out. These codes, norms, rules and values-for instance, cardinality and ordinality, counting and measuring, and sorting and comparing-take different forms according to the cultures in which they were generated, organized, and accepted. To recover these forms and behaviors in different cultural environments has been the main thrust of ethnomathematics... (p. xvi).

Ethnomathematics not only recognizes culturally shaped forms of cognition but also the notion that mathematical ideas develop from reflection on labor and other cultural activities. As human beings, we reflect on perceived quantitative, qualitative, and spatial forms in objects around us. Further, we attend to abstracted relations among objects and ideas that connect different relations. The Egyptian-born mathematician and educational psychologist, Caleb Gattegno, succinctly states that the elements of reality upon which mathematics is built are "objects, relations among objects, and dynamics linking different relations" (1987, p. 14). The objects, relations, and dynamics may be concrete and contextual or abstract and decontextual. Mathematical ideas may arise from context or apply to contextual situations not necessarily related to the origin of the ideas. Ethnomathematics notes that the power of mathematical ideas often becomes manifest once they transcend their physical, tangible origins.

Different cultures attend to different aspects of reality and express themselves differently. We interact with the world and attempt to contend with and give meaning to what we encounter and perceive. We try to comprehend, interpret, and explain the many aspects and challenges that reality presents. The means by which we comprehend, interpret, and explain, using number, logic, and spatial configurations are culturally shaped and are the ways we produce mathematical knowledge.

Gerdes examines culturally shaped products of mathematical knowledge expressed in African material culture as a search for meaning and beauty. In Chapter One, we learn that he is concerned with those spheres of African life in which "geometrical ideas, geometrical considerations, geometrical explorations, geometrical imagination are interwoven, interbraided, interplaited, intercut, intercoiled, intercised, interpainted...." He has found that frequently "geometrical exploration also is an expression of and develops hand in hand with artistical and esthetical exploration...." This widespread, seamless connection between beauty and geometrical explorations represents a cultural value common throughout Africa. Gerdes uncovers this idea through his methodological approach.

Consistent with the theoretical foundations of ethnomathematics, Gerdes has developed a research methodology. Gerdes (1986, [1988] 1997) explains an aspect of his research methodology in this way:

> We developed a complementary methodology that enables one to uncover in traditional, material culture some hidden moments of geometrical thinking. It can be characterized as follows. We looked to the geometrical forms and patterns of traditional objects like baskets, mats, pots, houses, fishtraps, and so forth and posed the question: why do these material products possess the form they have? In order to answer this question, we learned the usual production techniques and tried to vary the forms. It came out that the form of these objects is almost never arbitrary, but generally represents many practical advantages and is, quite a lot of times, the only possible or optimal solution of a production problem. The traditional form reflects accumulated experience and wisdom. It constitutes not only biological and physical knowledge about the materials that are used, but also mathematical knowledge, knowledge about the properties and relations of circles, angles, rectangles, squares, regular pentagons and hexagons, cones, pyramids, cylinders, and so forth (p. 227–228).

To uncover the "accumulated experience and wisdom" that indicate, among other things, the mathematical knowledge of a culture requires respectful and attentive focus on encoded ideas of the material culture. Gerdes's research perspective and outcomes inform his curriculum development and pedagogical work (Gerdes, 1998). This book represents the painstaking empirical—nearly archeological—spadework of a world-class ethnomathematician. Moreover, the empirical evidence within this book begets questions concerning whether creators of cultural artifacts and ornaments engage in mathematical thinking. Leaning on his methodology, Gerdes ([1988], 1997) answers these theoretical questions.

> Applying this method, we discovered quite a lot of 'hidden' or 'frozen' mathematics. The artisan, who imitates a known production technique, is, generally, doing some mathematics. But the artisans, who discovered the techniques, did and invented quite a lot of mathematics, were thinking mathematically (p. 228, original emphasis).

Importantly, the justification for his claim rests on both theoretical and empirical premises. Expressions of mathematical thinking vary, as do modes of communicating mathematical ideas. On this point, Marcia Ascher, an important ethnomathematician, states that

> [t]here has to be more understanding of their [people in particular fields] ideas, as they are culturally embedded, so that their mathematical aspects can be recognized. Take weaving for example. That surely involves geometric visualization. It not only requires the creation and conception of a pattern, but also requires knowing what moves to execute or colors to use to cause a pattern to emerge. In effect, the weaver is digitalizing the pattern. The weaver expresses the visualization through actions and materials (Ascher in Ascher and D'Ambrosio, 1994, p. 42).

The visualization and corresponding mental (and physical) actions of the weaver are akin to visual and mental functionings of the professional geometer. The weaver mathematizes her or his visual field and materials. These cognitive products and practices of a culture correspond to Gattegno's criteria and are mathematical. Furthermore, the mathematical attribution of the weaver's cognitive products and practices is independent of whether a professional mathematician expropriates them in their original form and transforms them into an academic, codified form. This, I believe, is an important point we reach from a reading of this book.

The mathematical and historical significance of this book proceeds from repercussions of Eurocentric projects predating modernity and ending hopefully with the start of the new millennia. African artisans, laborers, and professional mathematicians have much to learn from each other. Others can also learn from participating in this dialogue.

Arthur B. Powell
June 1998
Rutgers University, Newark, NJ
(1993–1994), Universidade Pedagogica, Mozambique

References

Ascher, M. and D'Ambrosio, U. (1994). *Ethnomathematics: A dialogue. For the Learning of Mathematics*, 14, 2: 36–43.

D'Ambrosio, U. (1997). Foreword. In A. B. Powell and M. Frankenstein (Eds.), Ethnomathematics: Challenging eurocentrism in mathematics education (pp. xv–xxi). New York: State University of New York.

——. (1990). *Etnomatematica: Arte ou tecnica de explicar e conhecer* [Ethnomathematics: Art or technique of explaining and knowing]. Sao Paulo: Atica.

Gattegno, C. (1987). *The science of education*: Part 1: Theoretical considerations. New York: Educational Solutions.

Gerdes, P. (1998). On culture and mathematics teacher education. *Journal of Mathematics Teacher Education*, 1: 33–53.

——. ([1988], 1997). On culture, geometrical thinking and mathematics education. In A. B. Powell and M. Frankenstein (Eds.), *Ethnomathematics: Challenging eurocentrism in mathematics education* (pp. 223–248). New York: State University of New York. [Original: in Educational Studies in Mathematics, Dordrecht, 19, 3: 137–162.]

——. (1995). *Ethnomathematics and education in Africa.* Stockholm: Institute for International Education.

——. ([1992], 1994). *African Pythagoras: A study in culture and mathematics education*. Maputo: Instituto Superior Pedagógico. [Original: Pitágoras africano: Um estudo em cultura e educacão matemática. Maputo: Instituto Superior Pedagógico.]

——. ([1985], 1992). *Sobre o despertar do pensamento geometrico* [The origins of geometric thinking]. Curitiba: Universidade Federal do Parana. [Original: Maputo: Instituto Superior Pedagógico. English translation: *Culture and the awakening of geometrical thinking: Anthropological, historical, and philosophical considerations: An ethnomathematical study.* Minneapolis: MEP, In press.]

——. (1986). On culture, mathematics and curriculum development in Mozambique. In S. Mellin-Olsen and M. Johnsen Hoines (Eds.), *Mathematics and culture, A seminar report* (pp. 15–42). Radal: Casper Forlag.

Contents

Preface

Geometrical and educational explorations inspired by African cultural activities

The peoples of Africa south of the Sahara desert constitute a vibrant cultural mosaic, extremely rich in its diversity. Among the peoples of the subsaharan region, interest in imagining, creating and exploring forms and shapes has blossomed in diverse cultural and social contexts with such an intensity that with reason, to paraphrase Claudia Zaslavsky's *Africa Counts*, it may be said that "Africa geometrizes".

In the first chapter of *Geometry from Africa*, I present, through some examples, an overview—necessarily incomplete—of geometrical ideas in African cultures, as manifested in the work of wood and ivory carvers, potters, painters, weavers, mat and basket makers, and of so many other laborious and creative African men and women alike.

In the second chapter of this book I show, using examples from Senegal in the west to Madagascar in the southeast, how diverse African ornaments and artifacts, varying from woven knots, to symmetrical designs, and to infinite decorative patterns, may be used as a starting point to create an attractive educational context in which students may be led to discover the Pythagorean Theorem and find proofs of it. I also explore connections to related geometrical ideas and propositions such as Pappus' Theorem, similar right triangles, Latin and magic squares, and arithmetic modulo n.

In the third chapter of this book, I analyze geometrical ideas inherient in various crafts and explore possibilities for their educational use. Chapters deal with topics, such as symmetrical wall decoration in Lesotho and South Africa; house building in Mozambique and Liberia; weaving pyramidal baskets in Congo/Zaire, Mozambique, and Tanzania; plaited strip patterns from Guinea, Mozambique, Senegal, and Uganda; finite geometrical designs from the Lower Congo region. Exploration of a hexagonal basket weaving technique from Cameroon to Kenya, Congo to Madagascar and Mozambique, will lead to attractive, and probably surprising connections between the underlying geometry and chemical models of recently discovered carbon molecules. Pentagrams will be discovered in knots. I develop alternative ways for rectangle constructions and for the determination of areas of circles and volumes of spatial figures, including a twisted decahedron. In various sections of this chapter, I discuss and/or propose diverse enumeration problems.

The theme of the fourth chapter of this book is the geometry of the southern-central African sand drawing tradition—the drawings are called *sona* in the Chokwe language (predominantly northeast Angola). As slavery and colonial domination disrupted and destroyed so many African traditions, the *sona* tradition with its strong geometrical component virtually disappeared. Consequently, I first present elements of a reconstruction of the sand drawing tradition. In particular, I discuss symmetry and monolinearity as cultural values, and analyze geometric algorithms and construction rules for monolinear drawings. Secondly, I present examples of an educational exploration of *sona*, varying from the primary to university levels, from arithmetical relationships and progressions, to enumeration problems, to the geometrical determination of the greatest common divisor of two natural numbers, to algorithms for systematically constructing monolinear designs. Finally, I end the book with an excursion, underscoring the mathematical potential of the *sona* tradition. Here I contribute to developing the geometry of Lunda-designs and Lunda-patterns, which I discovered in the context of analyzing a particular class of *sona*. In this sense, I hope that this book will contribute to a more widespread appreciation and valuation of the diversity, richness, and potential of African cultural, including mathematical, traditions.

The four chapters and the sections of which they are composed, may be read independently one of another. The bibliography is organized by chapter.

I thank Victor Katz (MAA and University of the District of Columbia) for suggesting that I write *Geometry from Africa* as a summary of ideas included in some of my earlier works. I tried to do so, but as always new ideas emerge, some of which have been incorporated. Furthermore, I would like to thank Andrew Sterrett, MAA Classroom Resource Materials Editor, and Elaine Pedreira, Editorial Manager, for the editorial guidance and production that they and their staff provided me.

I thank Gordan Patel, Dean of the Graduate School, Larry Hatfield, Chairman of the Mathematics Education Department and Jeremy Kilpatrick for the invitation to spend my sabbatical leave at the University of Georgia and for the working conditions that enabled me to write this book. I thank my colleagues at the University of Georgia for the intellectually stimulating environment, in particular, Thomas Cooney, Nicholas Oppong, James Wilson, and Patricia Wilson.

I am grateful to colleagues who invited me during my sabbatical leave to speak at their institutions for the interest in my work and for providing a forum for discussion: Gloria Emeagwali and Segun Odesina (Central Connecticut State University), Maurice Bazin (Exploratorium, San Francisco), Henri Polak, Bruce Vogeli, Joel Schneider, and Daniel Ness (Columbia University, New York), Allan Feldman, Steve Nathan and Mort Sternheim (University of Massachusetts, Amherst), Ismael Ramirez-Soto and Marilyn Frankenstein (University of Massachusetts, Boston), Robert Davis, Carolyn Maher and Gerold Golden (Rutgers University, New Brunswick), Colm Mulcahy and Malaika Moses (Spelman College, Atlanta), Mary Pratt-Cotter (Georgia College and State University, Milledgeville), Erwin Marquit (University of Minnesota), Israel Kleiner, Pinayur Rajagopal, and Lee Lorch (York University), Chandler Davis and Craig Fraser (University of Toronto), John Sims (Ringling School of Art and Design, Sarasota).

In the context of this MAA publication, I would like to thank those colleagues in the USA and Canada, who have been encouraging my research for several years, in particular, Sam Anderson, Marcia Ascher, Maurice Bazin, Donald Crowe, Chandler Davis, Marilyn Frankenstein, Gloria Gilmer, Beatrice Lumpkin, Lee Lorch, Arthur Powell, Dirk Struik, and Claudia Zaslavsky.

I thank Aarnout Brombacher (Cape Town, South Africa and 1996/1997 University of Georgia) and Donald Crowe (University of Wisconsin) for their comments on the manuscript, Donald Crowe for suggesting some photographs for inclusion in the book, and Arthur B. Powell (Rutgers University–Newark, and 1993/94 Visiting Professor at the Universidade Pedagógica in Mozambique) for his enthusiastic support of the idea of writing this book, his comments, and his preface.

The book is mostly based on research and experimentation I did in the context of Ethnomathematics Research Project at Mozambique's Universidade Pedagógica. I thank the SAREC Research Department of the Swedish International Development Agency for financially supporting this project. I thank my colleagues at the Universidade Pedagógica, in particular, Abdulcarimo Ismael, Abílio Mapapá, Daniel Soares, Marcos Cherinda, and Jan and Frouke Draisma, for stimulating and fruitful conversations. I owe a special and warm word of appreciation to my parents Otto and Marie-José, and to my wife Marcela Libombo, who always have supported me. I thank my mother-in-law Idalina for coming to the USA to help us care for our daughters Lesira and Likilisa. It is to them that I dedicate *Geometry from Africa.*

Athens GA, 11 November 1997
Paulus Gerdes

For
Lesira
and
Likilisa

1

On geometrical ideas in Africa south of the Sahara

Peoples of Africa south of the Sahara constitute a vibrant cultural mosaic, extremely rich in its diversity. Widespread interest in imagining, creating and exploring forms and shapes blossomed and continues to flourish in diverse cultural and social contexts. In this first chapter, I'll try to convey with examples spheres of African life where geometrical thinking has been active, often interwoven with artistic, artisanal, architectonic, ceremonial, educational, aesthetic, musical, religious, social and symbolic considerations and practices.

Historians generally assume that Africa, in particular Eastern or Southern Africa, constitutes the cradle of mankind. In a lecture at the University of the United Nations, the first president of the African Mathematical Union (1975–1986), Henri Hogbe-Nlend of Cameroon, author of several books on bornologies and functional analysis, asked whether Africa can also be considered the cradle of world mathematics. His response was an unequivocal yes. He gave particular attention to ancient Egyptian mathematics (Hogbe-Nlend, 1985). In a recent study, the Congolese egyptologist and historian of science, Théophile Obenga, stressed the philosophical and demonstrative character of ancient Egyptian geometry and suggested its relationship with geometrical ideas in Africa south of the Sahara (Obenga, 1995). In the journal *Historia Mathematica* I (1994) presented an overview of research findings and of sources on or related to the history of mathematics in Africa-south-of-the-Sahara, particularly studies that have appeared since the publication of Claudia Zaslavsky's classic study "*Africa Counts: Number and Pattern in African Culture*" (1973). She and Donald Crowe analyze geometric shape in architecture, and geometric form, pattern and symmetry in African art (pp. 153–196).

The development of geometrical thinking starts early in African history. In general, early humans learned to geometrize in the context of their labor activities (cf. Gerdes, 1990, 1998). The hunter-gatherers of the Kalahari desert in southern Africa learned to track animals, learned to recognize and interpret spoors. The shape of a spoor makes it possible to evaluate what animal passed by, how long ago, if it was hungry or not, etc. Liebenberg's study (1990) led him to conclude that "the critical attitude of contemporary Kalahari Desert trackers, and the role of critical discussion in

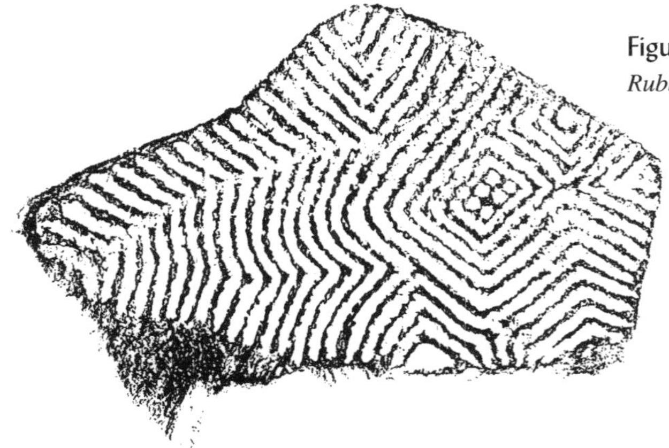

Figure 1.1a
Rubbing of a rock engraving (South Africa)

Figure 1.1b
Rock paintings at Chicolone (northwestern Mozambique)

tracking suggest ... that the rationalist tradition of science may well have been practised by hunter-gatherers long before the Greek philosophic schools were founded" (Liebenberg, p.45).

From all over Africa rock paintings and rock engravings have been reported (cf. map in Willcox, p.19). Some date from several hundreds years ago, others maybe several thousands. Often they have a geometric structure, as the examples presented in Figure 1.1.

Other archaeological finds that may give an indication of geometrical exploration by African hunters, farmers and artisans are stone and metal tools and ceramics. Figure 1.2 displays some examples of pot decorations.

Figure 1.3 presents some of the most common pot shapes found in Africa south of the Sahara.

Particularly exceptional are archaeological finds of perishable materials, like of baskets, textiles, and wooden objects. The finds from the Tellem are extremely important and provide us with some idea of earlier geometrical explorations.

Figure 1.1c
Petroglyphs from Calunda (extreme east Angola)

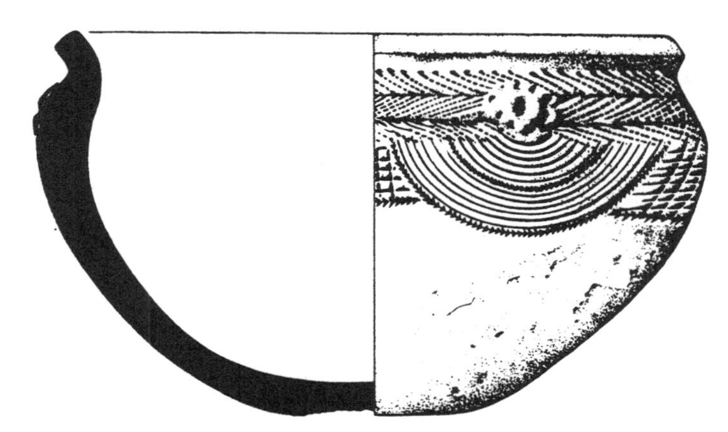

Figure 1.2a
Pot found at Batalino-Maluba, c. 400–200 BC (northeast Congo/Zaire)

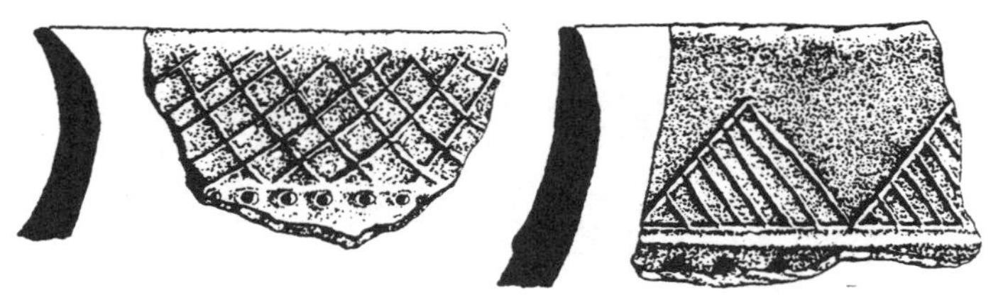

Figure 1.2b
Pieces of pots found near Monapo, c. AD 600–900 (northeast Mozambique)

Figure 1.2c
Pot from Igbo-Ukwu, c. AD 800–100 (southeast Nigeria)

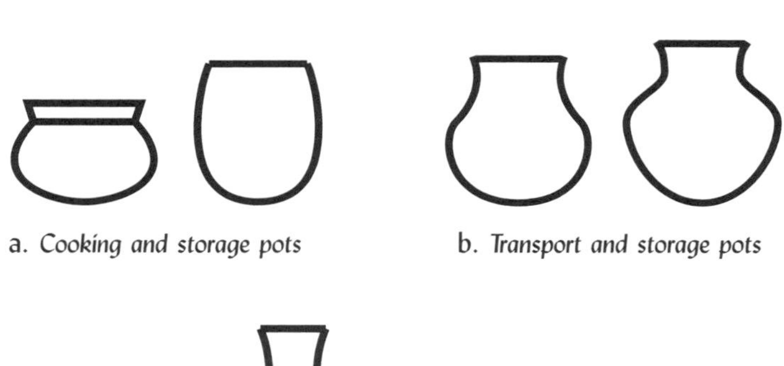

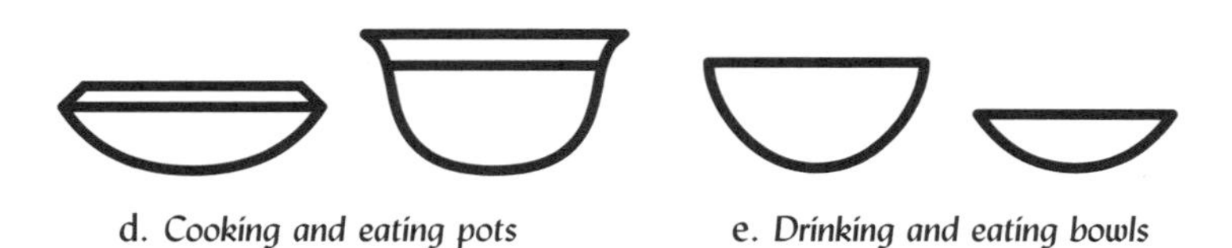

Figure 1.3. *Common pot shapes*

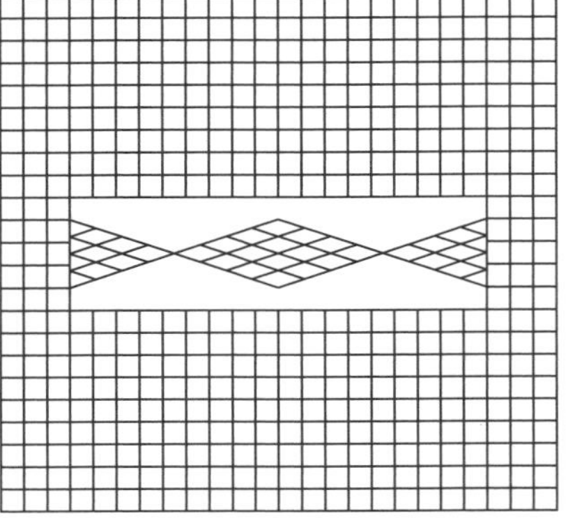

Figure 1.4a
Wooden headrest seen from above, 11th–12th century (left)
Structure of that design (above)

Geometrical patterns from the Tellem

The archaeological finds from caves in the Cliff of Bandiagara in the center of the Republic of Mali present clear evidence of exploration of forms, shapes and symmetries. The earliest buildings in the caves are cylindrical granaries made of mud coils. They date from the 3rd to 2nd century BC. In the 11th century, the now vanished people of the Tellem—as they are called by the Dogon who inhabit the region from the 16th century onwards—entered the area coming from the south, perhaps from the tropical rain forest. From the 11th up to the 15th century, the Tellem people deposited their dead in the remaining old granaries and in new buildings they made in the caves. The dead were buried with agricultural implements, hunting weapons, and household goods, like wooden headrests, bows, quivers, hoes and musical instruments, baskets, gourds, leather sandals, boots, bags and armlets, woolen and cotton blankets, coifs, caps, tunics, and fiber aprons (see the examples in Figure 1.4).

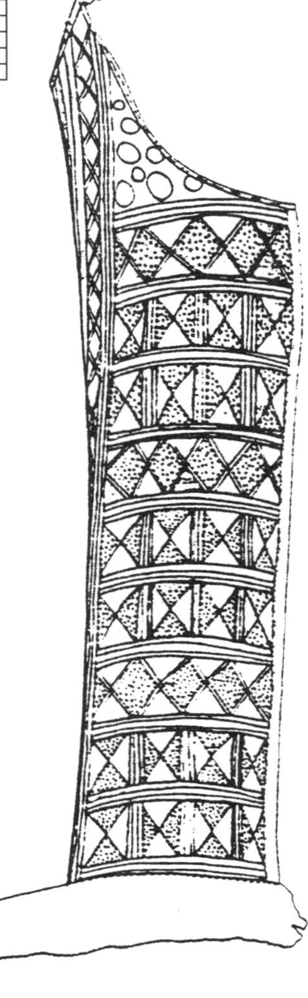

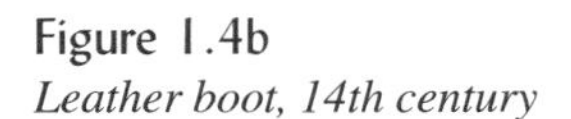

Figure 1.4b
Leather boot, 14th century

These objects of perishable materials found in a reasonably good state of preservation in the caves belong to the oldest objects that have been preserved from Africa south of the Sahara. Archaeologists and textile specialists who analyzed the Tellem textiles underscore that "... there is no other region in the world where such a great variety of linear and geometrical patterns has been obtained in cotton fabrics by means of a single color, the only one available locally, i.e., indigo.... The designs have indeed been the object of search for infinite combinations which have persisted until our own day" (Bolland, p. 50).

As an illustration of this search by Tellem weavers, let us examine some patterns found on preserved fragments of tunics, sleeves, coifs and caps, woven in plain weave, that is, the weave in which the horizontal and vertical threads cross each other one over, one under. The average width of the threads is 1 mm. The weavers alternated groups of natural white cotton threads with groups of blue, indigo-dyed, threads. Figure 1.5 gives an example: (from left to right) six vertical white threads are followed by four blue threads; (from top to bottom) three horizontal white threads are followed by three blue threads (see the basic woven rectangle in Figure 1.5a), leading to the plane pattern, a part of which is shown in Figure 1.5b). In this example, the basic rectangle has dimensions ten ($=6+4$) by six ($=3+3$), or $(6+4)\times(3+3)$.

In general, the dimensions are $(m+n)\times(p+q)$, where m, n, p, and q are natural numbers. The Tellem weavers experimented with dimensions and discovered relationships between the dimensions and the (symmetry) properties of the resulting patterns. In particular, the variation among the found plain weave fragments gives the impression that they knew the effect on the patterns of the selection of even and odd dimensions, as well as how these dimensions $(m+n)$ and $(p+q)$ are built up. The reader is invited to discover these relationships. The Tellem patterns from the 11th and 12th centuries in Figures 1.6 to 1.16 (woven rectangles followed by fragment of respective plane pattern) may facilitate this analysis. The plane patterns in Figures 1.12 to 1.16 are two-color patterns, in the sense that for each there exist a rigid motion of the plane (translation, rotation, reflection) that reverses the colors (white and blue).

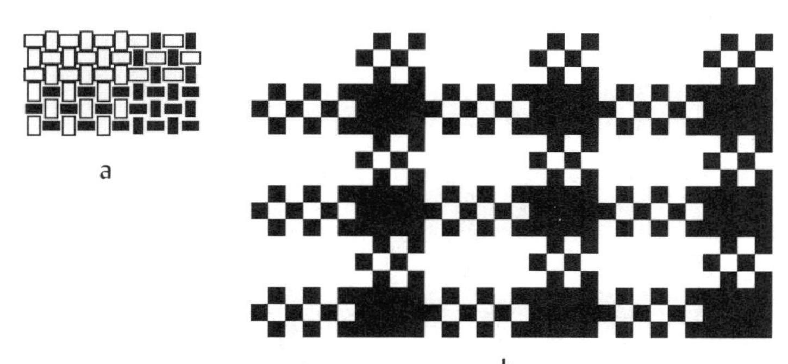

Figure 1.5
Tellem textile.
Dimensions $(6+4)\times(3+3)$
[14th century]

Figure 1.6
Tellem textile.
Dimensions $(2+1)\times(2+1)$
[11th/12th century]

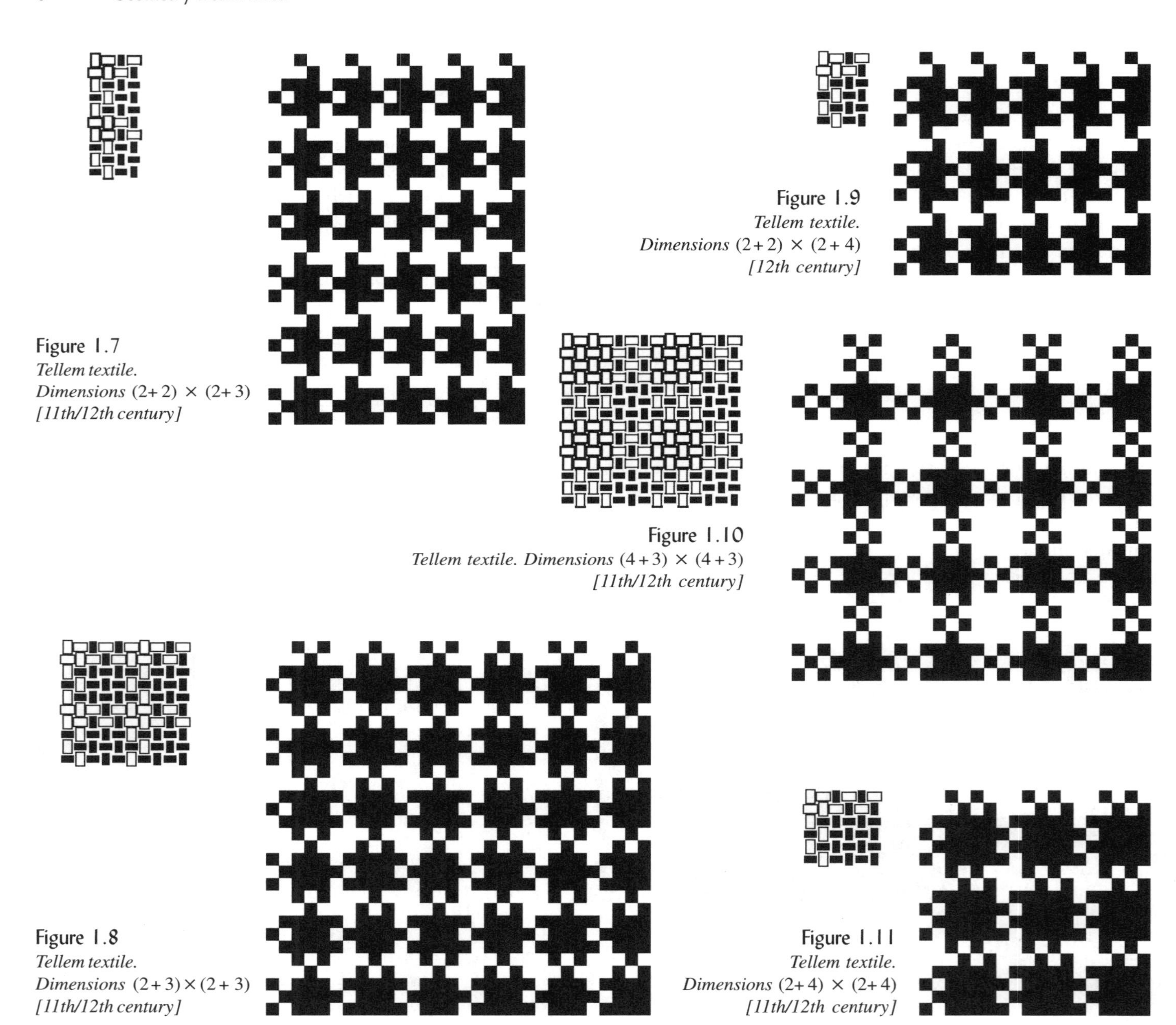

Figure 1.7
Tellem textile.
Dimensions (2+2) × (2+3)
[11th/12th century]

Figure 1.9
Tellem textile.
Dimensions (2+2) × (2+4)
[12th century]

Figure 1.10
Tellem textile. Dimensions (4+3) × (4+3)
[11th/12th century]

Figure 1.8
Tellem textile.
Dimensions (2+3) × (2+3)
[11th/12th century]

Figure 1.11
Tellem textile.
Dimensions (2+4) × (2+4)
[11th/12th century]

Figure 1.12
Tellem textile.
Dimensions (2+2) × (1+1)
[11th/12th century]

Figure 1.13
Tellem textile.
Dimensions (2+2) × (2+2)
[11th/12th century]

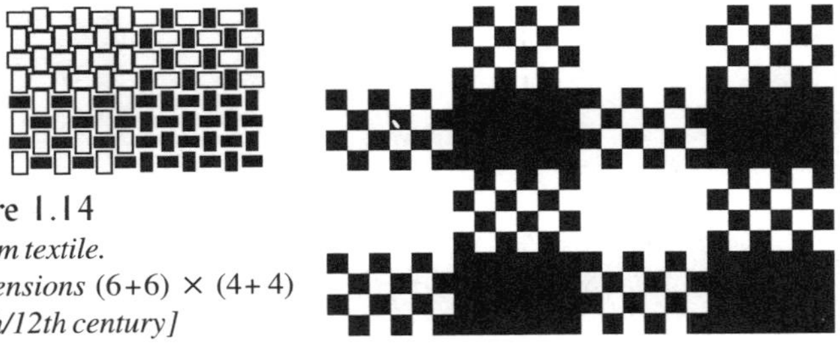

Figure 1.14
Tellem textile.
Dimensions (6+6) × (4+4)
[11th/12th century]

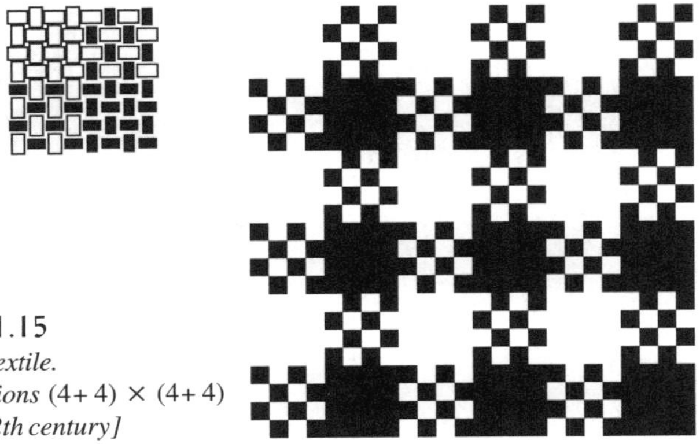

Figure 1.15
Tellem textile.
Dimensions (4+4) × (4+4)
[11th/12th century]

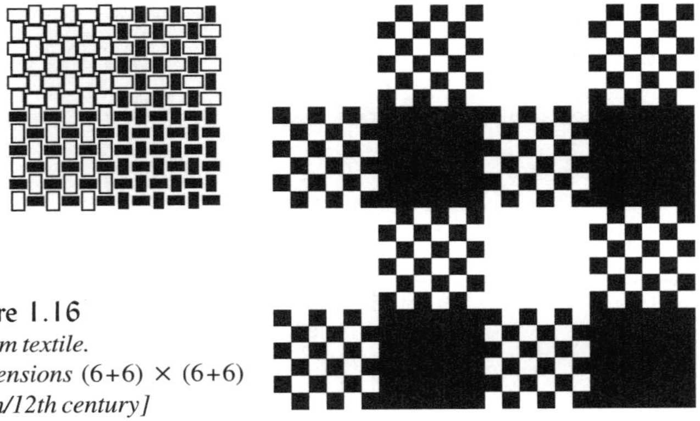

Figure 1.16
Tellem textile.
Dimensions (6+6) × (6+6)
[11th/12th century]

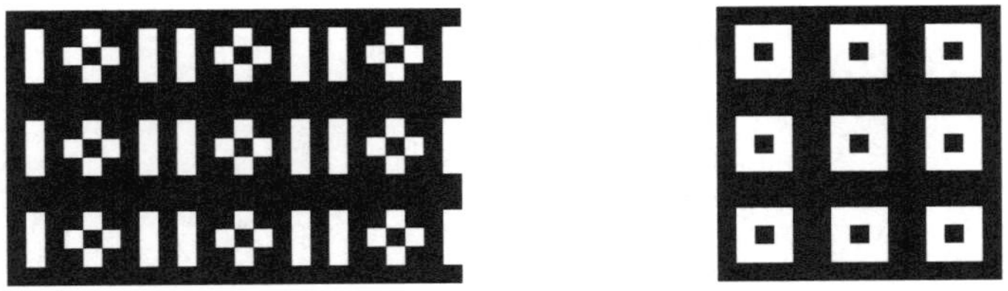

Figure 1.17
Plane Patterns.
[11th/12th cenury]

The Tellem weavers also used a variant of the plain weave, whereby in one direction double threads are employed instead of single threads. In this way they were able to weave cloths with decorative plane patterns like the ones presented in Figure 1.17, and strip patterns as the example in Figure 1.18 illustrates.

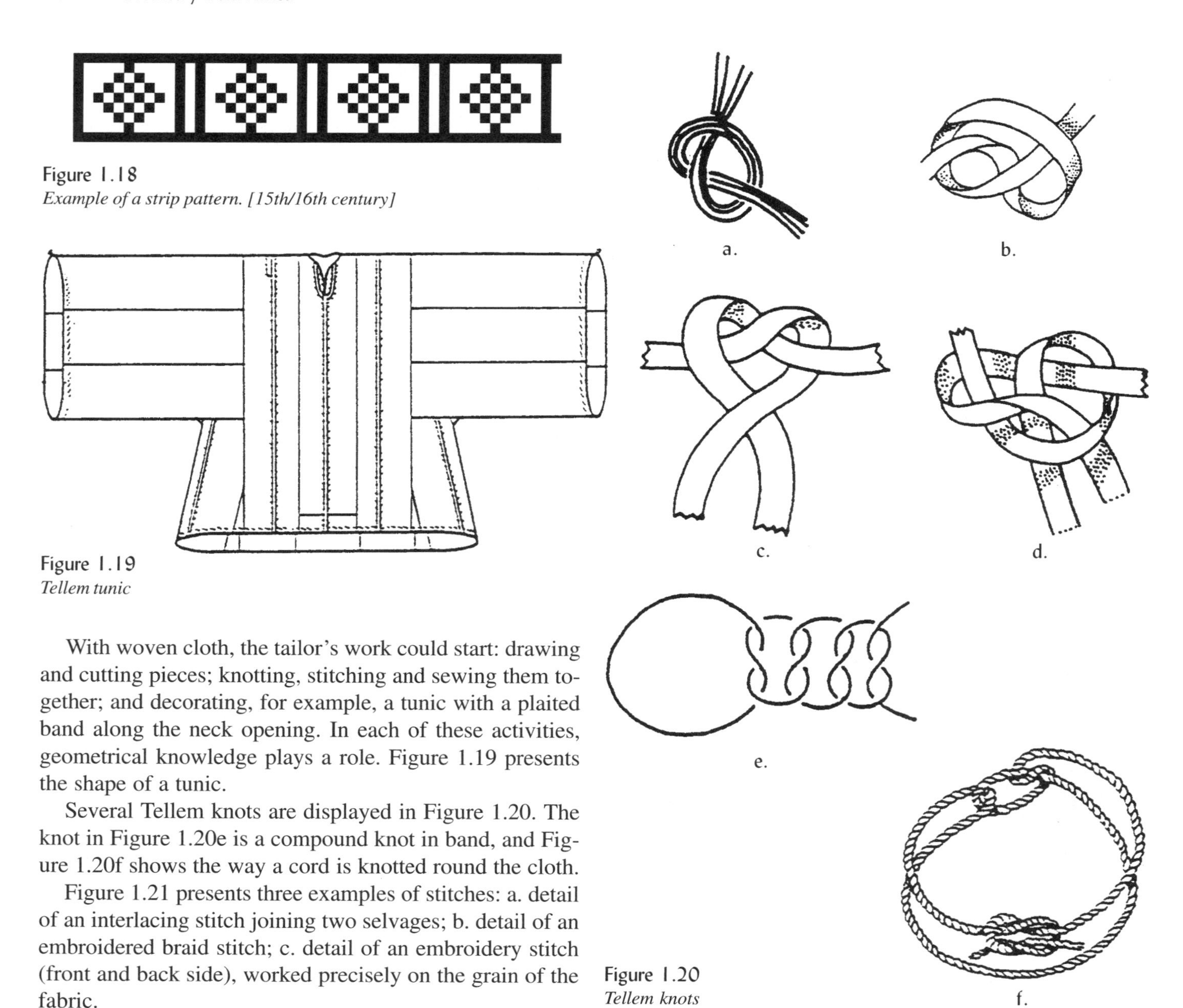

Figure 1.18
Example of a strip pattern. [15th/16th century]

Figure 1.19
Tellem tunic

Figure 1.20
Tellem knots

With woven cloth, the tailor's work could start: drawing and cutting pieces; knotting, stitching and sewing them together; and decorating, for example, a tunic with a plaited band along the neck opening. In each of these activities, geometrical knowledge plays a role. Figure 1.19 presents the shape of a tunic.

Several Tellem knots are displayed in Figure 1.20. The knot in Figure 1.20e is a compound knot in band, and Figure 1.20f shows the way a cord is knotted round the cloth.

Figure 1.21 presents three examples of stitches: a. detail of an interlacing stitch joining two selvages; b. detail of an embroidered braid stitch; c. detail of an embroidery stitch (front and back side), worked precisely on the grain of the fabric.

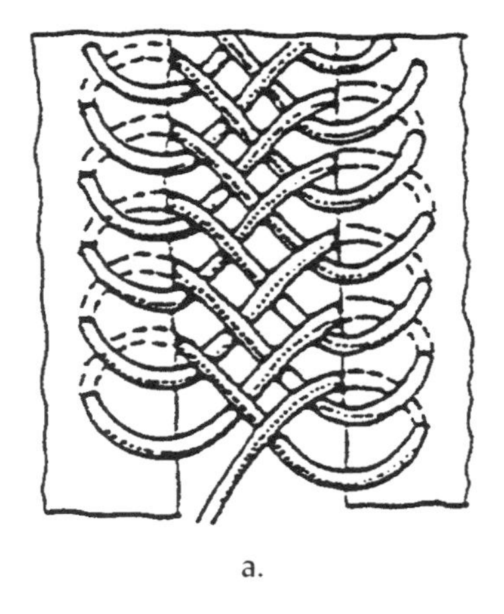

a.

b.

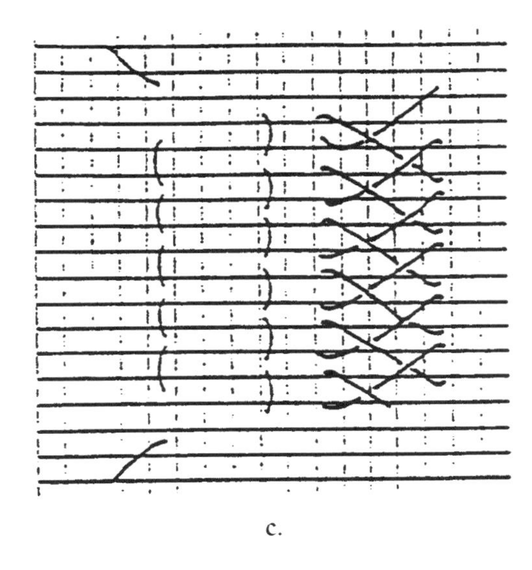

c.

Figure 1.21
Examples of Tellem stitches

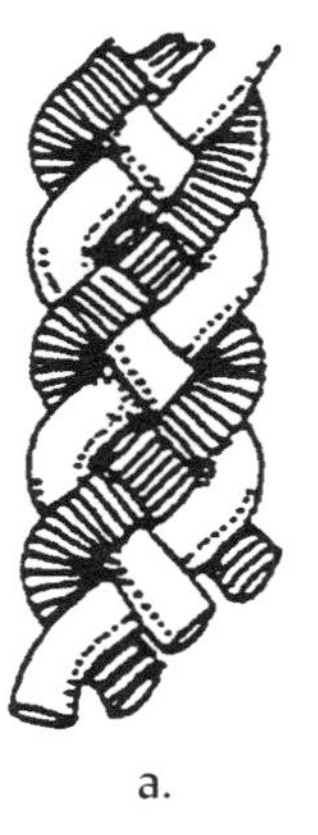

a.

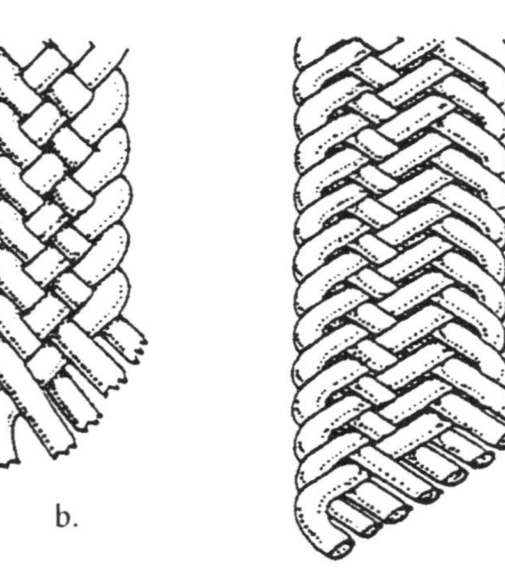

b.

c.

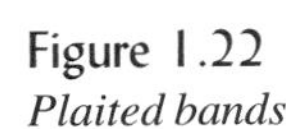

Figure 1.22
Plaited bands

Decorative bands were plaited both with even and odd numbers of strings. Among the plaited bands found in the caves, there are on the one hand bands made out of 4 (see Figure 1.22a), 6, 8 (Figure 1.22c), 14, and, on the other hand, out of 5 (Figure 1.22b), 7, and 9 strings. Both the selection of an even or odd number of strings and the weave (plain weave or not) has implications for the visible decorative patterns.

In addition to cotton textiles, the Tellem weavers also produced woolen goods such as blankets. Figure 1.23a shows a plane pattern on a tapestry-woven woolen textile, and Figure 1.23b gives an idea of the structure of the weaving process involved.

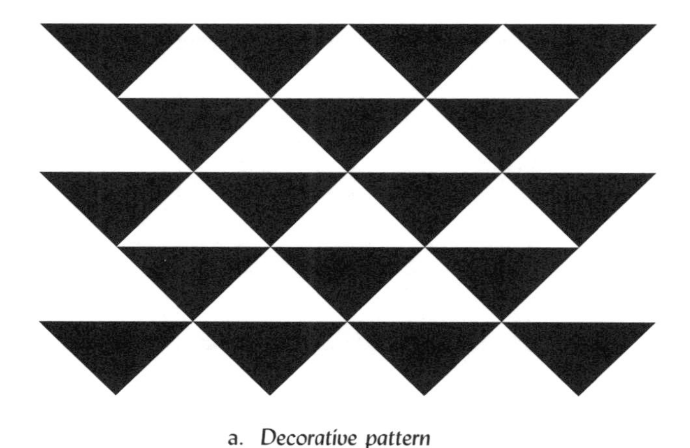

a. *Decorative pattern*

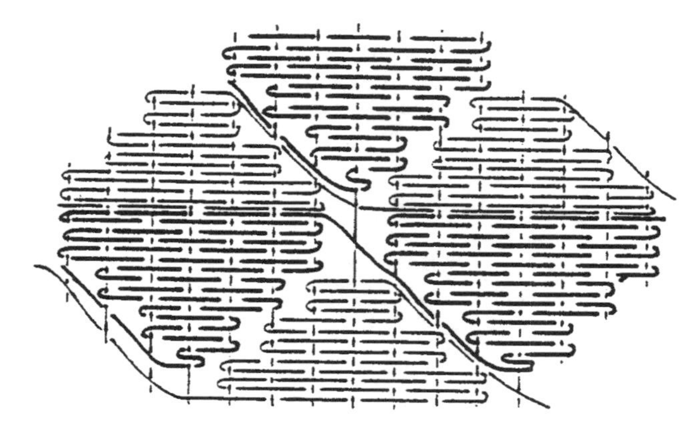

b. *Weaving structure*

Figure 1.23
Tellem woolen textile

Geometric design in the Kuba Kingdom

The Kuba Kingdom in the eastern Kasai region in central Congo/Zaire is one of the African cultural contexts that particularly has stimulated design with a strong geometric accent. The Kuba people are an association of various culturally related ethnic groups, among them the Bushoong, the Ngongo, the Shoowa, and the Ngeende. The reign of Shyaam a-Mbul (c. 1600–1620) contributed to the intensification of industrious and artistic activities: copper and iron work, basket and mat weaving, construction of palaces, wood carving, embroidery, pottery, beadwork. All manufactured objects are decorated with geometric design (cf. Meurant, p. 135).

Figure 1.24 presents examples of Bushoong plane patterns on raffia cloth. First the men weave the plain light tan raffia cloth on a single heddle loom. This cloth is decorated with geometric designs by women who embroider plush motifs on the plain cloth (cf. Washburn, p. 23).

Figure 1.24a
Bushoong Raffia cloth design

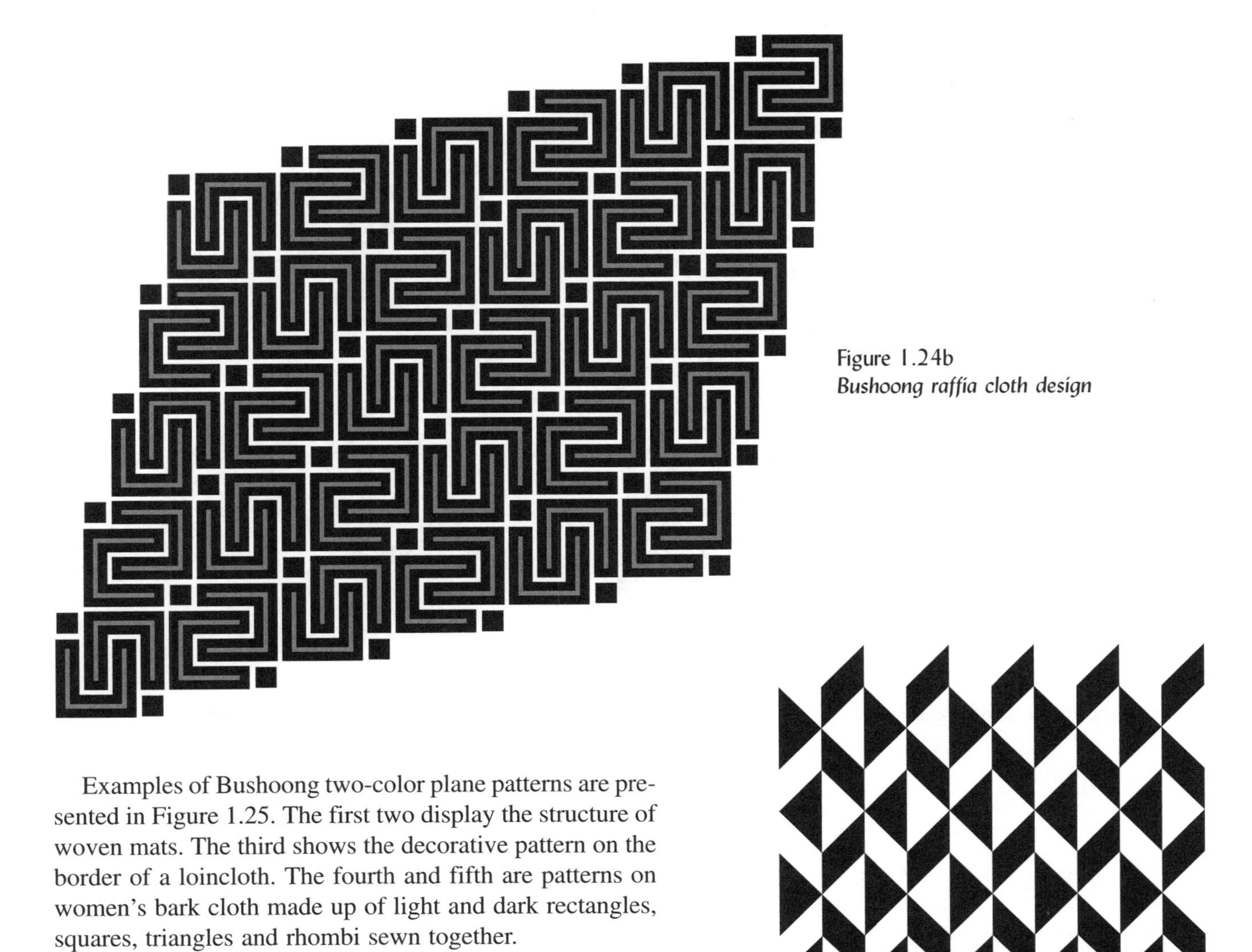

Figure 1.24b
Bushoong raffia cloth design

Figure 1.25a
Example of a Bushoong two-color pattern *(continued on next page)*

Examples of Bushoong two-color plane patterns are presented in Figure 1.25. The first two display the structure of woven mats. The third shows the decorative pattern on the border of a loincloth. The fourth and fifth are patterns on women's bark cloth made up of light and dark rectangles, squares, triangles and rhombi sewn together.

Among the Ngongo the decoration of the walls of houses and palaces with matwork is widespread. A collection of such architectural mats observed by the Hungarian ethnographer Torday was published in 1910. Figure 1.26 presents examples. The plane patterns have various symmetries.

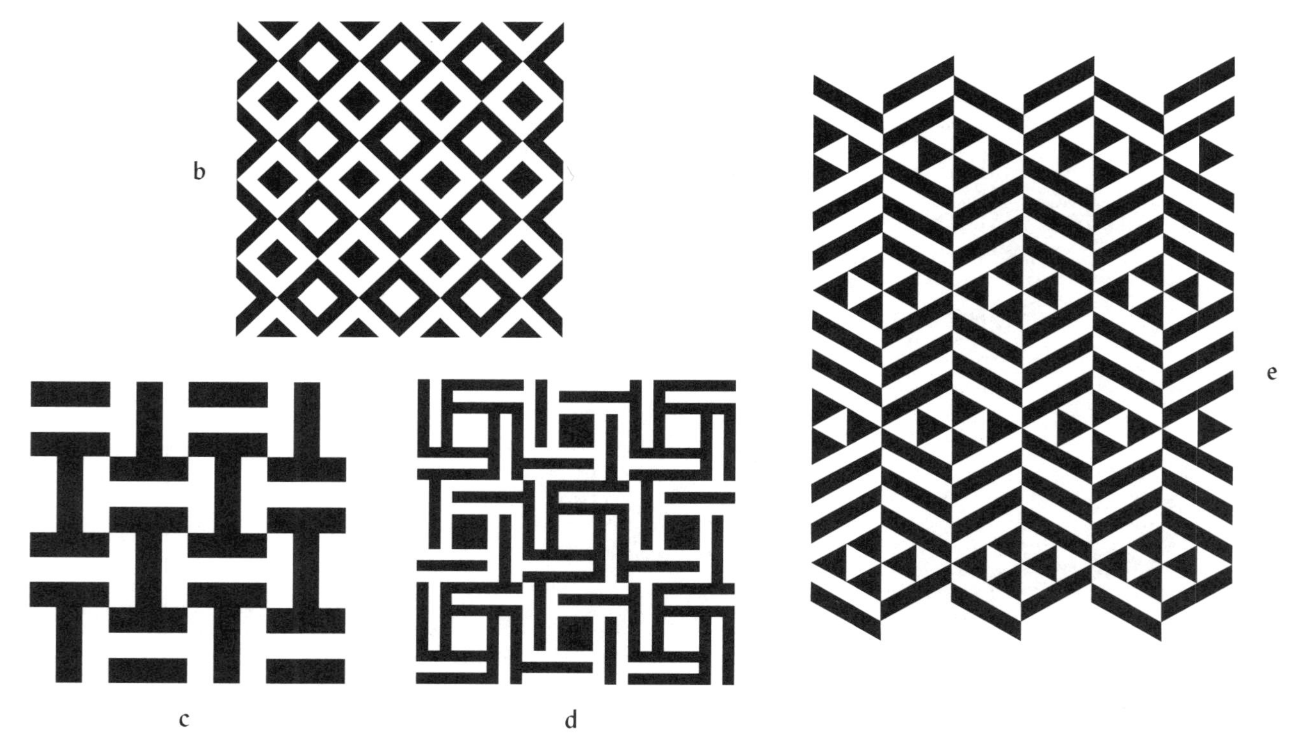

Figure 1.25 (b–e)
Examples of Bushoong Two-color patterns

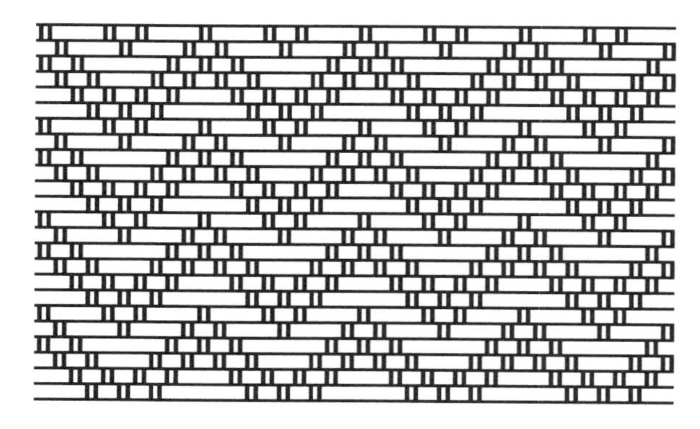

Figure 1.26a
Example of a Bushoong architectural mat

Horizontally one sees the sticks which are woven together by the vertical lianas.

Figures 1.27a, b, c, and d present wooden cups decorated withsymmetrical patterns. Examples of Kuba wood carving motifs with rotational symmetry are presented in Figures 1.27e, f, and g.

Figure 1.28 presents an example of a double axial symmetric decoration of copper on an iron knife.

Plaited band motifs occur frequently as decoration on wooden and metal objects, carved on gourds, as well as a

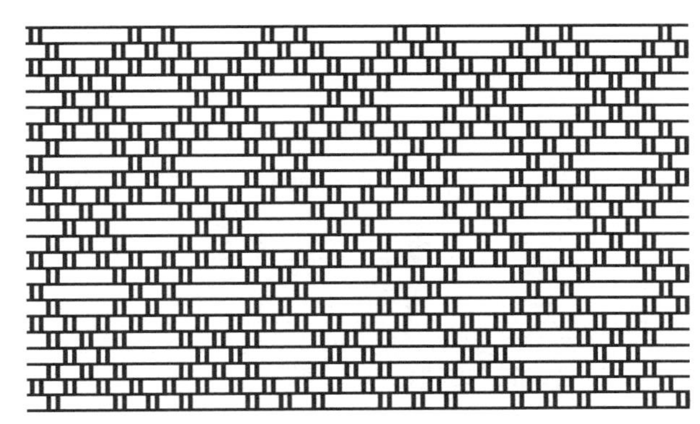

Figure 1.26b

Example of a Bushoong architectural mat

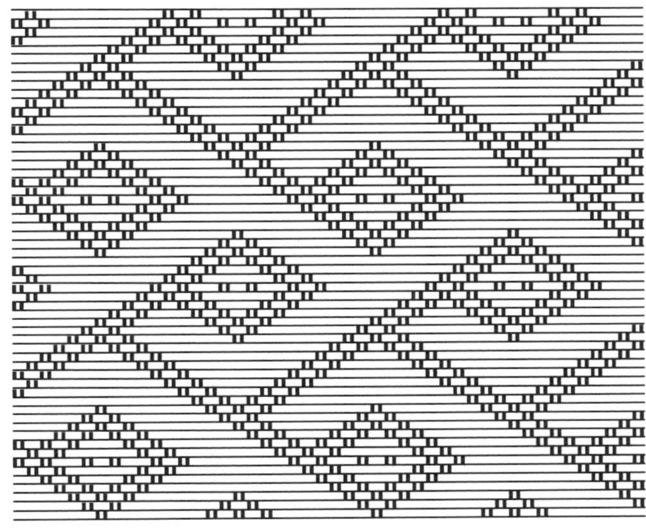

Figure 1.26c

Example of a Bushoong architectural mat

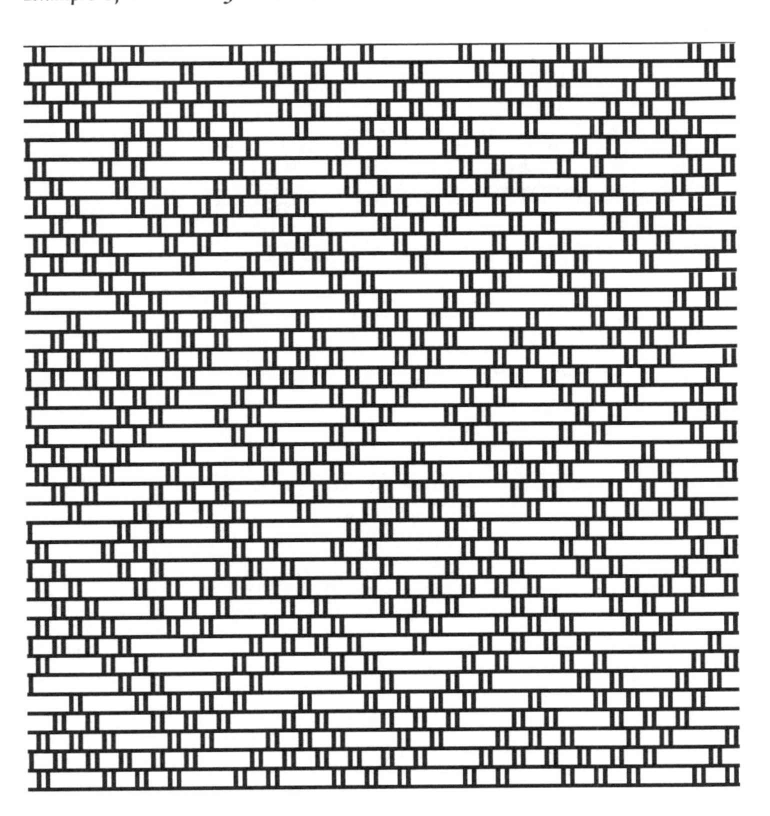

Figure 1.26d

Example of a Bushoong architectural mat

Figure 1.26e
Example of a Bushoong architectural mat

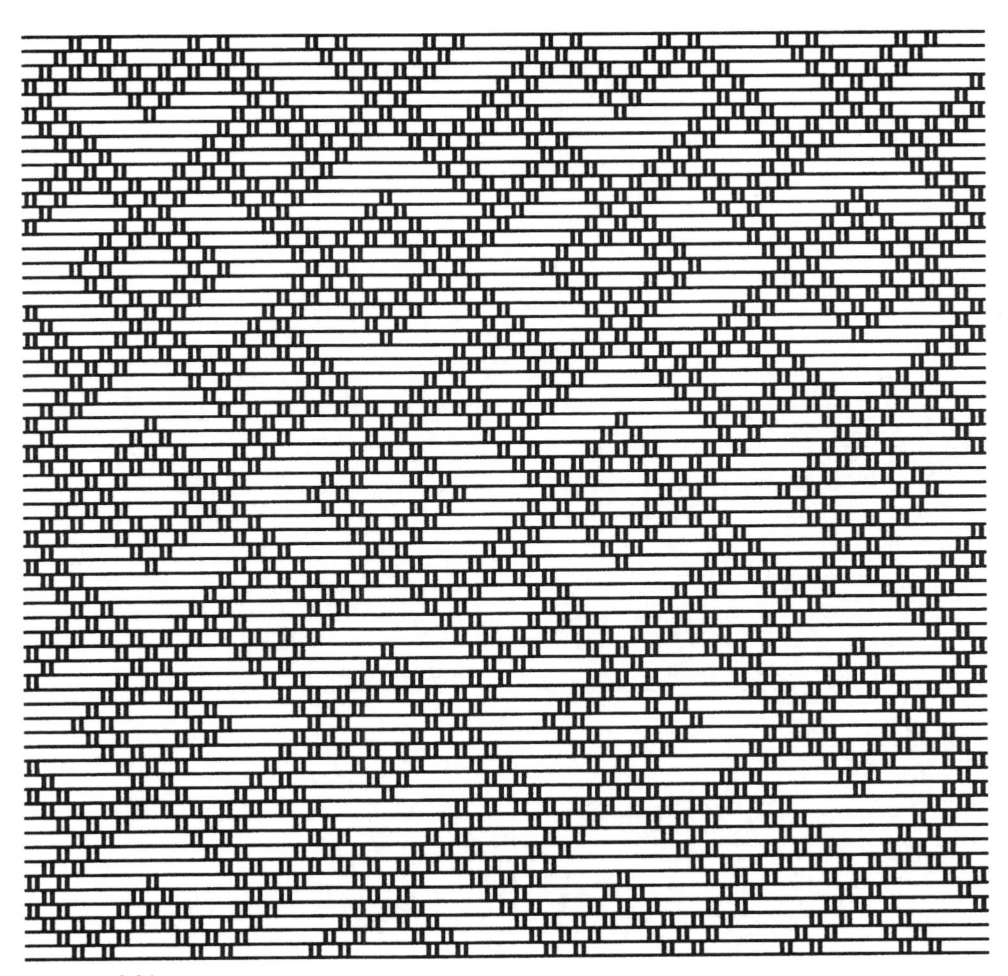

Figure 1.26f

Example of a Bushoong architectural mat

Figure 1.26g
Example of a Bushoong architectural mat

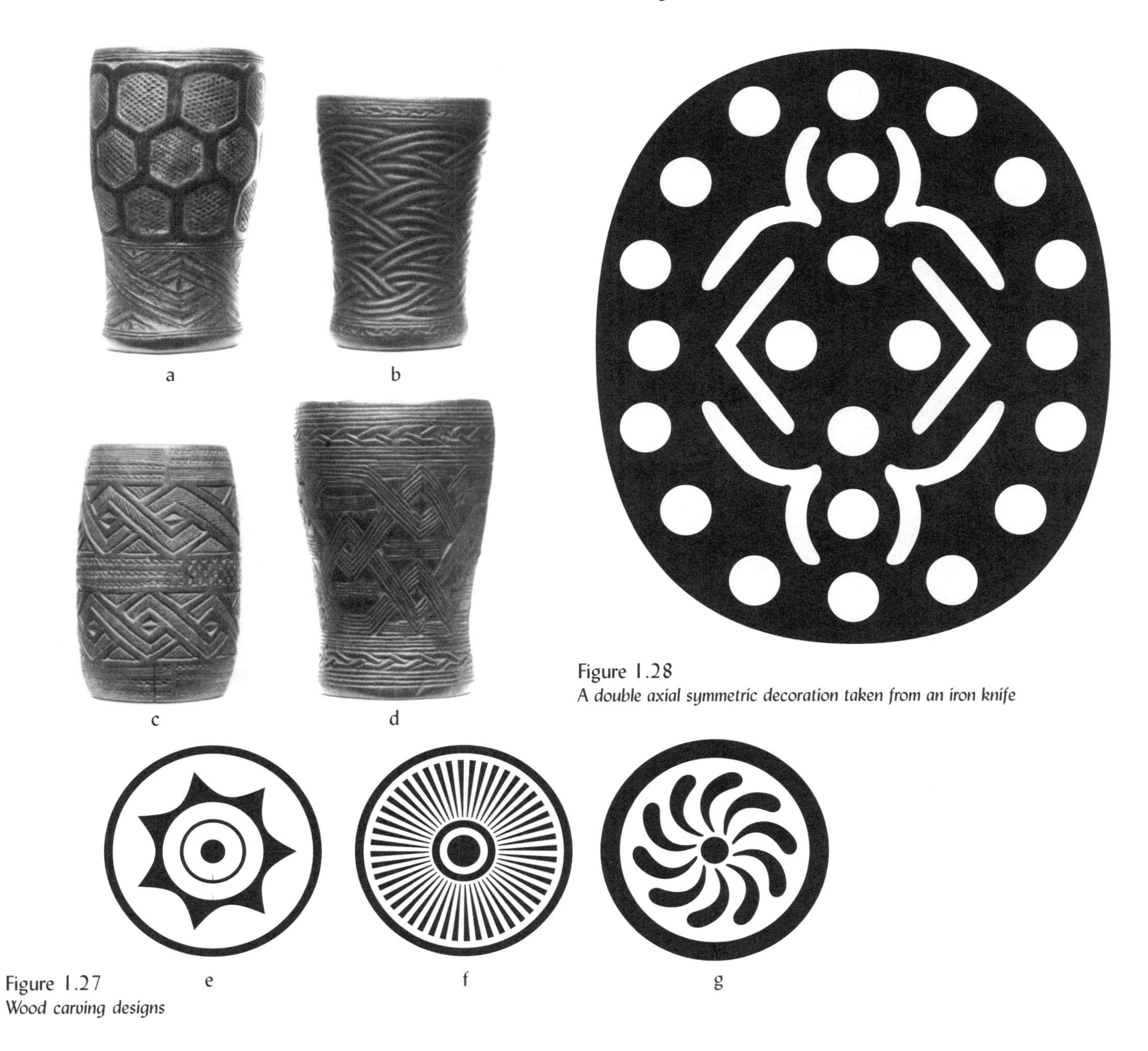

Figure 1.28
A double axial symmetric decoration taken from an iron knife

Figure 1.27
Wood carving designs

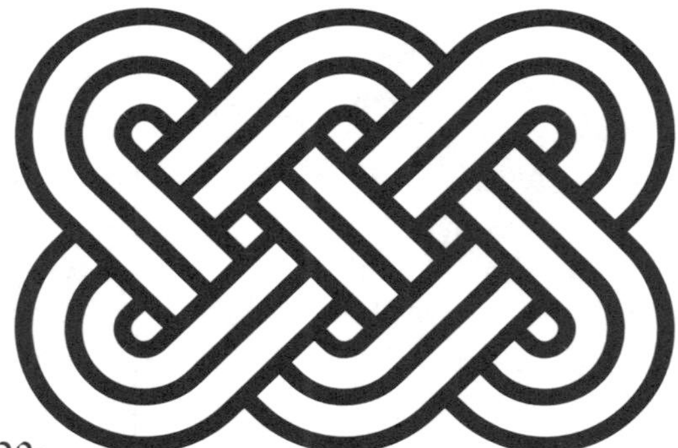

Figure 1.29a
Bushoong tattoo design

female Bushoong tattoo, like the example presented in Figure 1.29. Plaited band motifs made out of only one continuous line are drawn by boys as an amusement in the sand (see the example in Figure 1.30; cf. Section 4 on the Chokwe sand drawings). Examples of the use in wood carving of other geometrical line algorithms, leading to strip patterns, are presented in Figure 1.31 (cf. Crowe, 1971, 1973).

Torday and Joyce (1910, p. 90, 96) recorded a 'topological' puzzle Torday saw in the Kuba capital. It consists of two pieces of calabash and a string arranged as shown in Figure 1.32. The player has to separate one of the calabash pieces from the string without cutting or untieing the string.

Geometric design among the Ashanti

Among the Akan peoples in the forests of today's southern Ghana and eastern Côte d'Ivoire, the Ashanti Kingdom with its capital in Kumasi became dominant in the beginning of the 18th century. Before, from the 14th century onwards, the city of Begho, just north of the forest had been an important center of commerce and craft. Among the archaeological finds of Begho are the smoking pipes, for which Crowe (1982) analyzed the symmetries.

Figure 1.29b
Plaited band motifs

Ashanti women brought pottery to a high level of perfection, richly decorated with geometric designs. Figure 1.33 presents a design with 8-fold symmetry and a strip pattern that decorate the same jar.

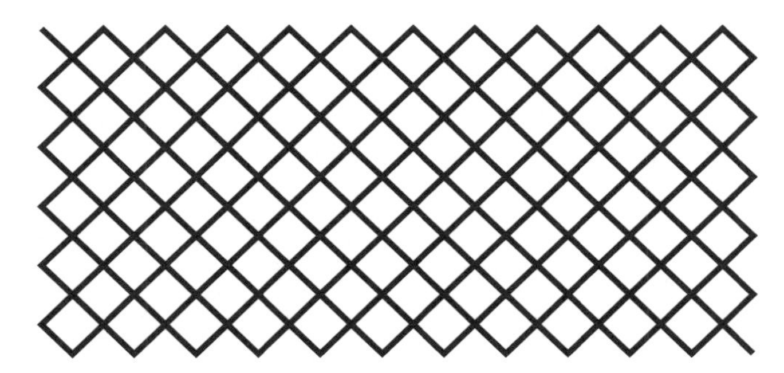

Figure 1.30
Drawing in the sand

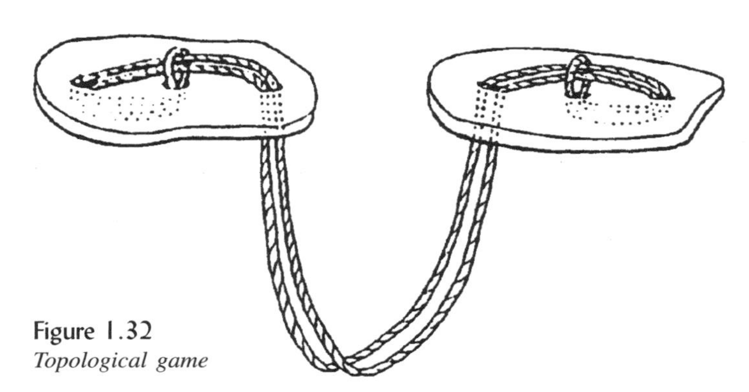

Figure 1.32
Topological game

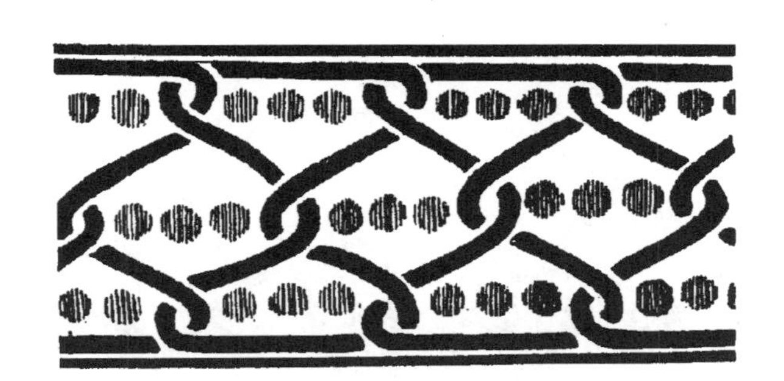

Figure 1.31
Line figures on wood carvings

Figure 1.33
Ashanti ceramic decoration

Figure 1.34
Plaster work wall decoration

Figure 1.35
Metal decoration

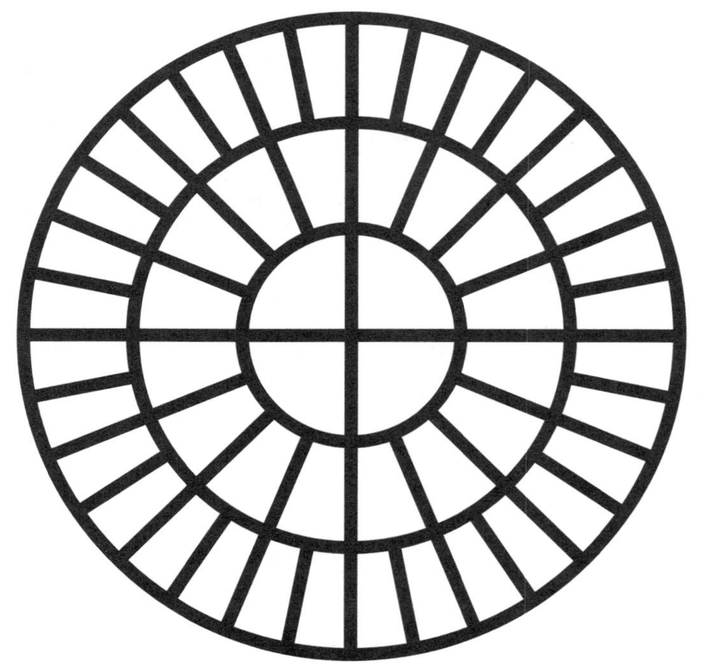

Figure 1.36
Goldweight motif

Among other Ashanti cultural activities interwoven with geometrical ideas and experimentation, are plaster and smith work and the weaving and decoration of cloth, as the following examples illustrate. Figure 1.34 presents an example of an Ashanti plaster work wall decoration. Similar designs are also seen in the traditional metalwork on the decorated blades of the state swords, made of iron, which are used by chiefs in ceremonies. Figure 1.35 presents an example of two designs on a sword blade. To weigh gold dust the Ashanti have used

Figure 1.37a
Adinkra stamp motifs

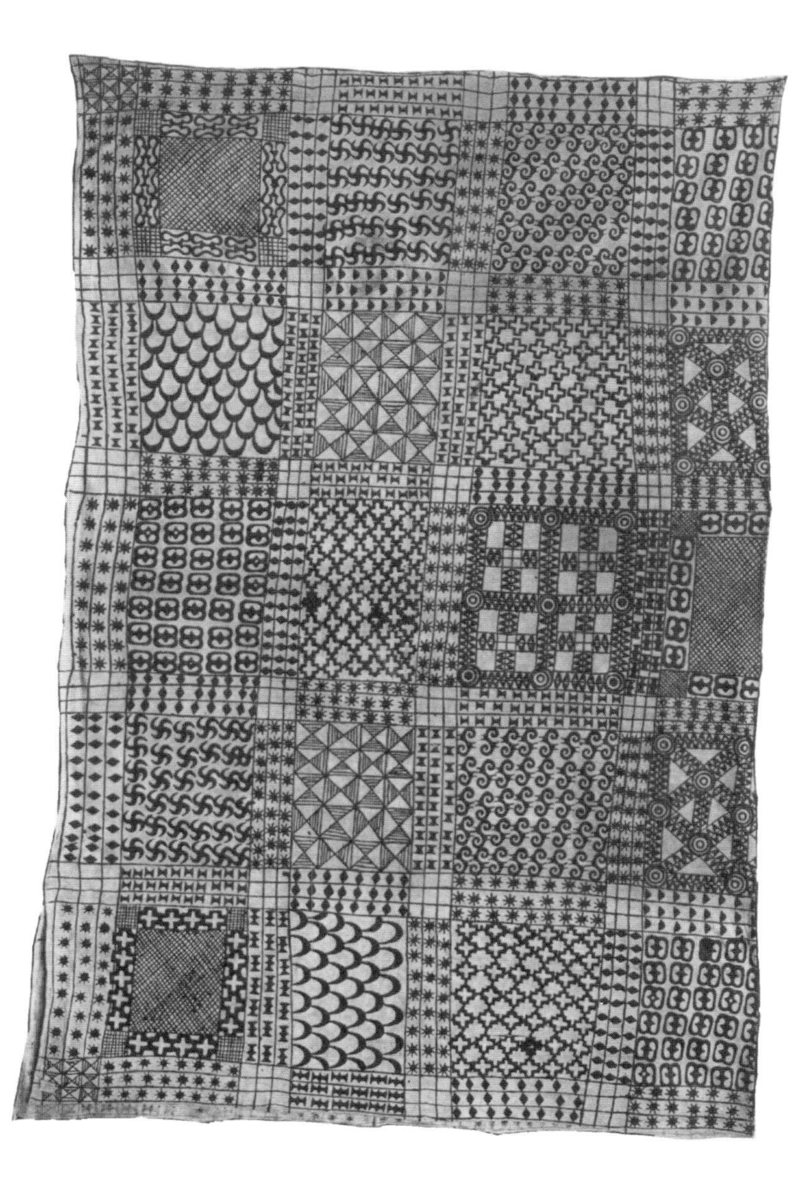

Figure 1.37b
Adinkra cloth

brass weights often with animal or geometric forms. The first volume of Niangoran-Bouah's book *The Akan World of Gold Weights* displays beautifully many of these goldweights with various geometrical shapes and forms. Figure 1.36 presents the bottom face of a cylindrical goldweight.

Among the decorated cloths produced by the Ashanti is hand-stamped cloth called *adinkra*. White cotton fabric is divided in rectangles which are filled up by a series of copies of the same stamp. Each stamp is made from a piece of calabash, cut in a design (Figure 1.37a presents three examples), with sticks glued to its back as a handle. Figure 1.37b presents an *adinkra* cloth from Bonwire, near Kumasi, acquired before 1934 by the British Museum.

The best known west-African fabric is the multi-colored *kente* cloth, woven by the Ashanti and by the Ewe in Togo. Weavers use horizontal looms to produce long and narrow strips of cloth, which then may be sewn together to form square or rectangular pieces used to make robes. Adler and Barnard's book *African Majesty. The Textile Art of the Ashanti and Ewe* displays beautiful examples of *kente* cloth (cf. also Lamb). Figure 1.38 presents several *kente* strip patterns. These are examples of two-color patterns in the sense that there exist translations, reflections or rotations which reverse the colors of the pattern. Figure 1.39 presents an example where three colors are used. Figure 1.40 presents two finite designs on a *kente* cloth.

Figure 1.38a–g
Kente cloth patterns

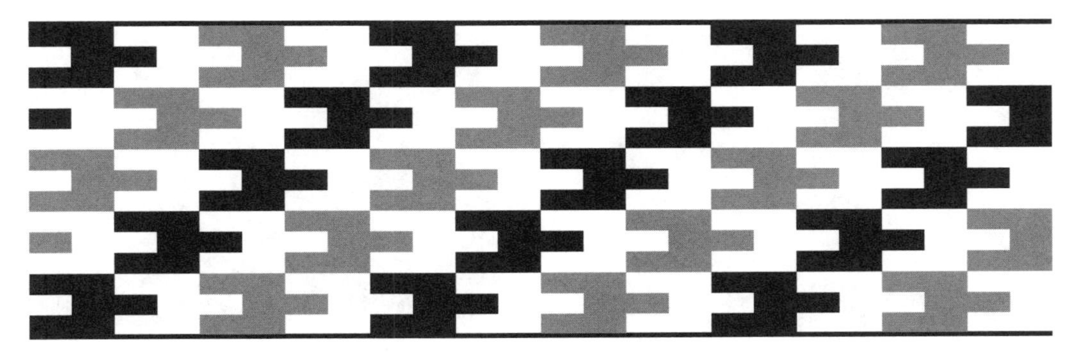

Figure 1.39
Kente strip pattern with three colors

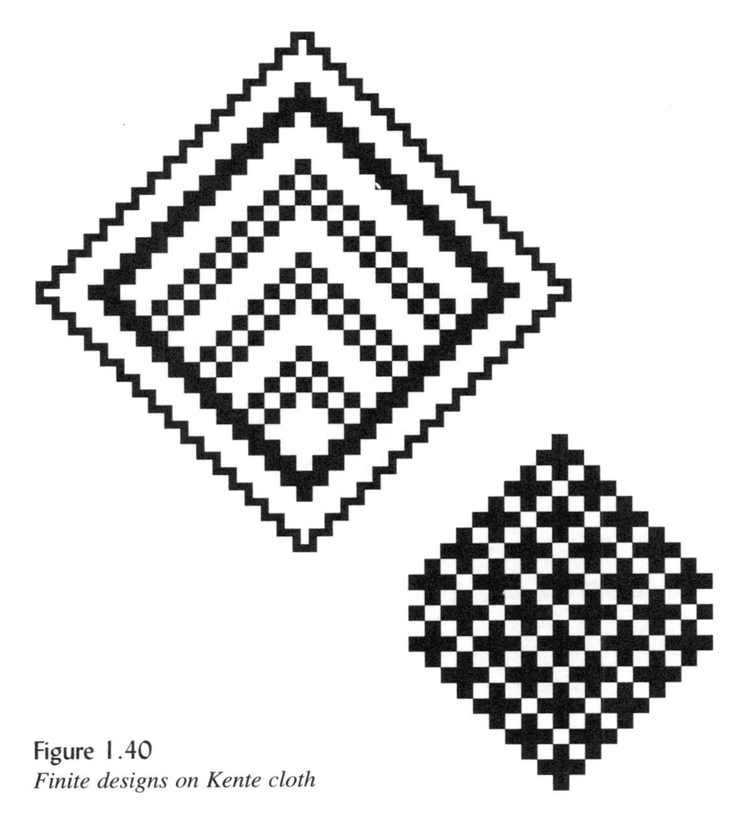

Figure 1.40
Finite designs on Kente cloth

Further African textile designs

Lamb & Lamb's *Au Cameroun: Weaving - Tissage,* Lamb & Holmes' *Nigerian Weaving,* Sieber's *African Textiles and Decorative Arts,* Picton & Mack's *African Textiles: Looms, Weaving and Design,* Picton's *The Art of African Textiles: Technology, Tradition, and Lurex,* and Clarke's *The Art of African Textiles* give information about weaving traditions in Africa south of the Sahara and present colorful photographs of fabrics, many with geometric designs and patterns. As an illustration of these rich and varied traditions, some further examples of such designs and patterns are presented in Figure 1.41.

Mangbetu geometric design in northeastern Congo/Zaire

According to Schildkrout & Keim (p.133) the Mangbetu ideal of feminine beauty included body-painting in geometric patterns. The women of the ruling class painted their bodies with *bianga,* a black juice made from the gardenia

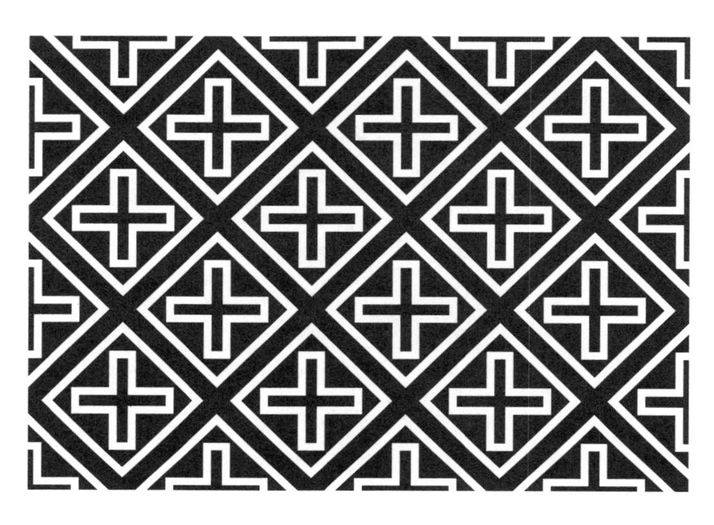

Figure 1.41b
Plane pattern on a Bambara painted shirt (Mali)

Figure 1.41a
Two plane patterns on Yoruba painted-resist-dyed cloth shirts (Nigeria)

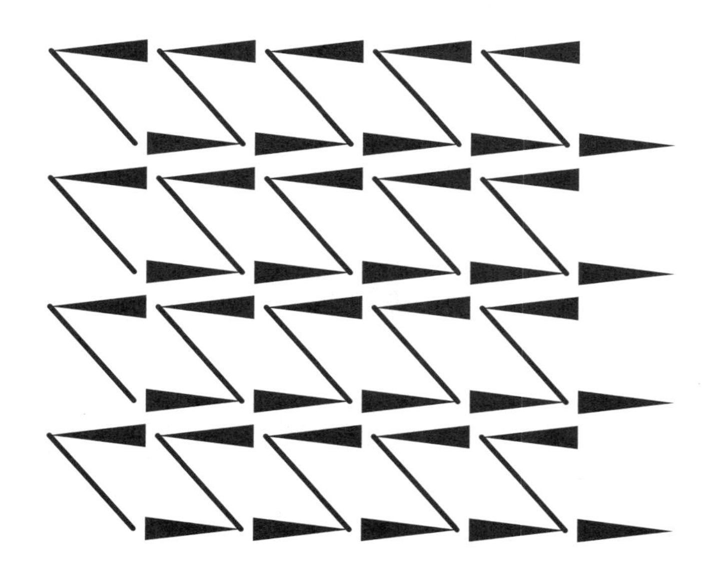

Figure 1.41c
Mende woven cloth pattern (Sierra Leone)

Figure 1.41d
Cloth pattern (western grasslands, Cameroon)

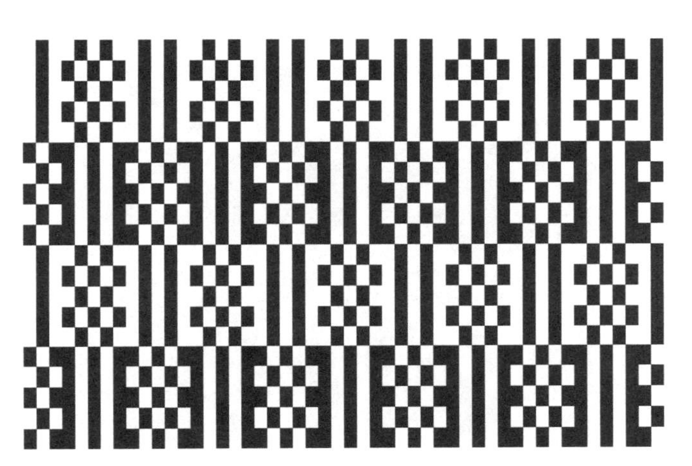

Figure 1.41f
Two-color pattern on cloth woven by Ibo women (Nigeria)

Figure 1.41e
Detail of a blue-and-white dyed display cloth from Bamum (northwest Cameroon)

Figure 1.41g
Strip patterns on a woven cloth from the Mende (Sierra Leone)

Figure 1.41h
Design on a dyed royal display cloth from Bamum (northwest Cameroon)

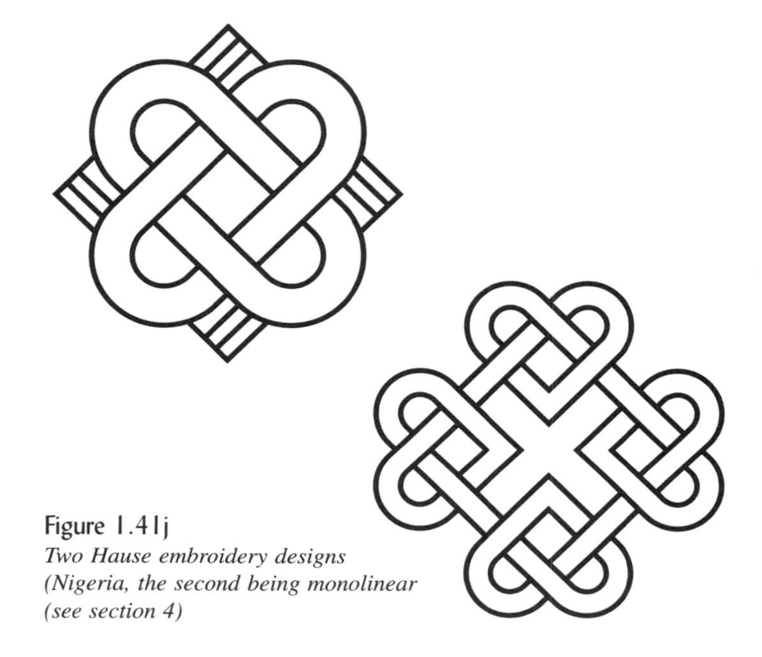

Figure 1.41j
Two Hause embroidery designs (Nigeria, the second being monolinear (see section 4)

Figure 1.41i
Example of a stenciled adire eleko design from Oshogbo, Nigeria

Figure 1.41k
Embroidered design on a woven raffia fibre cloth bag, collected in Bamum in 1905 (northwestern Cameroon)

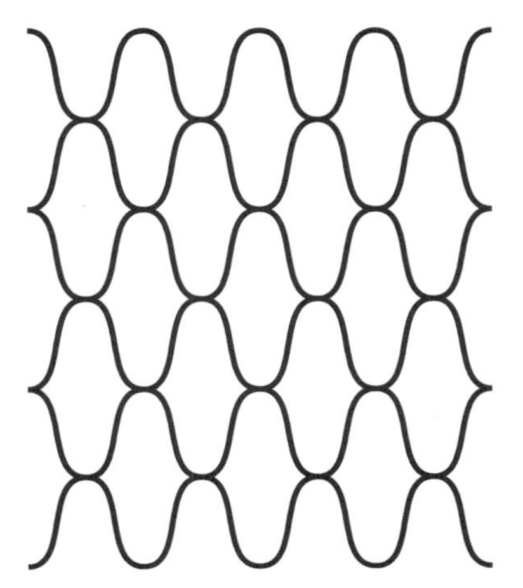

Figure 1.42
Plane Pattern present in a body painting, reproduced on a wooden sculpture collected in Niangara in 1910

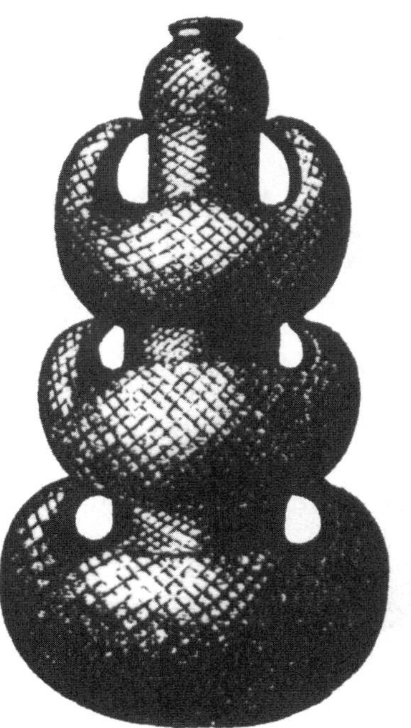

Figure 1.44
Mangbetu water bottles

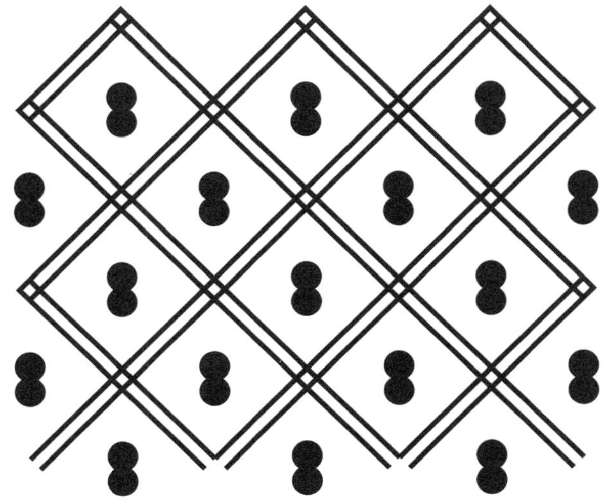

Figure 1.43
Plane pattern present on a barkcloth, collected in 1910

plant. Figure 1.42 presents an example of a Mangbetu body painting plane pattern. Mangbetu women wore pieces of barkcloth upon their right shoulder and whenever a woman sat down she laid it first upon the stool before sitting down. Figure 1.43 presents an example of a painted pattern on such a barkcloth.

Figure 1.44 presents two Mangbetu ceramic water bottles from the 19th century. Examples of useful ivory carvings, collected in 1918, are displayed in Figure 1.45. Schweinfurth's 19th century engraving in Figure 1.46 represents the great hall of the court of King Mbunza, giving

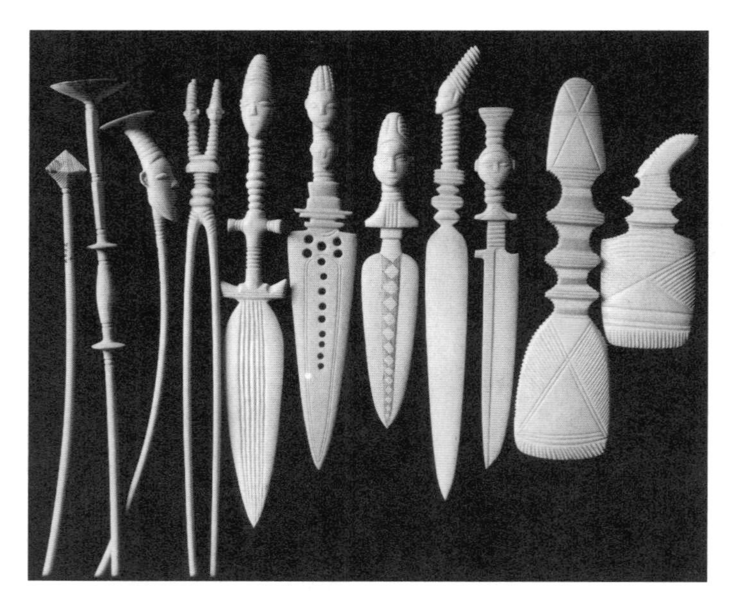

Figure 1.47
Rattles

Figure 1.45
Four hat and hairpins (at the left) are followed by five show knives and by two line gauges used to trace designs on mud walls.

Figure 1.46
Great hall of King Munza

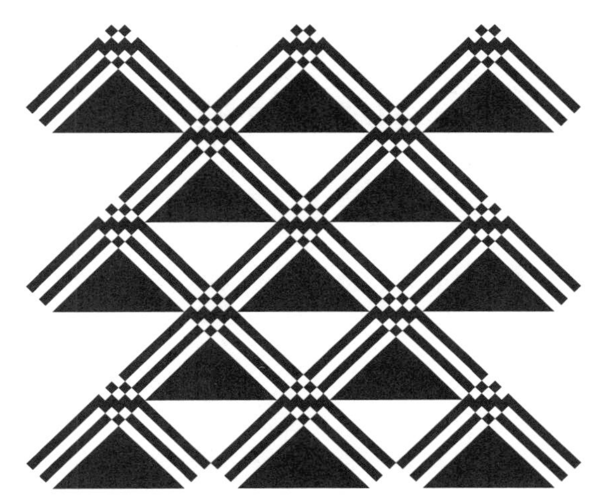

Figure 1.48a
Mangbetu mural painting

an idea of its geometrical structure. This hall was used for dancing, and Figure 1.47 shows rattles, made from strips of palm fibers and filled with seeds, used during dances. Figure 1.48a presents the plane pattern structure of a Mangbetu mural painting photographed before 1915 in Niangara.

Mural decoration

We encountered already examples of mural decoration among the Kuba, Ashanti, and Mangbetu. Mural decoration is one of the cultural spheres most used for geometrical exploration in Africa. Courtney-Clarke's *Ndebele: The art of an African tribe* and Changuion's *The African mural* describe house decoration and mural painting in Southern Africa, in particular among the Ndebele. Two chapters of my book *Women, Art, and Geometry in Southern Africa* analyze mural decoration among the Sotho (cf. section 3) and the Ndebele. Figure 1.48b gives an example of an Ndebele mural painting in South Africa.

Figure 1.48b
Ndebele mural painting

In West Africa, as in southern Africa, it is mostly the women who decorate the walls of their houses with geometrical figures. Each year after the harvest, they gather to restore and paint their mud dwellings which have been washed clean by the rains of the wet season. Courtney-

Figure 1.49
Drawing of a house front in Zinder (Niger)

Figure 1.50
Zande mural paintings

Clarke's *African Canvas* presents beautiful examples of geometrical mural decoration in West Africa. Figure 1.49 presents a drawing of a house front in Zinder (Niger).

Figure 1.50 presents two central African examples of mural paintings. They are from the Zande people in north-eastern Congo/Zaire. An example of a Swahili plaster work design from the northern Kenyan coastal region is given in Figure 1.51, and an example of a Nubian stencil wall painting is presented in Figure 1.52.

Figure 1.51
Swahili plaster work design

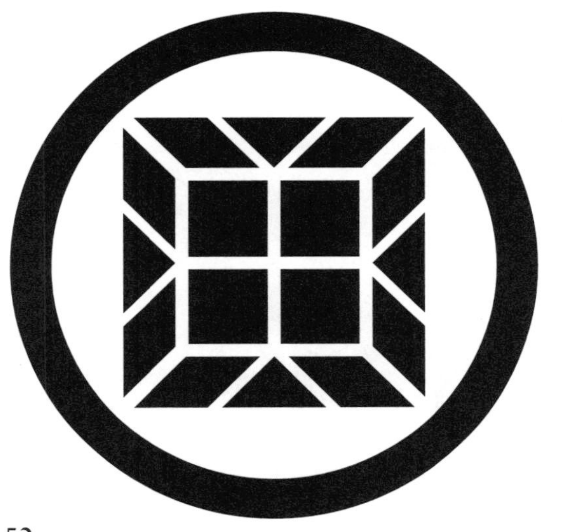

Figure 1.52
Nubian stencil wall painting

Figure 1.53 presents a detail of a beautifully plaited strip design (the black strands are the horizontal ones), that decorates part of the wall above the door of the house of a Bamileke chief in Cameroon. Figure 1.54 presents a wall decoration motif from Ghana.

Mural paintings and architecture in Burkina Faso are analyzed in Boudier & Minh-Ha's *African Spaces: Designs for Living in Upper Volta*. Oliver's *Shelter in Africa* and Denyer's *African traditional architecture* analyze architecture in general with many attractive illustrations. Figure 1.55 sketches the roof structure, as seen from below, of a Fulani house in Cameroon. Figure 1.56 presents a schematic plan of a large Massa farmhouse enclosure in Cameroon. Figure 1.57 reproduces a 19th century drawing of the circular settlement structure of Zulu cattle keepers in South Africa. The relationship between such enclosures and fractal geometry is one of the themes of Eglash's book *African Fractals*.

Figure 1.53
Bamileke wall decoration

Figure 1.54
Example of a wall decoration motif among the Nabdam in north Ghana

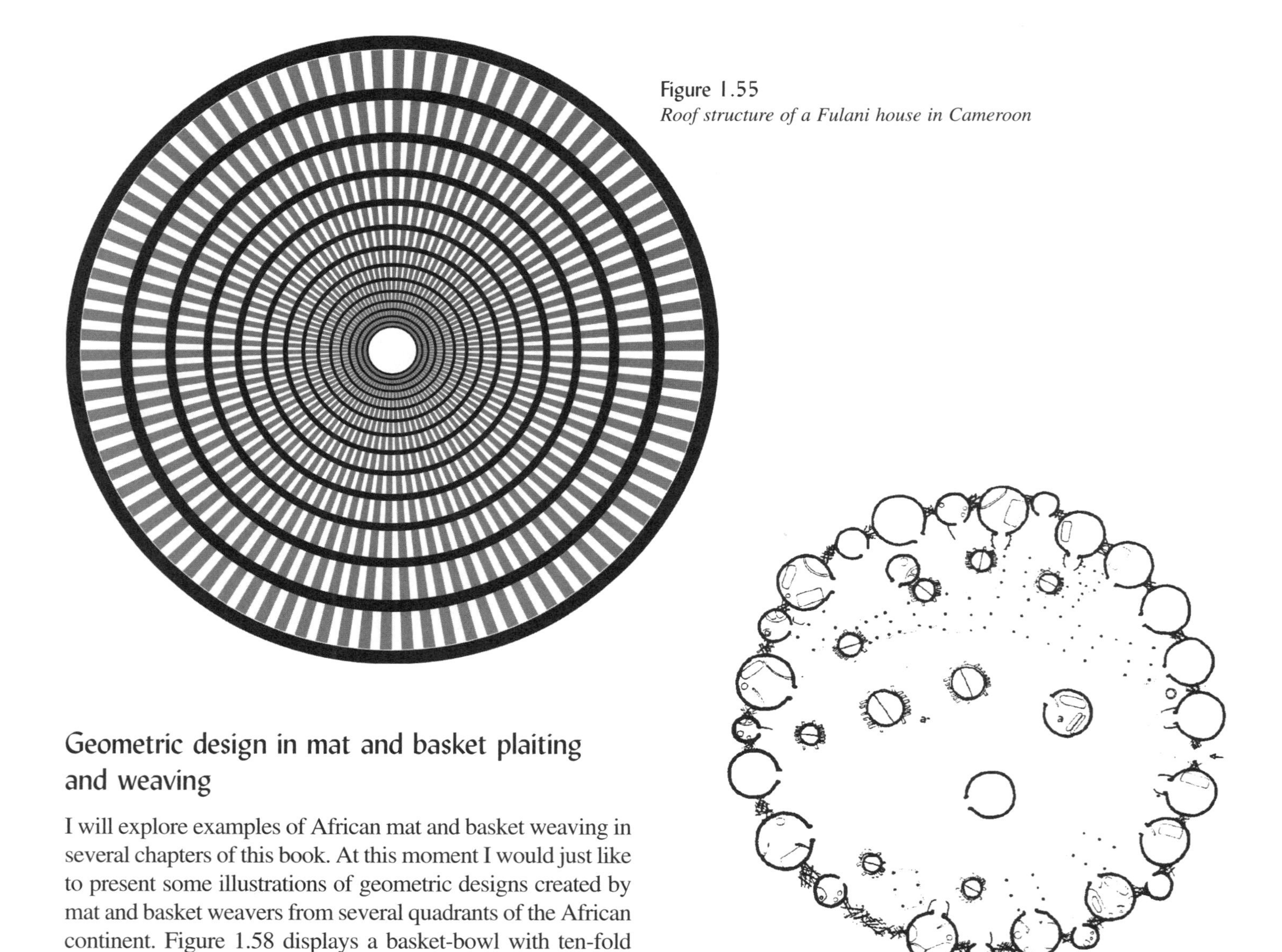

Figure 1.55
Roof structure of a Fulani house in Cameroon

Figure 1.56
Massa farmhouse enclosure (Cameroon)

Geometric design in mat and basket plaiting and weaving

I will explore examples of African mat and basket weaving in several chapters of this book. At this moment I would just like to present some illustrations of geometric designs created by mat and basket weavers from several quadrants of the African continent. Figure 1.58 displays a basket-bowl with ten-fold and a small pendant with five-fold axial symmetry produced by Swazi women using the coiled basket making technique (see the coil structure in Figure 1.59).

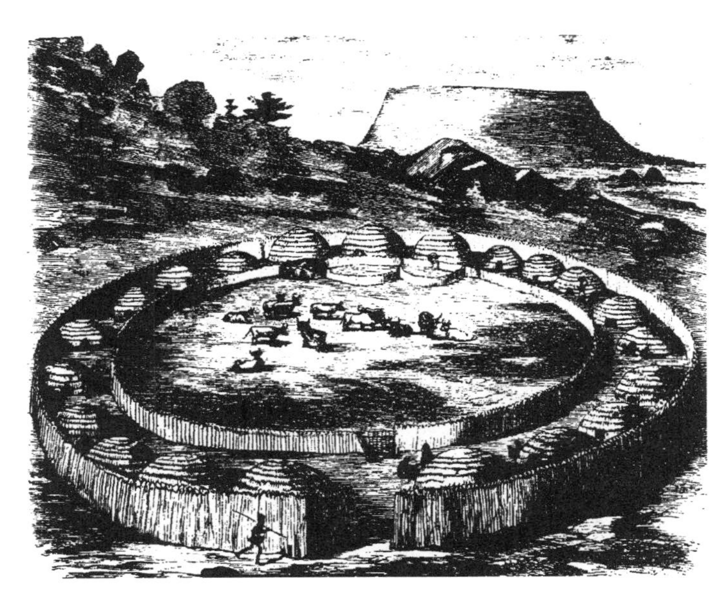

Figure 1.57
Nineteenth century Zulu settlement

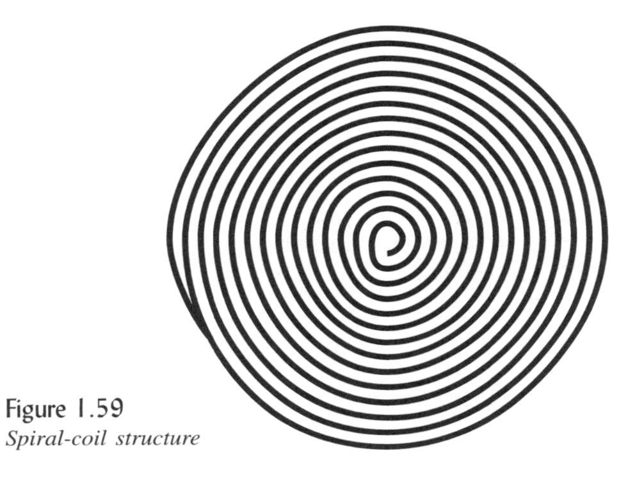

Figure 1.59
Spiral-coil structure

Figure 1.60a
Ovimbundu motif: bird's feet

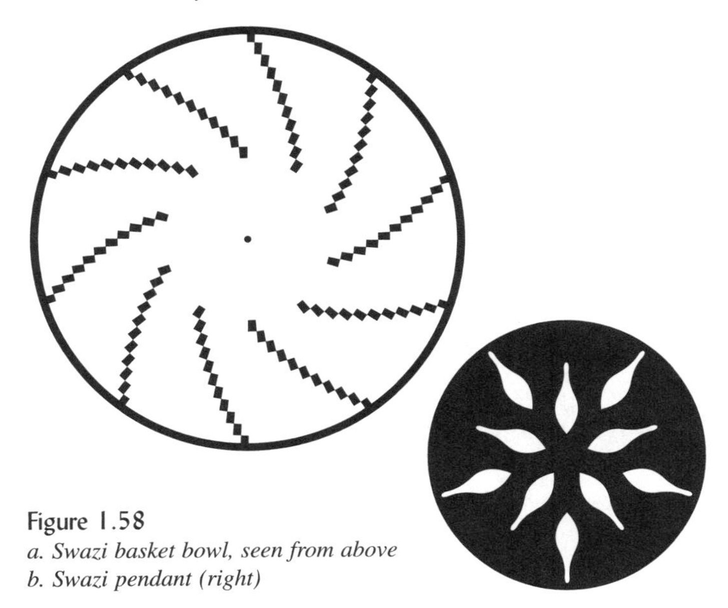

Figure 1.58
a. Swazi basket bowl, seen from above
b. Swazi pendant (right)

Figure 1.60 presents two coiled basketry motifs from Ovimbundu women in Angola, and Figure 1.61 a Somali coiled basketry motif.

Returning to the Mangbetu in northeastern Congo/Zaire, they are among the many African peoples who produce beautiful plaited mats and baskets. As an example, a detail of the wall of a plaited basket collected in 1910 in Niangara

Figure 1.60b
Ovimbundu motif: leaves of a dried tree

Figure 1.62
Detail of a Mangbetu basket

Figure 1.61
Somali decorative motif

is presented in Figure 1.62. The black strands are horizontal. Both the experimentation with producing and joining woven "toothed squares" and with two-color symmetry are noteworthy.

Figure 1.63 presents some strip patterns on baskets from the Tabwa people who live in southeastern Congo/Zaire.

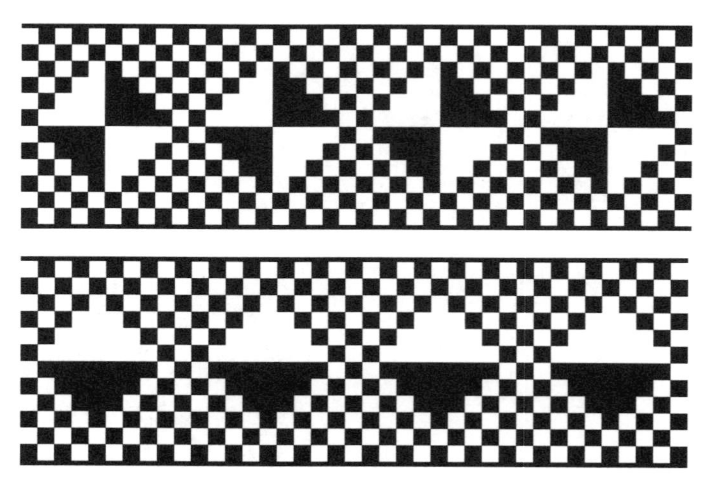

Figure 1.63a and b
Strip patterns on baskets from the Tabwa (southeastern Congo/Zaire)

Figure 1.63c and d
Strip patterns on baskets from the Tabwa (southeastern Congo/Zaire)

Leather work

Frequently designs from mat and basket weaving are applied and further developed in other contexts, using different materials. As an example, Figure 1.64 displays part of the cover of a bottle from Senegal. Sections of leather (black) have been slit and are woven with dried palm fibres (white).

Similarly, Figure 1.65 illustrates the decoration of a leather money purse from the the Mali / Côte d'Ivoire border region and of a leather bolster from Korhogo (Côte d'Ivoire). Another type of leather work is illustrated in Figure 1.66: examples of leather appliqués with which Tuareg women in Niger decorate camel saddle bags. Figure 1.67 presents a painted strip pattern on a leather bracelet from Tamale (Ghana).

Female Hausa artisans in Nigeria produce parchment boxes of dyed skin, richly decorated with geometric designs as Figure 1.68 illustrates.

Figure 1.64
Part of a bottle decoration (Senegal)

Figure 1.69 displays several leather shields. Those at the top are (left to right) from the Nandi and Kipsiki peoples in Kenya, the Lukha of Sudan and the Zulu of South Africa; those at the bottom are from the Embu and Masai (twice) of Kenya and the Sonjo of Tanzania.

Figure 1.65
Decorations in leather

Figure 1.67
Painted strip pattern

Figure 1.66
Tuareg leather appliqués (Niger)

Figure 1.68
Examples of Hausa parchment boxes

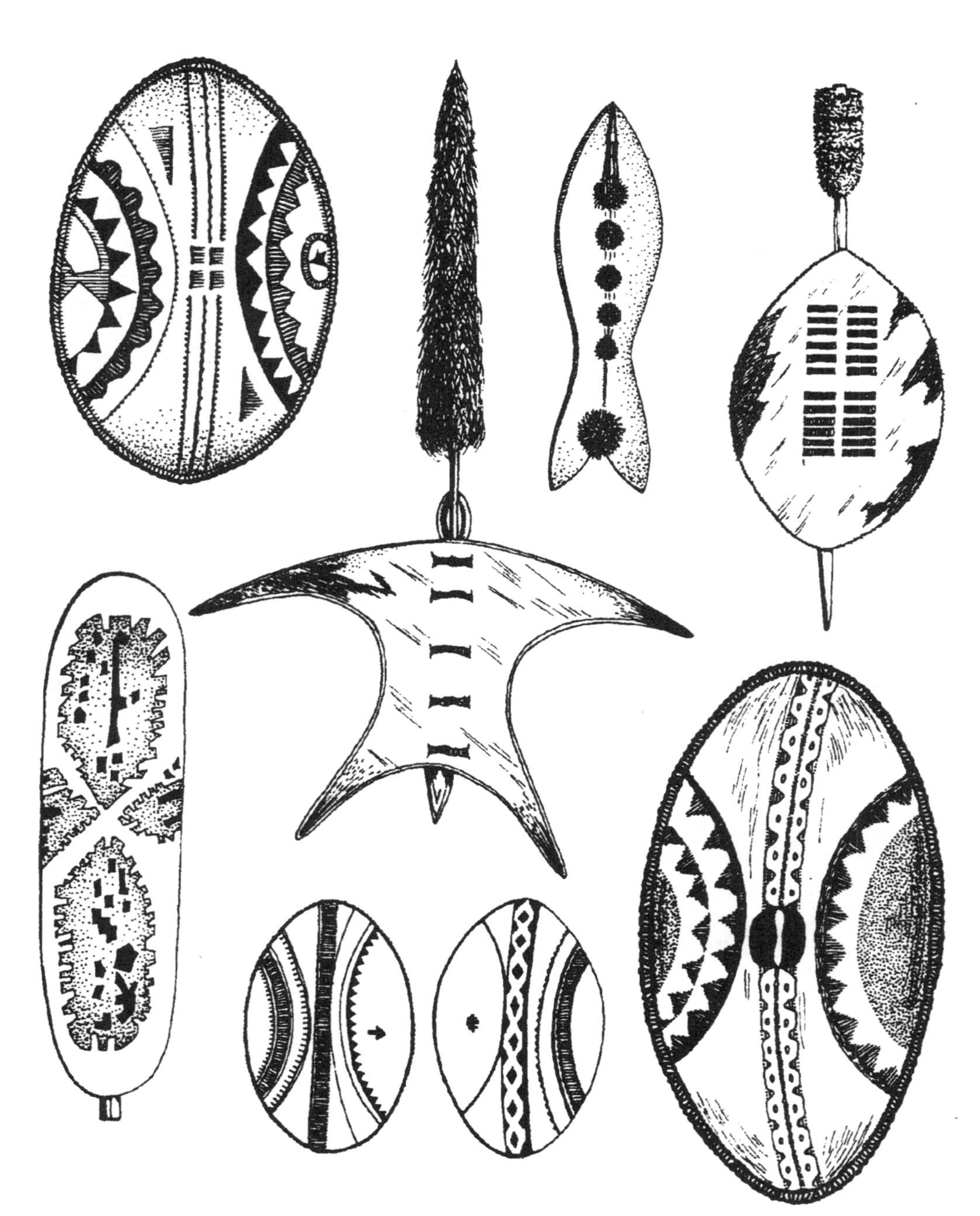

Figure 1.69
Examples of leather shields. Those at the top are (left to right) from the Nandi and Kipsiki peoples in Kenya, the Lukha of Sudan and the Zulu of South Africa; those at the bottom are from the Embu and Masai (twice) of Kenya and the Sonjo of Tanzania.

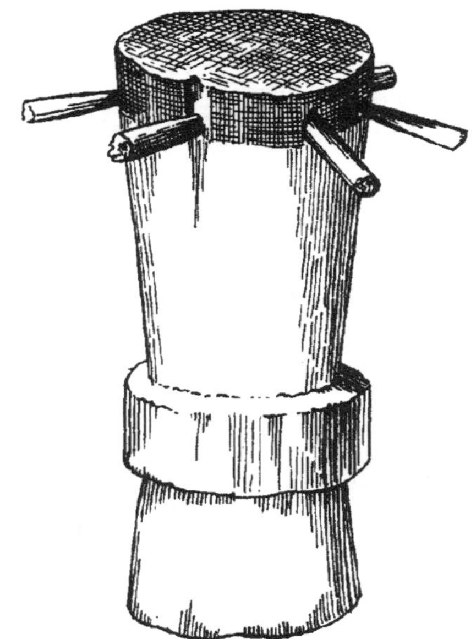

Figure 1.70
Bassari drum from Senegal

Figure 1.71
Decorative design on a semi-spherical calabash from Kano, Nigeria

To make a good sounding drum it is important to fix the covering skin evenly to the drum's wall. This means that the pins have to be equally spaced. In other words, the holes in which the pins are nailed constitute a regular polygon. Figure 1.70 illustrates a drum from Senegal.

Calabash and gourd decoration

A particularly apt domain for geometrical expression and exploration is the decoration of calabashes or gourds. The calabashes may be cut open and emptied, their shells being dried in the sun until hard. The yellow surface maybe darkened or colored, and then designs may be applied by incision and/or, for example, by burning lines into it with a sharp hot metal object, like a knife or a nail. Figure 1.71 presents a design incised and burned onto a semi-spherical calabash. Its provenance is Kano in Nigeria. Chappel's *Decorated gourds in north-eastern Nigeria* analyzes calabash decoration by Fulani, Yungur and Bata craftsmen and women in north-eastern Nigeria, displaying the richness and variety of forms, the creativity of their producers. The decorated gourds are used as containers, but also, as those made by Bura and Waja women (Nigeria), as sun hats for babies (cf. Arnoldi & Kreamer, p. 39). Decorated gourds are also produced in other parts of Nigeria, Benin, Cameroon, Kenya, Uganda, Congo/Zaire (cf. Trowell, pl. XLVI–LIII). Figure 1.72 presents a burnt-in decoration on a palm wine calabash collected in 1898 in northwestern Cameroon and Figure 1.73 some gourd decorations from Nigeria.

Figure 1.72
Decorative pattern on a calabash from Cameroon

Figure 1.73b
Calabash decoration: Hona calabash bowl seen from above (Nigeria)

Figure 1.73a
Calabash decoration: Fulani water bottle calabash (Nigeria)

Wood carving and geometric design

Another activity that invites geometrical exploration is that of wood carving. African wooden boxes, stools, seats (cf. Bocola), headrests, doorposts, pilasters, canoes and boats, spears, drums, pestles, spoons, cups, masks, combs, ... often have symmetric shapes, as well as being often decorated with geometric designs. Figure 1.74 presents some examples.

Figure 1.73c
Calabash decoration:
Yungur calabash bowl seen from above (Nigeria)

Figure 1.73d
Calabash decoration: Bata calabash bowl seen from above (Nigeria)

Figure 1.73e
Calabash decoration: Painted interior of a Fulani calabash bowl (Nigeria)

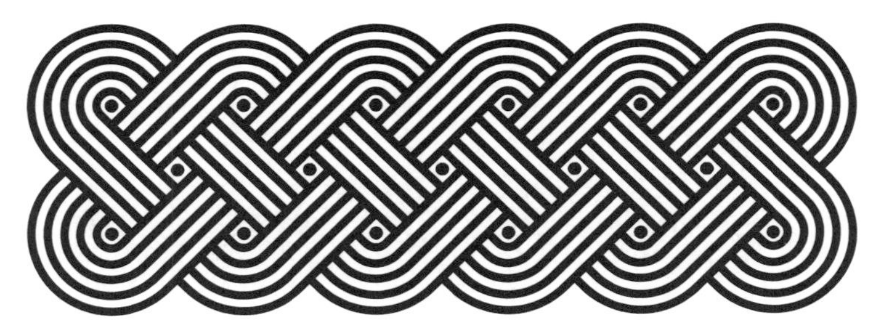

Figure 1.74a
Woven band carved on a wooden stool from Benin

Figure 1.74b
Carved hardwood combs from the upper Côte d'Ivoire

Figure 1.74e
Tabwa stool decoration (southeastern Zaire/ Congo, ca. 1884)

Figure 1.74c
Design on a wooden Ligbe mask (Côte d'Ivoire)

Figure 1.74d
Design on a wooden Dogon pilaster (Mali)

Figure 1.74f
Plane pattern on Igbo/Ibo mask (Nigeria)

Figure 1.74g
Plane pattern on a Bamileke mask (Cameroon)

Figure 1.74h

Plane pattern (frog motif on Babanki mask (Cameroon, 19th Century)

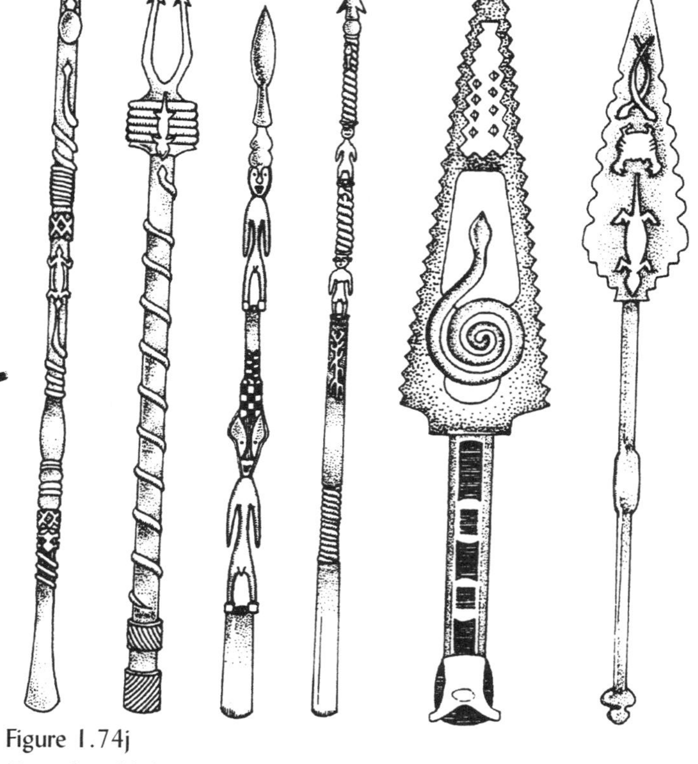

Figure 1.74j

Examples of fishing spears and canoe paddles of the Kalabari (Nigeria)

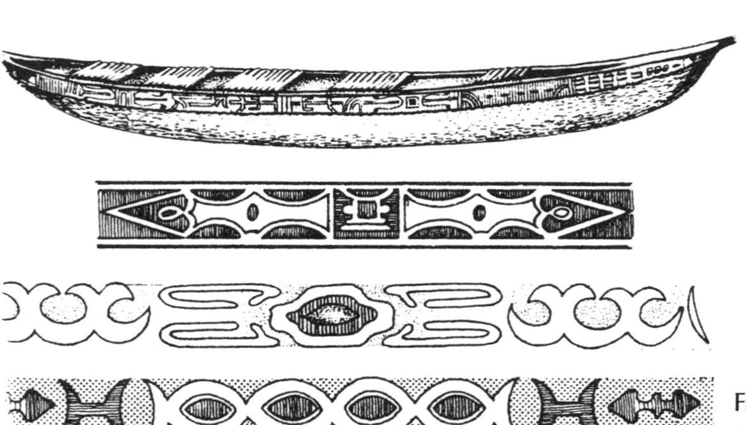

Figure 1.74i

Designs for canoes (Ghana)

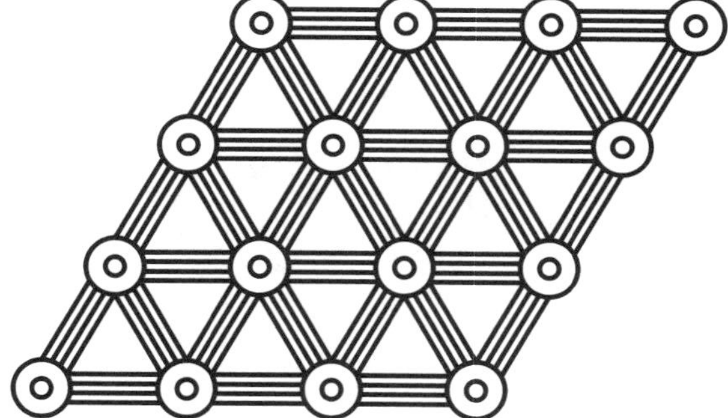

Figure 1.74k

Plane structure of decorative pattern on a Bamum mask (Cameroon)

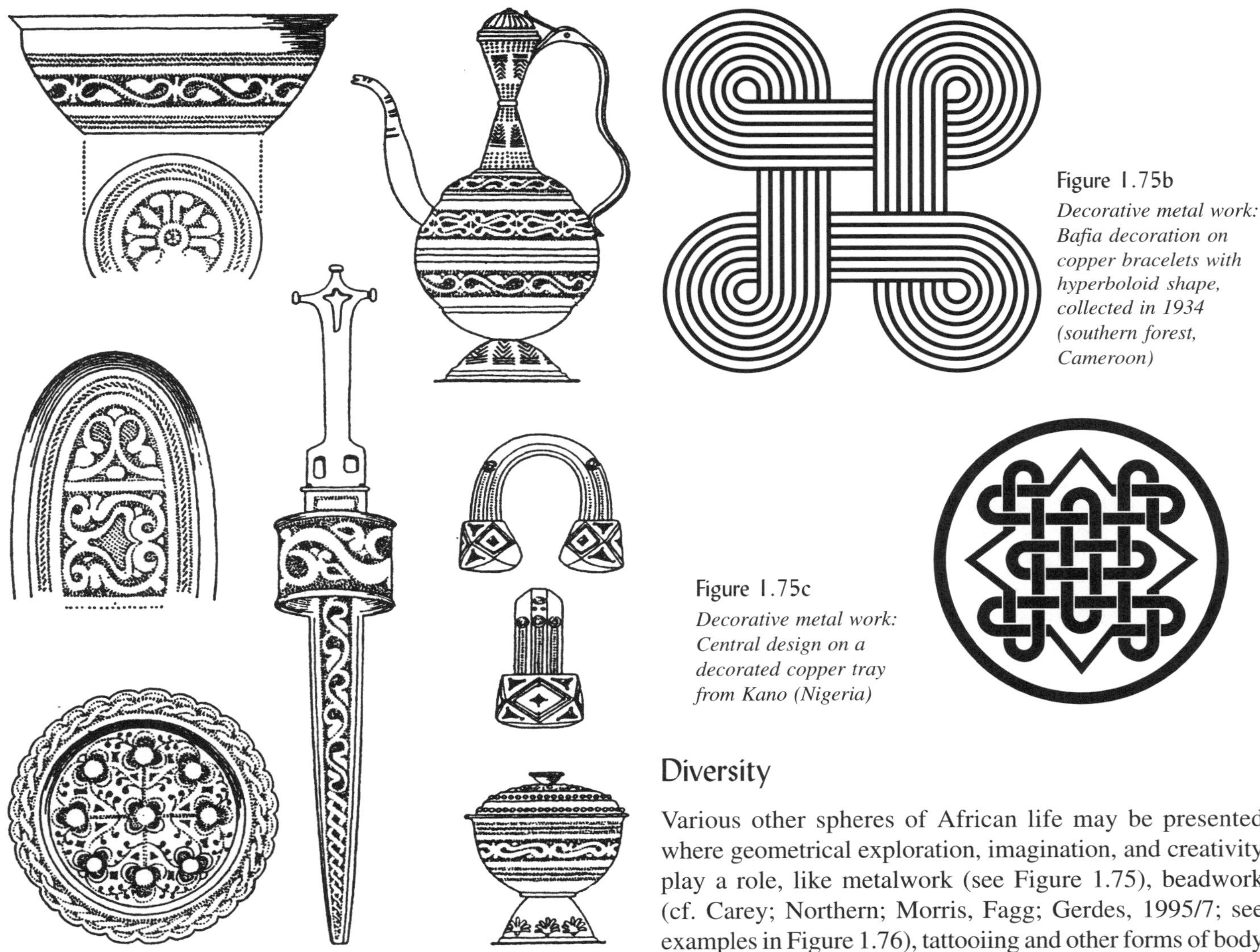

Figure 1.75b
Decorative metal work: Bafia decoration on copper bracelets with hyperboloid shape, collected in 1934 (southern forest, Cameroon)

Figure 1.75c
Decorative metal work: Central design on a decorated copper tray from Kano (Nigeria)

Figure 1.75a
Decorative metal work: Dagger, anklets, jug, trays, and containers made of beaten brass from Bida (Nigeria)

Diversity

Various other spheres of African life may be presented where geometrical exploration, imagination, and creativity play a role, like metalwork (see Figure 1.75), beadwork (cf. Carey; Northern; Morris, Fagg; Gerdes, 1995/7; see examples in Figure 1.76), tattooiing and other forms of body decoration (cf. e.g. Fischer's *Africa Adorned*; Gerdes, 1995/7; see examples in Figure 1.77), hairstyles (cf. Arnoldi & Kreamer; see examples in Figure 1.78), string figure and other games.

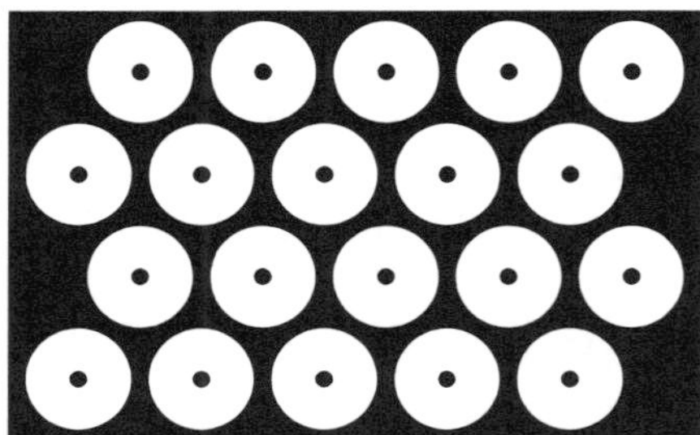

Figure 1.76a

Bamileke bead pattern on a hat for an elephant mask (northwest Cameroon, 1913)

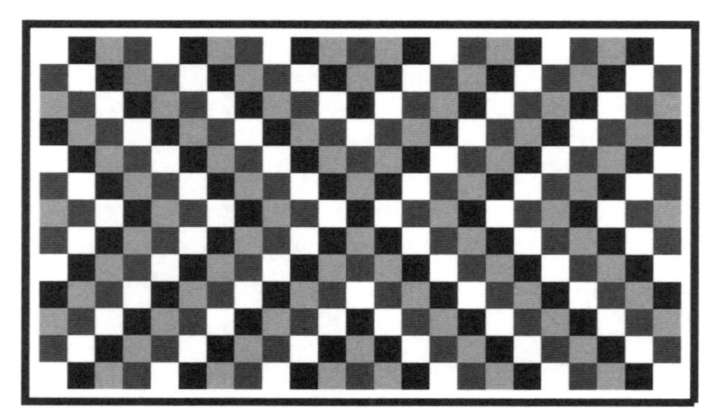

Figure 1.76b

Examples of Zulu beaded aprons (South Africa)

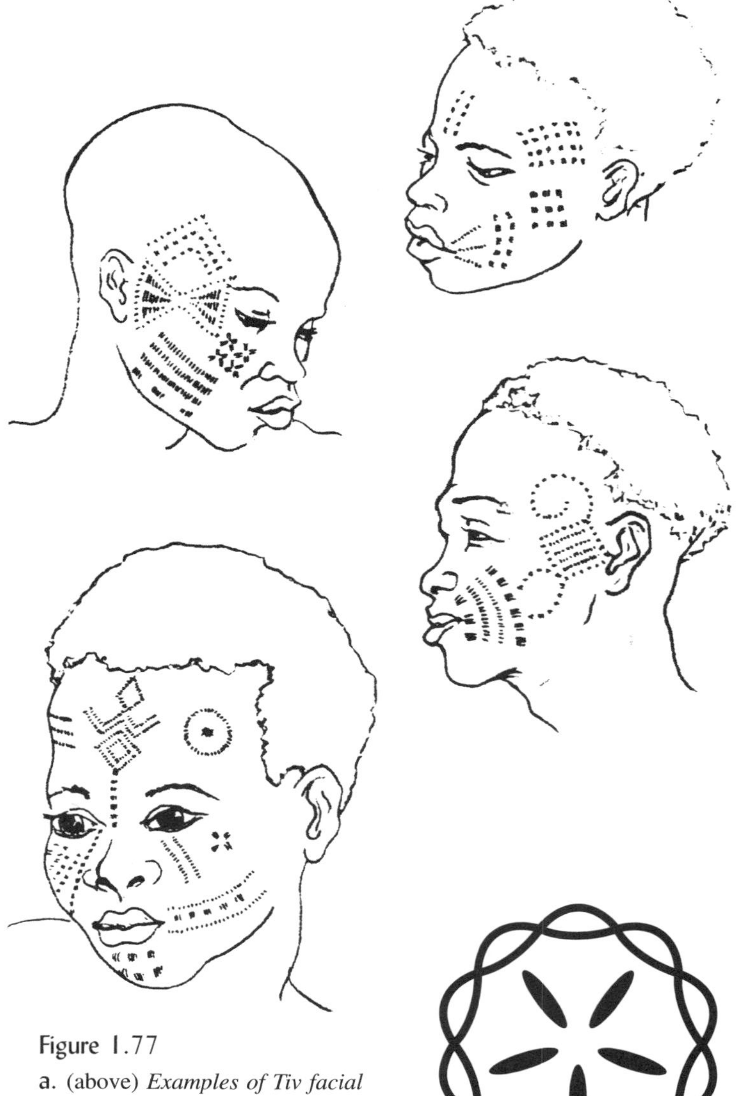

Figure 1.77

a. (above) *Examples of Tiv facial tattoos (South Africa)*

b. (right) *Detail of a facial tattoo among Makhuwa women (Mozambique)*

Figure 1.78a

Woman with bamboo hooped coiffure, 19th century (Congo/Zaire)

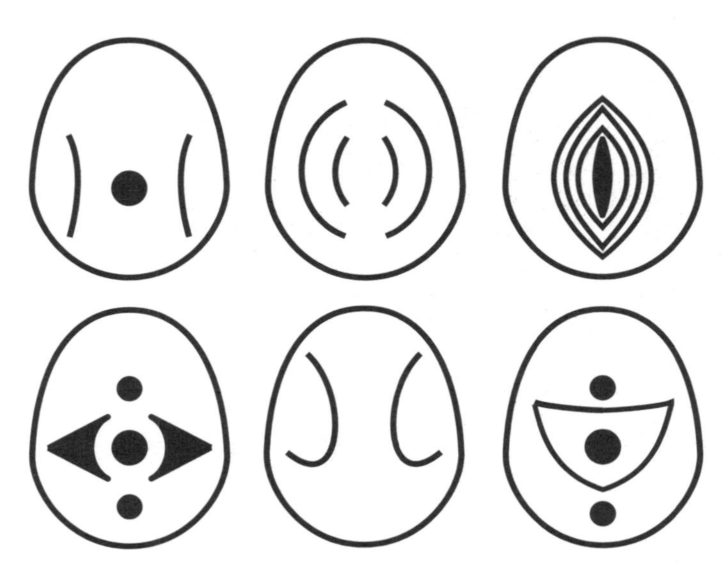

Figure 1.78b

Twa designs from the banks of Lake Kivu, shaved into the hair (c. 1900, Ruanda)

Figure 1.78c

Nineteenth century African hair styles

In any region or country, the diversity of activities in which geometrical considerations are involved, may be extremely great. Let's take the example of Kenya: The Turkana and the el-Molo weave semi-spherical basket fish traps; Wata woodcarvers produce containers, fashioned from raw logs, decorating them with strip patterns; Kamba women decorate gourds; Luyia on the eastern shore of Lake Victoria build and decorate boats; Luyia basket makers weave conical quail baskets with a hexagonally woven bottom; Meru woodcarvers on the eastern slopes of Mount Kenya shape symmetrical spoons and Meru drummers use log-drums painted with symmetrical designs; the Pokomo dugout canoes; Boran and Taita women weave mats out of strips; and Sakuye weave hats with a decorative motif with fourfold symmetry (cf. photographs in Amin & Moll).

In all spheres of African life considered in this chapter—as in the rest of this book—geometrical ideas, geometrical considerations, geometrical explorations, geometrical imagination are interwoven, interbraided, interplaited, intercut, intercoiled, intercised, interpainted, ... Often geometrical exploration also is expression of and develops hand in hand with artistical/aesthetical exploration, looking for symmetry and beauty. The Cameroonean mathematician Georges Njock once characterized the relationship between African art and mathematics as follows: "Pure mathematics is the art of creating and imaginating. In this sense black art is mathematics" (Njock, p.8).

May the examples given in this chapter convey to the reader an idea for how women and men all over Africa south of the Sahara, in diverse historical and cultural contexts, traditionally have been geometrizing.

Bibliography

Adler, Peter and Nicholas Barnard (1992), *African Majesty: The Textile Art of the Ashanti and Ewe,* Thames and Hudson, London.

Amin, Mohamed and Peter Moll (1983), *Portraits of Africa* [Kenya], Harvill Press, London.

Arnoldi, Mary and Christine Kreamer (1995), *Crowning Achievements: African Arts of Dressing the Head,* Fowler Museum of Cultural History, Los Angeles.

Ben-Amos, Paula (1980), *The Art of Benin,* Thames and Hudson, London.

Bocola, Sandro (Ed.) (1995), *African Seats,* Prestel, Munich/New York.

Bolland, Rita (1991), *Tellem Textiles: Archaeological finds from burial caves in Mali's Bandiagara Cliff,* Tropenmuseum, Amsterdam/Musée National, Bamako.

Bourdier, Jean-Paul and Trinh Minh-Ha (1985), *African Spaces: Designs for Living in Upper Volta,* Africana Publishing Company, New York.

Carey, Margret (1986), *Beads and Beadwork of East and Southern Africa*, Shire, Bucks.

Changuion, P., T. Matthews, and A. Changuion (1989), *The African Mural*, Struik, Cape Town.

Chappel, T.J. (1977), *Decorated Gourds in North-Eastern Nigeria,* Ethnographica, London/The Nigerian Museum, Lagos.

Clarke, Duncan (1997), *The Art of African Textiles*, Thunderday Press, San Diego.

Courtney-Clarke, Margaret (1986), *Ndebele: The Art of an African Tribe*, Rizzoli, New York.

—— (1990), *African Canvas*, Rizzoli, New York.

Crowe, Donald (1971), The geometry of African art I. Bakuba art. *Journal of Geometry*, Vol. 1, 169–182.

—— (1973), Geometric symmetries in African art. In *Africa Counts*, C. Zaslavsky, pp. 190–196.

—— (1975), The geometry of African art II. A catalog of Benin patterns. *Historia Mathematica*, Vol. 2, 253–271.

—— (1982), The geometry of African art III. The smoking pipes of Begho. In *The geometric vein, the Coxeter Festschrift*,

C.Davis, B. Grünbaum, and F. Sherk Eds., Springer Verlag, New York, pp. 177–189.

Denyer, Susan (1978), *African Traditional Architecture*. Heinemann, London.

Dowson, Thomas (1992), *The Rock Engravings of Southern Africa*, Witswatersrand University Press, Johannesburg.

Eglash, Ron (1998), *African Fractals* (in preparation).

Etienne-Nugue, Jocelyne (1982), *Crafts and the Arts of Living in the Cameroon*, Louisiana State University Press, Baton Rouge.

Fagg, William (1980), *Yoruba Beadwork — Art of Nigeria*. Rizzoli, New York.

Fischer, Angela (1984), *Africa Adorned: A Panorama of Jewelry, Dress, Body Decoration, and Hairstyle*, Harry N. Abrams, New York.

Gardi, René (1969), *African Crafts and Craftsmen*, Van Norstrand Reinhold, New York.

Gerdes, Paulus (1990), *Ethnogeometrie. Kulturanthropologische Beiträge zur Genese und Didaktik der Geometrie*, Verlag Franzbecker, Bad Salzdetfurth (Germany).

—— (1994), On mathematics in the history of subsaharan Africa, *Historia Mathematica*, New York, Vol. 21, 345–376.

—— (1995), *Women and Geometry in Southern Africa*, Universidade Pedagógica, Maputo.

—— (1998), *Women, Art and Geometry in Southern Africa*, Africa World Press, Trenton NJ and Asmara, Eritrea.

—— (1999), *On Culture and the Awakening of Geometrical Thinking*, MEP - University of Minnesota, Minneapolis.

Ghaidan, Usam (1973), Swahili Plasterwork, in: *African Arts*, Vol. VI, No. 2, 46–49.

Gilfoy, Peggy (1987), *Patterns of Life: West African Strip-Weaving Traditions*, National Museum of African Art, Washington.

Guidoni, Enrico (1987), *Primitive Architecture*, Electa/Rizzoli, New York.

Hambly, Wilfrid (1945), *Clever Hands of the African Negro*, The Associated Publishers, Washington.

Hauenstein, Alfred (1988), *Examen de motifs décoratifs chez les Ovimbundu et Tchokwe d'Angola*, Instituto de Antropologia, Universidade de Coimbra, Coimbra.

Hogbe-Nlend, Henri (1985), *L'Afrique, berceau de la mathématique mondiale?*, United Nations University, Nairobi (mimeo).

Jefferson, Louise (1974), *The Decorative Arts of Africa*, Collins, London.

Lamb, Venice (1975), *West African Weaving*, Duckworth, London.

Lamb, Venice and Judy Holmes (1980), *Nigerian Weaving*, Lamb, Roxford.

Lamb, Venice and Alastair Lamb (1981), *Au Cameroon: Weaving — Tissage*, Lamb, Roxford.

Lestrange, Marie-Thérèse and Monique Gessain (1976), *Collections Bassari du Musée de l'Homme*, Musée de l'Homme, Paris.

Liebenberg, Louis (1990), *The Art of Tracking: The Origin of Science*, David Philip Publ., Claremont (South Africa).

Maurer, Evan and Allen Roberts (1985), *Tabwa: The Rising of a New Moon, a Century of Tabwa Art*, University of Michigan Muse, Ann Arbor.

Meurant, Georges (1986), *Shoowa design: African Textiles from the Kingdom of Kuba*, Thames and Hudson, London.

Meurant, Georges and Robert Thompson (1995), *Mbuti Design: Paintings by Pygmy Women of the Ituri Forest*, Thames and Hudson, London.

Morris, Jean (1994), *Speaking with beads — Zulu Arts from Southern Africa*, Thames and Hudson, New York.

Newman, Thelma (1974), *Contemporary African Arts and Crafts*, Crown Publishers, New York.

Niangoran-Bouah, N. (1984), *L'univers Akan des poids a peser l'or, / The Akan world of gold weights*, Vol. 1: *Les poids non figuratifs/Abstract design weights*, Les Nouvelles Editions Africaines, Abidjan (Côte d'Ivoire).

Njock, George (1985), Mathématiques et environnement socio-culturel en Afrique Noire, *Présence Africaine*, Vol. 135, 3–21.

Nooter, Mary (1993), *Secrecy: African Art that Conceals and Reveals*, The Museum for African Art, New York.

Northern, Tamara (1975), *The Sign of the Leopard: Beaded Art of Cameroon*, The William Benton Museum of Art, Storrs.

Obenga, Théophile (1995), *La Géométrie Égyptienne — Contribution de l'Afrique antique à la Mathématique mondiale*, L'Harmattan, Paris.

Oliver, Paul (Ed.) (1971), *Shelter in Africa*, Barrie & Jenkins, London.

Picton, John (1995), *The Art of African Textiles: Technology, Tradition, and Lurex*, Barbican Art Gallery, London.

Picton, John and John Mack (1989), *African Textiles: Looms, Weaving and Design*, British Museum Publications, London.

Pruitt, Sharon (1990), *Art of the Cameroon: Selections from the Spellman College Collection of African Art*, University of Georgia, Athens.

Robbins, Warren and Nancy Nooter (1989), *African Art in American Collections*, Smithsonian Institution Press, Washington.

Schildkrout, Enid and Curtis Keim (Eds.) (1990), *African Reflections: Art from Northeastern Zaire*, American Museum of Natural History, New York.

Schweinfurth, Georg (1874), *The heart of Africa. Three years' travels and adventures in the unexplored regions of central Africa from 1868 to 1871*, Harpers & Brothers, New York, 2 volumes.

Shaw, Thurstan, Paul Sinclair, Bassey Andah, and Alex Okpoko (Eds.) (1993), *The Archaeology of Africa. Food, Metals and Towns*, Routledge, New York / London.

Sieber, Roy (1972), *African Textiles and Decorative Arts*, The Museum of Modern Art, New York.

—— (1980), *African Furniture and Household Objects*, The American Federation of Arts, New York.

Stössel, Arnulf (Ed.) (1984), *Afrikanische Keramik. Traditionelle Handwerkskunst südlich der Sahara*, Himmer Verlag, München.

Torday, Emil and T. A. Joyce (1910), *Notes ethnographiques sur les peuples communément appelés Bakuba, ainsi que sur les peuplades apparantées. Les Bushongo*, Musée Royal du Congo Belge, Bruxelles.

Trowell, Margaret (1960), *African Design*, Frederick Praeger, New York.

Washburn, Dorothy (1990), *Style, classification and ethnicity: design categories on Bakuba raffia cloth*, American Philosophical Society, Philadelphia.

Wenzel, Marian (1972), *House Decoration in Nubia*, University of Toronto Press, Toronto

Willcox, A. R. (1984), *The Rock Art of Africa*, Holmes & Meier, New York.

Zaslavsky, Claudia ([1973] 1990), *Africa Counts: Number and Pattern in African culture,* Lawrence Hill Books, Brooklyn, NY.

Part of a fence around a church. Both fence and church were decorated by Ndebele women (South Africa).
Photograph by Marilyn Frankenstein (University of Massachusetts Boston), 1991.

2

From African designs to discovering the Pythagorean Theorem

One of the most interesting and attractive theorems of early geometry is certainly the so-called 'Pythagorean Theorem.' Although legend has ascribed the theorem to Pythagoras of Samos (sixth century BC), the Senegalese Egyptologist Cheik Anta-Diop has suggested that Pythagoras may have learned the theorem during his long stay in Egypt (Diop, p. 436, 479).

In this section, I will show how diverse African ornaments and artifacts may be used to create an attractive educational context for the discovery of the Pythagorean Theorem and for finding proofs of it. I will also explore connections with related ideas and propositions, such as Pappus' Theorem, similar right triangles, Latin and magic squares, and arithmetic modulo n. This section reproduces several examples given in *African Pythagoras: A Study in Culture and Mathematics Education* (Gerdes, 1992; 1994).

1. From woven buttons to the Theorem of Pythagoras

In the south of Mozambique, the following technique is used to fasten a cover to a basket. Two plaited strips with little lassos at their ends are permanently fixed to the cover, and two cubic buttons are fastened to opposite sides of the basket. In order to close the basket each of the lassos is pulled around the corresponding button (see Figure 2.1).

A button is woven out of two strips of a palm leaf. Figure 2.2 shows how to make the initial knot, and Figure 2.3 displays this knot seen from the front and seen from the back. Once this knot is made, we continue, on the back side, to weave 'one-over-one-under,' building up successive layers of the button, until it has become more or less cubic.

Figure 2.4a shows the front side of the resulting button. If we make a drawing of it, rectifying the slightly curved lines and turning visible the hidden lines, we obtain the design illustrated in Figure 2.4b. In its middle a second square appears.

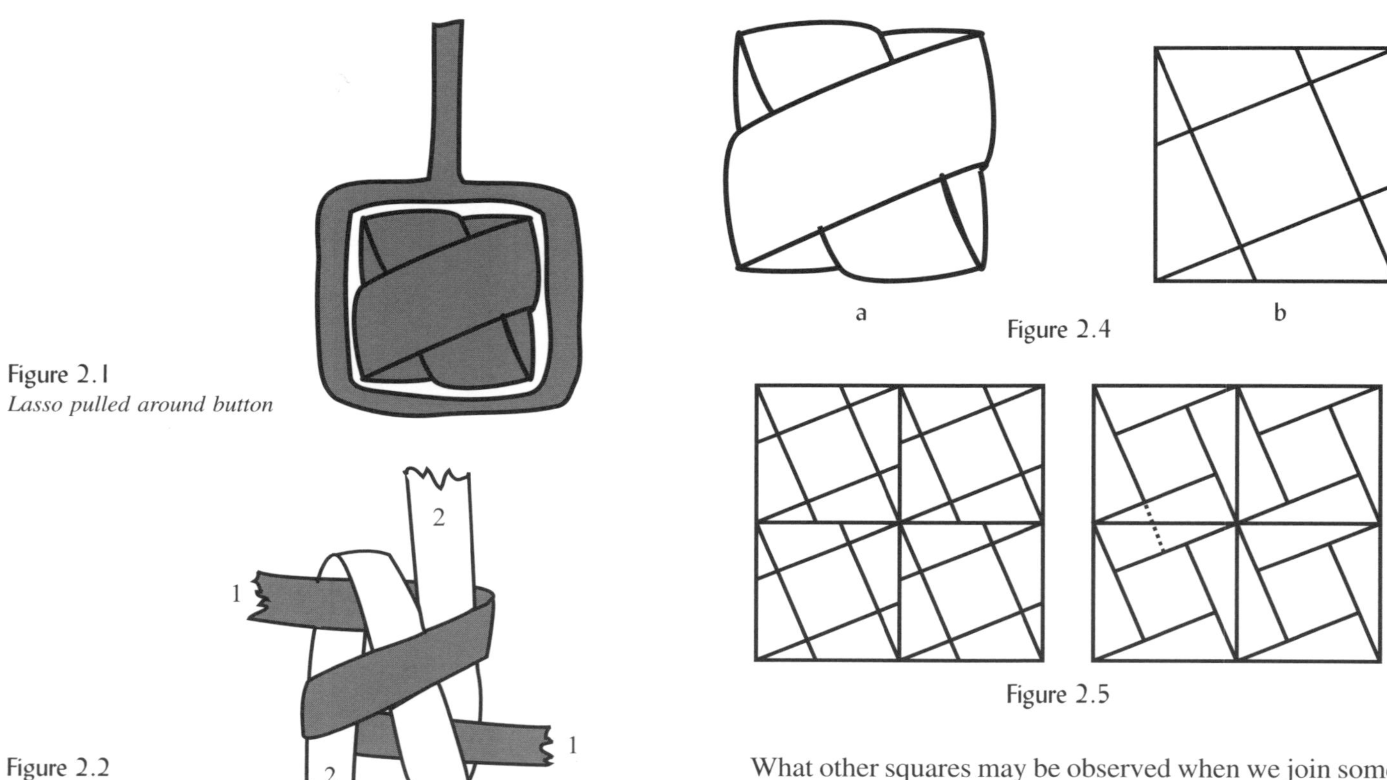

Figure 2.1
Lasso pulled around button

Figure 2.2
Making a knot

Figure 2.3

Figure 2.4

Figure 2.5

What other squares may be observed when we join some of these button inspired designs together? Do other figures with the same area as the face of the cubic button appear? If you like, you may extend some of the line segments or rub out some others (Figure 2.5). What may be observed? Equality in areas? (cf. Figure 2.6)

Students may arrive at the conclusion that

$$C = A + B$$

that is, we arrive at the Theorem of Pythagoras (Figure 2.7). Other students may join the square designs to form a pattern in such a way that the oblique lines are continued as shown in Figure 2.8. Comparing the areas, they may find (see Figure 2.9):

$$S + T = U\,.$$

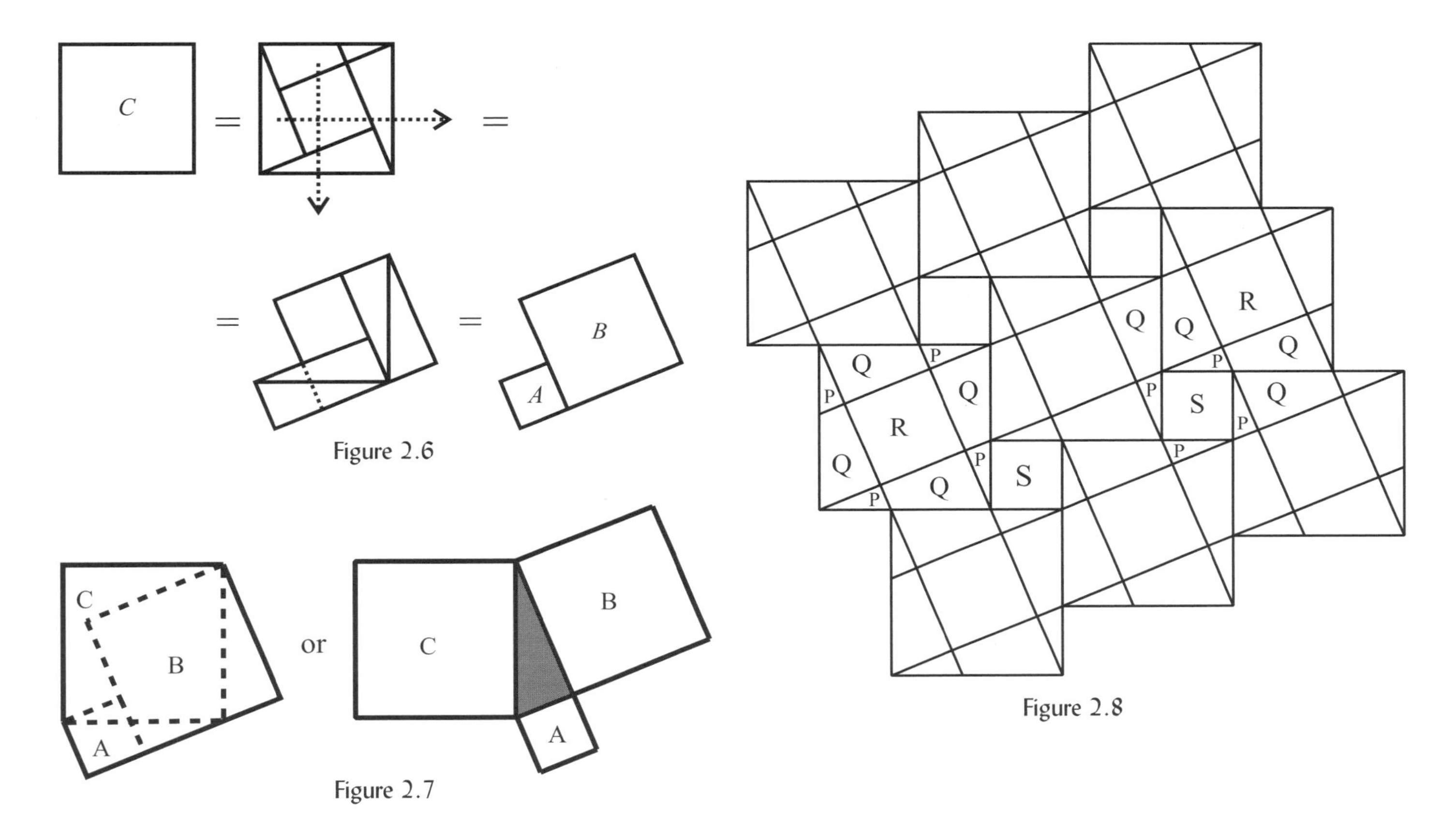

Figure 2.6

Figure 2.7

Figure 2.8

2. From decorative designs with fourfold symmetry to Pythagoras

Designs or pattern details that display a fourfold symmetry, that is a rotational symmetry of 90°, occur frequently in African decoration (see the examples in Figure 2.10). In this chapter, I will explore the idea, that four points which correspond to each other under a fourfold symmetry always constitute the vertices of a square.

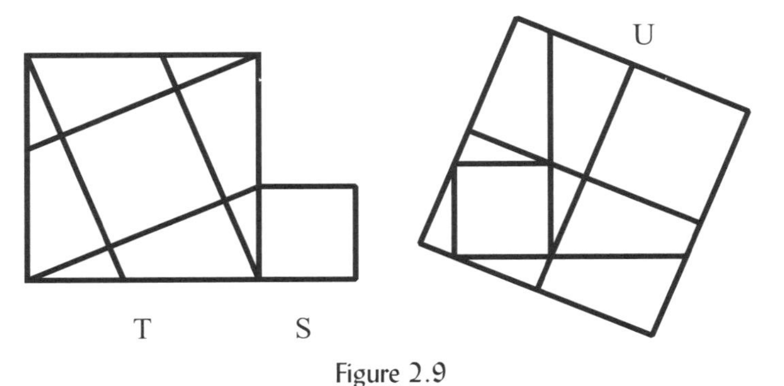

Figure 2.9

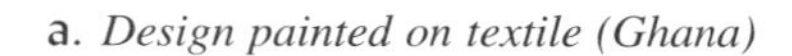

a. *Design painted on textile (Ghana)*

b. *Decorative design on a mat from the Lower-Congo area*

c. *Design on textile from the Upper-Senegal area*

d. *Examples of body stamps from the Ibo (Nigeria)*

e. *Ashanti bronze weight (Ghana)*

f. *Sudan*

g. *Weaving design, tattoo motif, decoration (Cameroon, Côte d'Ivoire, Congo/Zaire, Angola)*

h. *Madagascar*

Figure 2.10

Decorative motifs with fourfold symmetry *(continued on p. 59)*

i. *Design on a mat from Zanzibar Island (Tanzania)*

j. *Mosaic design (Madagascar; Lesotho)*

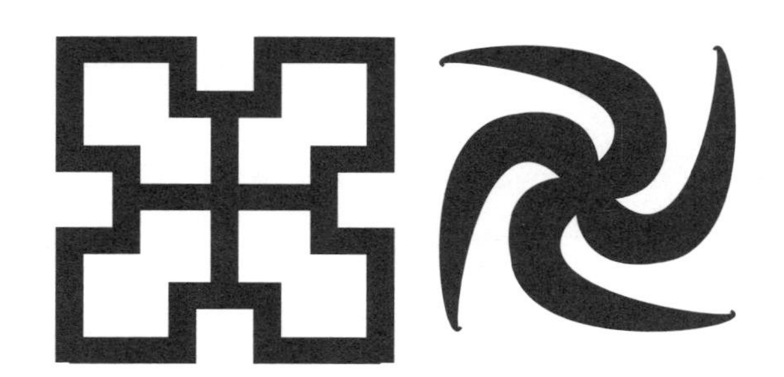

k. *Examples of adinkra stamps of the Ashanti (Ghana)*

l. *Motif engraved on wooden doors (Yoruba, Nigeria)*

m. *Detail of a design on Kuba plaited raffia clothes (Congo/Zaire)*

n. *Decorative motif from Angola*

o. *Letter of the Njoya alphabet (Cameroon)*

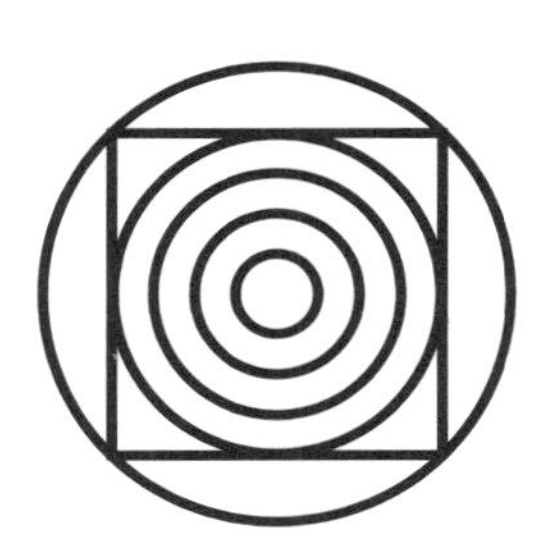

p. *Congo/Zaire*

q. *Angola*

r. *Benin*

A Mozambican decorative design as first example

We will take a traditional decorative design from Mozambique, shown in Figure 2.11, as our starting point. Any of the designs presented in Figure 2.10 could serve as an alternative starting point.

When we link four corresponding points on the circles (see the example in Figure 2.12) by straight line segments, we obtain a square (2.12b). The corresponding points of intersection of these segments with the circles are the vertices of a square too (2.12c), inscribed in the first square (2.12d). If, instead of linking the four neighboring points of intersection as in Figure 2.12d, we link the opposed intersection points as in Figure 2.12e, we obtain a cross that divides the first square into four congruent parts (2.12f). Both designs produced in this way (Figures 2.12d and f) lead easily to the Pythagorean proposition, as will be shown in the following. In this sense both designs may be called 'pythagorasable.' If, alternatively, we had started with linking the circle centers, we would have obtained the sequence in Figure 2.13, being the two designs (2.13d and f) similar to the ones in Figure 2.12.

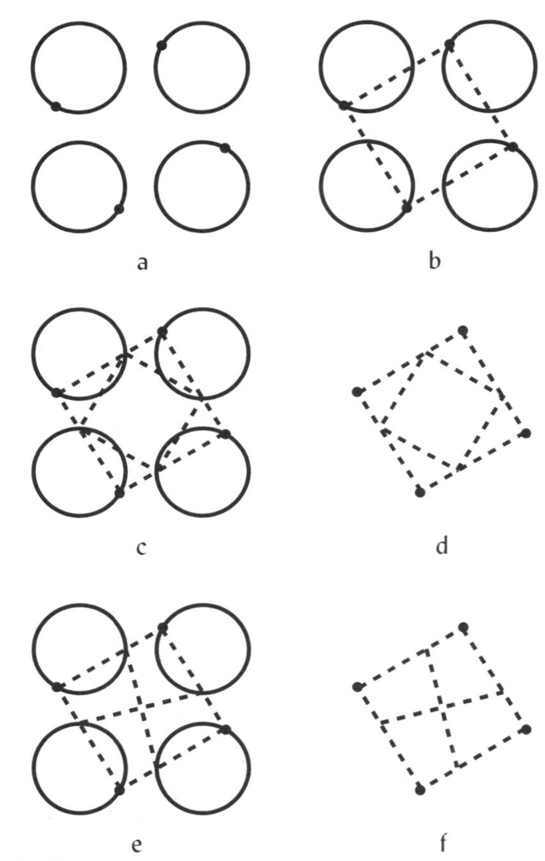

Figure 2.12

Figure 2.11
Headrest decoration (Mozambique)

Let a, b, and c denote the sides of the congruent right triangles in Figure 2.14 (Cf. Figures 2.12d and 2.13d). Students may be asked to compute the areas of the squares and triangles and analyze their relationships. As the side of the large square measures $a+b$, its area is $(a+b)^2$. The area of this square is equal to the sum of the areas of the four right triangles ($4ab/2$) plus the area of the inscribed square (c^2).

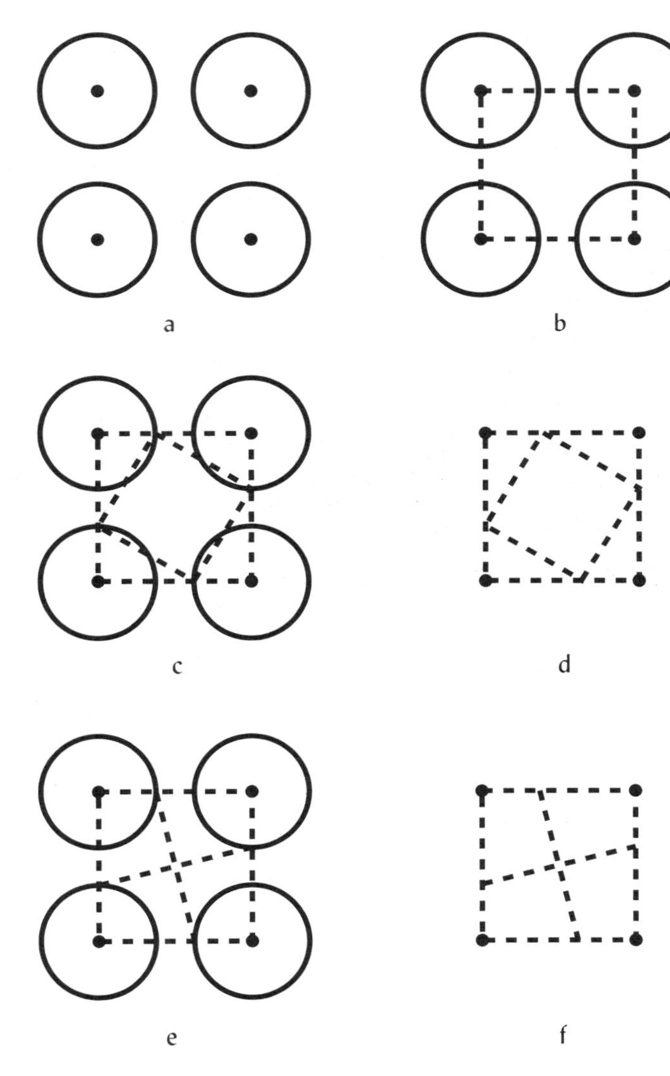

Figure 2.13

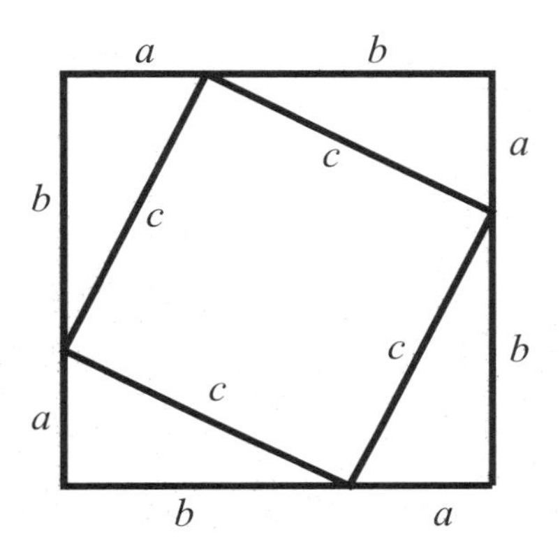

Figure 2.14

Therefore

$$(a+b)^2 = 2ab + c^2.$$

Taking into account the equality

$$(a+b)^2 = 2ab + a^2 + b^2,$$

the students may find

$$a^2 + b^2 = c^2.$$

In effect, they arrive at the Pythagorean proposition.

Geometrically we may arrive at this result in other ways too. For instance, rotate the upper-left right triangle clockwise 90° about it lowest vertex. Similarly, rotate the upper-right right triangle counterclockwise 90° about it lowest vertex. See Figure 2.15. We observe that the large square may be considered as composed of one square with side a, one square with side b and four right triangles with sides a, b and c. Initially the same large square was composed of a square of side c circumscribed by four right triangles with sides a, b and c. Therefore, the area of the square with side c is equal to the sum of the areas of the squares with sides a and b.

Let us now return to Figure 2.12f (or 2.13f). The cross (see Figure 2.16) divides the initial square into four quad-

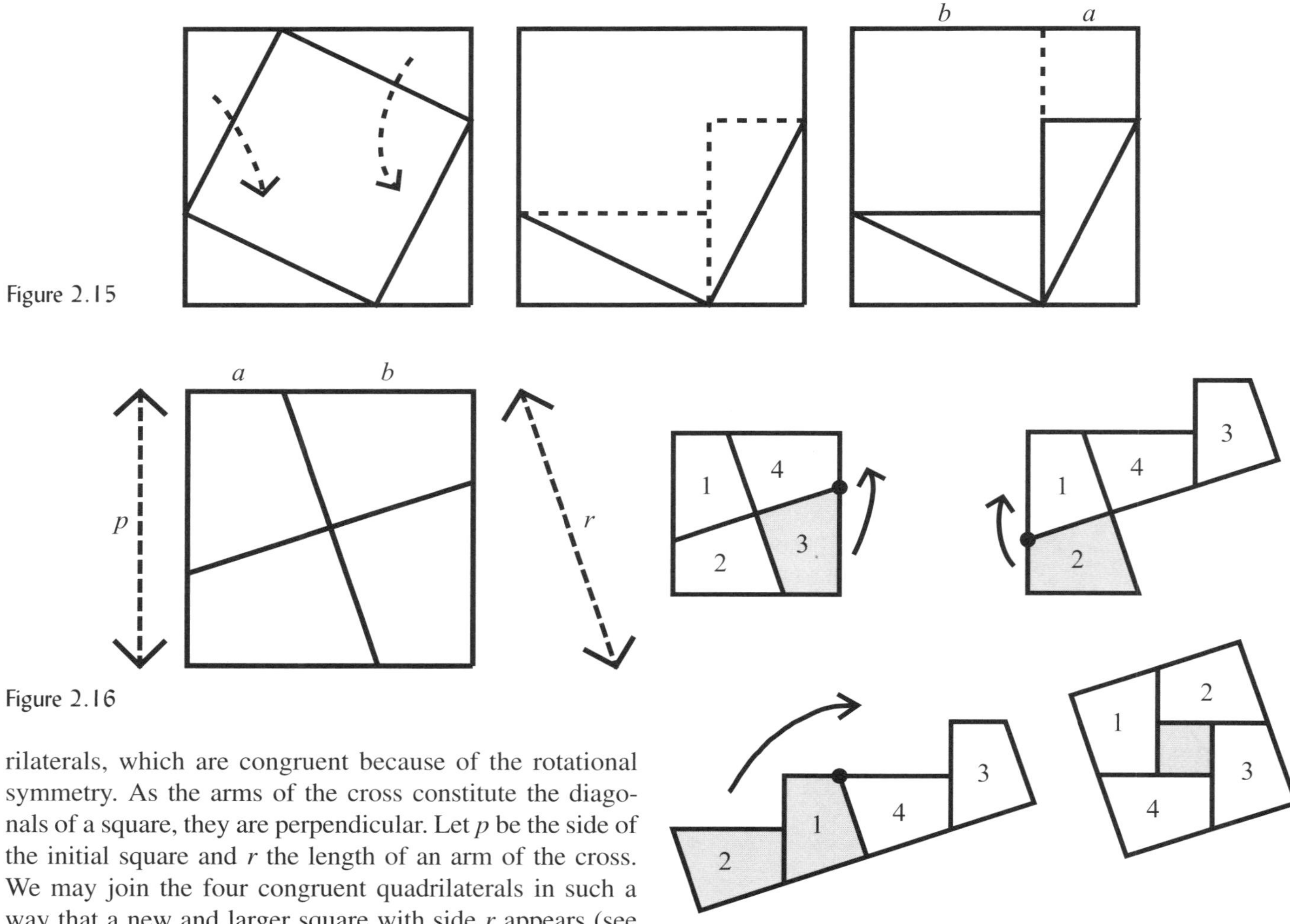

Figure 2.15

Figure 2.16

Figure 2.17

rilaterals, which are congruent because of the rotational symmetry. As the arms of the cross constitute the diagonals of a square, they are perpendicular. Let p be the side of the initial square and r the length of an arm of the cross. We may join the four congruent quadrilaterals in such a way that a new and larger square with side r appears (see Figure 2.17). At its center a square hole appears. Let q be its side. From the construction of the new square it follows immediately that

$$r^2 = p^2 + q^2.$$

As $q = b - a$, we see that p, q and r constitute the sides of a right triangle (see Figure 2.18).

This reasoning may be used to arrive at a 'dissection' proof for the Pythagorean proposition. This proof was found — for the first time? — by Perigal as late as 1873 (Cf. Loomis, p.104).

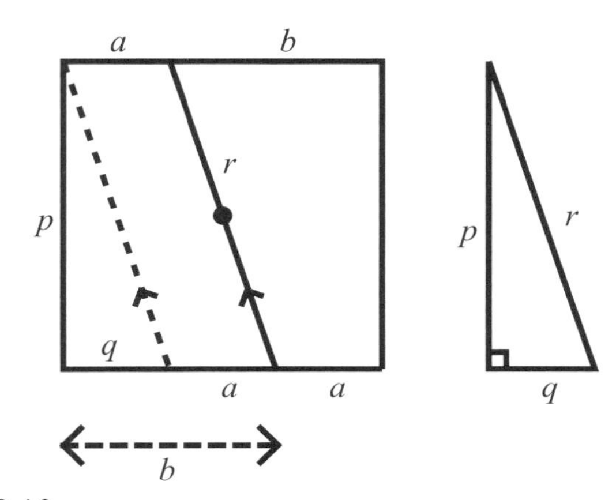

Figure 2.18

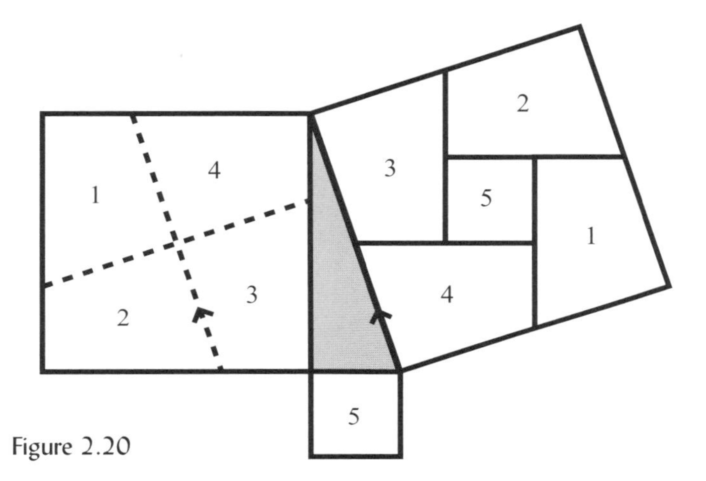

Figure 2.20

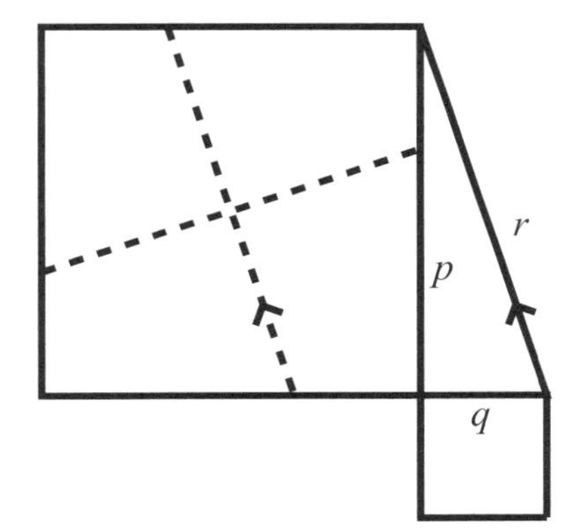

Figure 2.19

Consider an arbitrary right triangle with sides p, q and r. Let p be larger than q. Draw squares on the legs. Through the center of the square of side p, construct a cross with one of its arms parallel to the hypotenuse of the right triangle (see Figure 2.19).

As seen before, the four pieces into which the p-square has been dissected may be joined together with the q-square to obtain the r-square (see Figure 2.20). Therefore:

$$r^2 = p^2 + q^2.$$

Chokwe sand drawing designs (cf. Section 4)

Figure 2.21a shows some Chokwe sand drawings with fourfold symmetry. We can easily transform each drawing into one of the aforementioned 'Pythagorasable' designs (see Figure 2.21b).

A beautiful Kuba tiling

The Kuba people inhabit the central part of the Congo basin (in Congo/Zaire), living in the savannah south of the dense equatorial forest. Their metal products, such as weapons and jewelry, are famous. The villages themselves had specialized in certain types of craft, like the production of ornamented wooden boxes and cups, velvet carpets, copper pipes, raffia cloths, etc. (cf. Section 1).

Figure 2.22 illustrates an interesting tessellation of the plane composed of 'hooks' and squares. This design appears traditionally on woven mats and on embroidered raf-

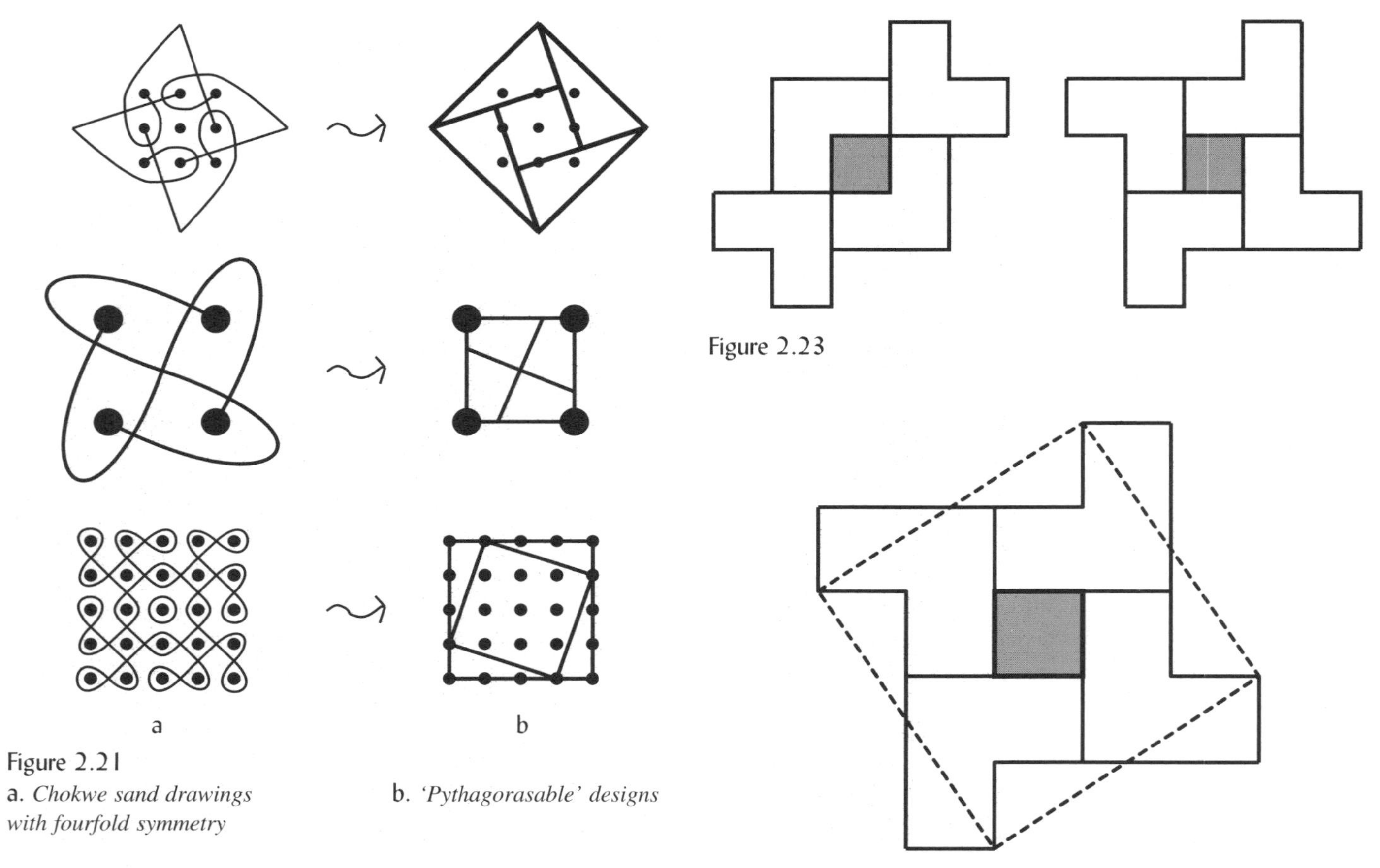

Figure 2.21
a. *Chokwe sand drawings with fourfold symmetry*
b. *'Pythagorasable' designs*

Figure 2.23

Figure 2.24

Figure 2.22
Kuba Tiling

fia textiles, and is called 'Mikope Ngoma', i.e., the drums of King Mikope.

Each square is embraced by four hooks as shown in Figure 2.23. The second design displays a fourfold symmetry. When we link the corresponding vertices as indicated in Figure 2.24, we obtain a square that has the same area as the design.

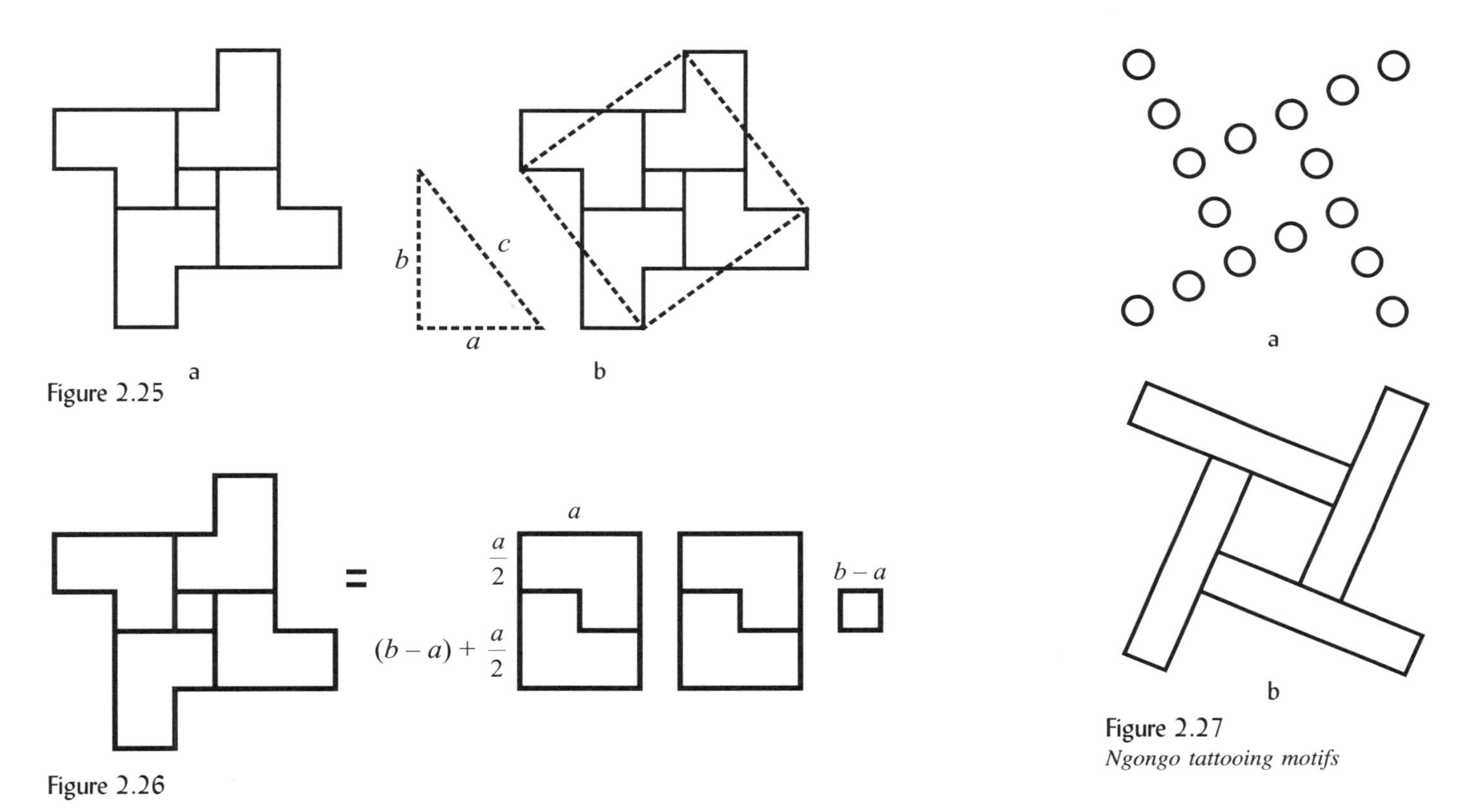

Figure 2.25

Figure 2.26

Figure 2.27
Ngongo tattooing motifs

We may generalize this design in such a way (see Figure 2.25a) that the new square still has the same area as the design (see Figure 2.25b) and that the corresponding right triangle is arbitrary with sides a, b and c. In this way, we find (see Figure 2.26) that the area of the large square (c^2) is equal to the area of the small square plus two times the area of the rectangle formed by joining the hooks. As the small square has side $b - a$, and the rectangles have sides a and $(a/2) + [(b - a) + (a/2)]$, or b, we may conclude that

$$c^2 = 2ab + (b - a)^2 = a^2 + b^2,$$

i.e., we prove the Theorem of Pythagoras.

Exploring the "elephants' defense" designs of the Kuba (Congo/Zaire)

Figure 2.27 shows two tattooing motifs formerly used by the Ngongo, one of the ethnic groups that belonged to the Kuba kingdom. When we draw lines between the ends of these tattooing motifs, we obtain a design similar to the Kuba engravings illustrated in Figure 2.28. The Bushongo—the dominant people in the old Kuba kingdom—call the engraving design in Figure 2.28a 'mwoong,' i.e. "elephants' defense", and the motif in Figure 2.28b

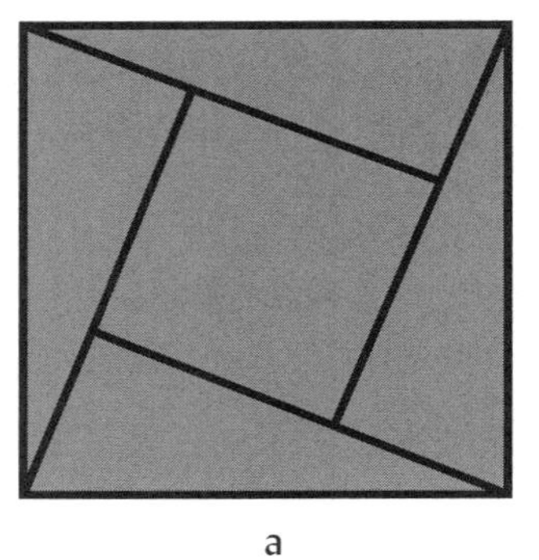

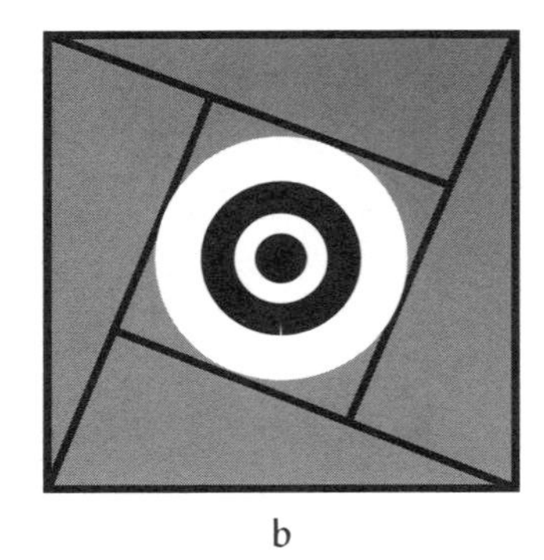

a b

Figure 2.28
"Elephants' defense" motif

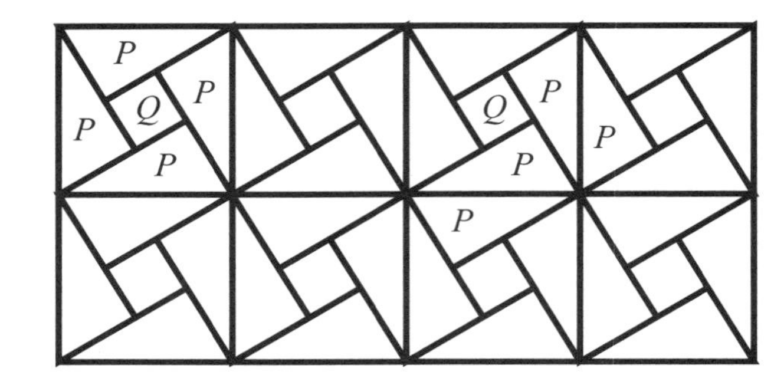

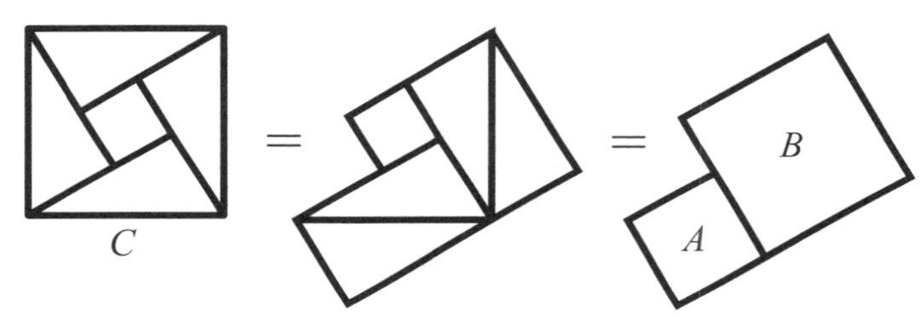

Figure 2.29

'ikwaakl'imwoong', i.e. "deformed elephants' defense". The design in Figure 2.28a is well known from the history of mathematics in Asia (China, India). When we place several copies of the design together, we may discover, 'reinvent' and even prove, easily and geometrically, the Pythagorean Theorem, as Figure 2.29 illustrates.

On the other hand, knowing the identity $(b - a)^2 = b^2 - 2ba + a^2$, and the formulas for the determination of the areas of squares and right triangles, using the following algebraic-geometric reasoning, some students may arrive at the same conclusion. The area of the small central square is equal to $(b - a)^2$ and the combined areas of the four neighboring right triangles are $2ab$. Therefore

$$c^2 = (b - a)^2 + 2ab = a^2 + b^2. \tag{1}$$

An ornamental variant and a generalization of the Pythagorean Theorem

Figure 2.30a shows a Kuba variant of the elephants' defense pattern. The big rectangle is composed of two pairs of similar triangles (see Figure 2.30b) and a little rectangle

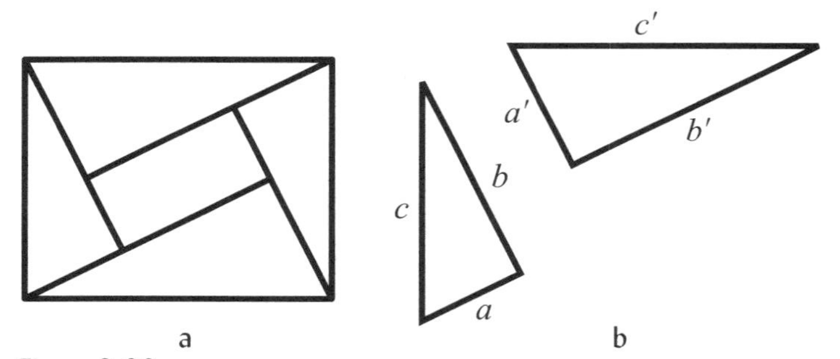

Figure 2.30
a. *Variant of the "elephants' defense" motif*

at the center. When we place several copies of this variant together (see Figure 2.31), we may discover/invent and prove the following generalization of the Pythagorean proposition:

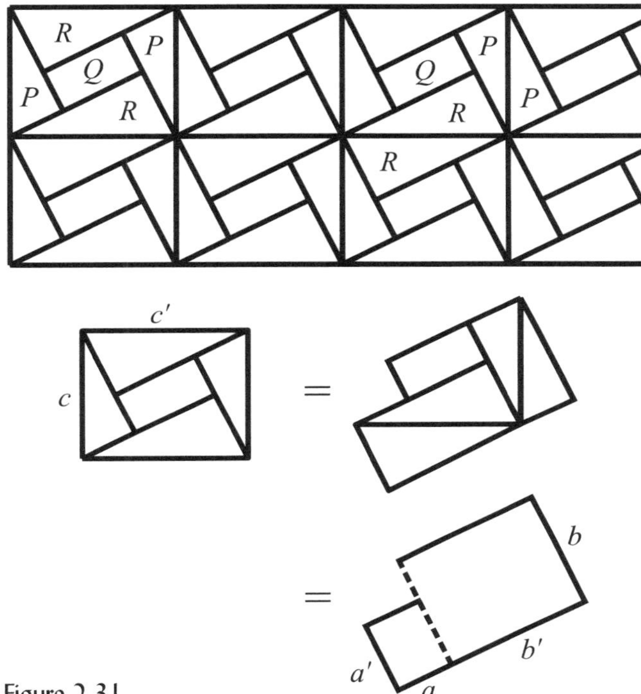

Figure 2.31

(2) $$cc' = aa' + bb'$$

On the other hand, students may be stimulated to arrive at the same conclusion by algebraic-geometrical reasoning. The area of the central rectangle is equal to $(b' - a)(b - a')$ and the areas of the four neighboring right triangles are together $ab + a'b'$. Therefore

$$c\,c' = (b' - a)(b - a') + a\,b + a'b' = a\,a' + b\,b'.$$

A combination of the Pythagorean Proposition with its generalization

What result may be obtained when we combine the Pythagorean theorem with its generalization?

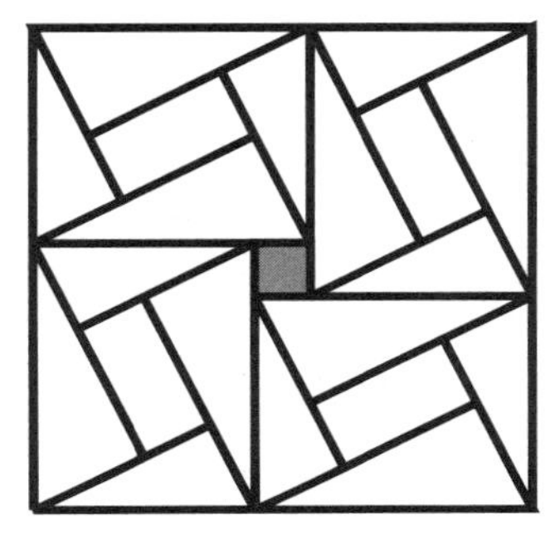

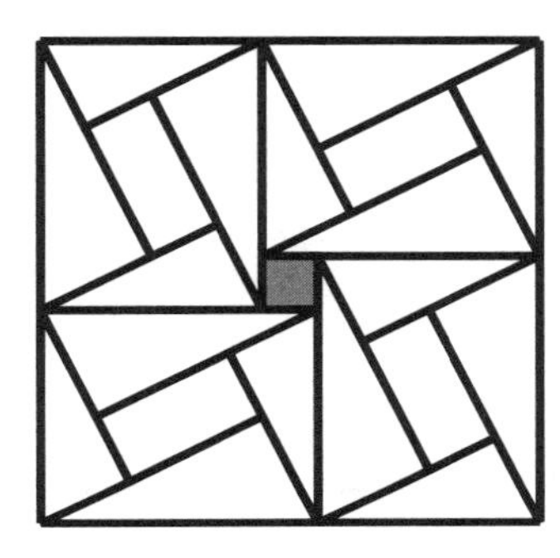

Figure 2.32

From $c^2 = a^2 + b^2$, $(c')^2 = (a')^2 + (b')^2$, and $cc' = aa' + bb'$, it follows algebraically easily that

(3) $$ab' = ba',$$

which implies that

(4) $$a : b = a' : b'.$$

In other words, the ratios of the sides (taken in the same order) of similar right triangles are equal. Does there exist a (purely) geometrical alternative to deduce this result?

Let us return to the rectangular Kuba variant of the elephants' defense pattern (Figure 2.30a). How can we join some of these ornamental rectangles in order to obtain a square?

Figure 2.32 shows two possibilities of obtaining a square of side $c + c'$. In both cases, a square 'hole' appears in the center, whose sides measure $c' - c$. When we extend the sides that end at the vertices of the 'hole' until they encounter other sides, we obtain Figures 2.33a and b. In both cases, four right triangles appear around the square 'hole' (see Figure 2.34) and, together with the square 'hole', they form a new square of side $(a' + b') - (a + b)$.

In both cases, around this new central square appear four rectangles which constitute patterns that are similar (see

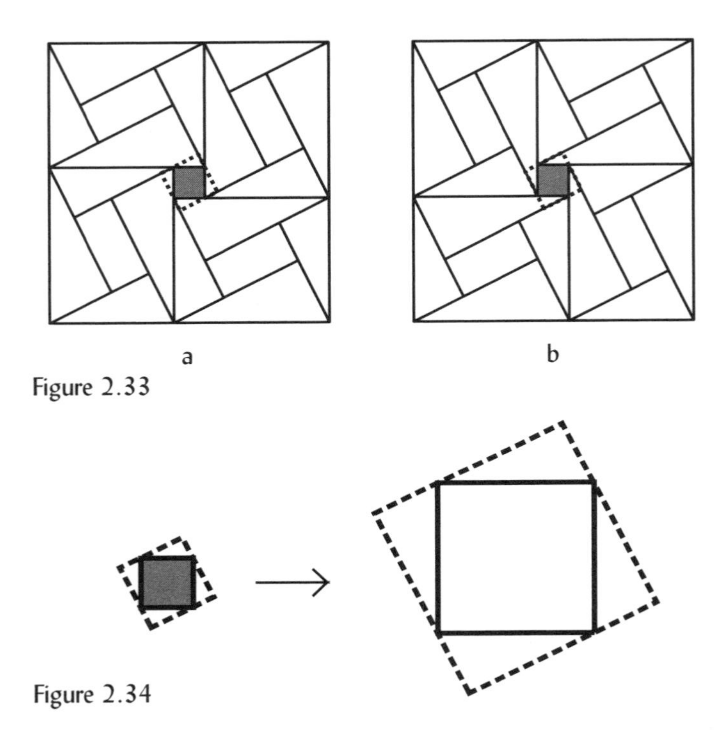

Figure 2.33

Figure 2.34

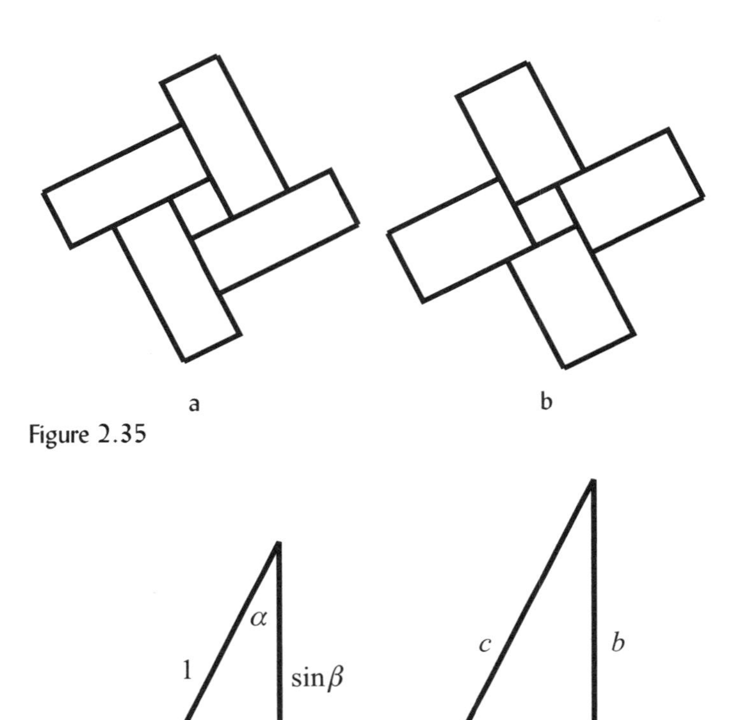

Figure 2.35

Figure 2.36

Figure 2.35) to the Kuba tattooing displayed in Figure 2.27b. As the large squares are congruent and the small squares are also congruent, we arrive at the conclusion that both designs themselves (see Figure 2.35) are also equal in area. Both are composed of four rectangles. Therefore, the area (ab') of one of these rectangles of the first design (Figure 2.35a) is equal to the area ($a'b$) of one of the rectangles of the second design (Figure 2.35b), that is

$$ab' = a'b.$$

In other words,

$$a : b = a' : b'.$$

Hence we have arrived geometrically at the conclusion that the ratios of the sides (taken in the same order) of similar right triangles are equal.

It turns out to be easy to prove the Fundamental Theorem of Similar Triangles on the basis of the foregoing result. Within the same context, we may also introduce the concepts of tangent, sine and cosine of an acute angle. Application of the generalization $cc' = aa' + bb'$ in the case of the triangles in Figure 2.36 leads to the trigonometric formula

$$c = a\sin\alpha + b\sin\beta.$$

3. From a widespread decorative motif to the discovery of Pythagoras' and Pappus' theorems and an infinity of proofs (cf. Gerdes, 1988)

Figure 2.37 displays variations of a decorative motif with a long tradition all over Africa and the world. It was already known in ancient Egypt. It appears on a painted basket in the tomb of Rekhmire (Thebes, 18th dynasty, c. 1475–1420 BC) (See Figure 2.38). Some other examples from distinct parts from Africa are shown in Figures 1.17, 1.18, 1.61, 1.64, 1.65). Among the Chokwe (Angola) the design is known as *manda a mbaci*, that is tortoise-shield (Cf. Bastin, p. 114, 116). The Ovimbunda (Angola) call it *olombungulu* (star), *ononguinguinini* (ants), or *alende* (clouds) (cf. Hauenstein, p. 39, 50, 54). In particular, it is frequently seen on baskets from all over the continent.

Discovering the Theorem of Pythagoras

Looking at the number of unit squares on each row of the 'star' or 'toothed square' in Figure 2.39, it is easy to see that the area of a 'star' is equal to the sum of areas of the 4×4 shaded square and 3×3 unshaded square.

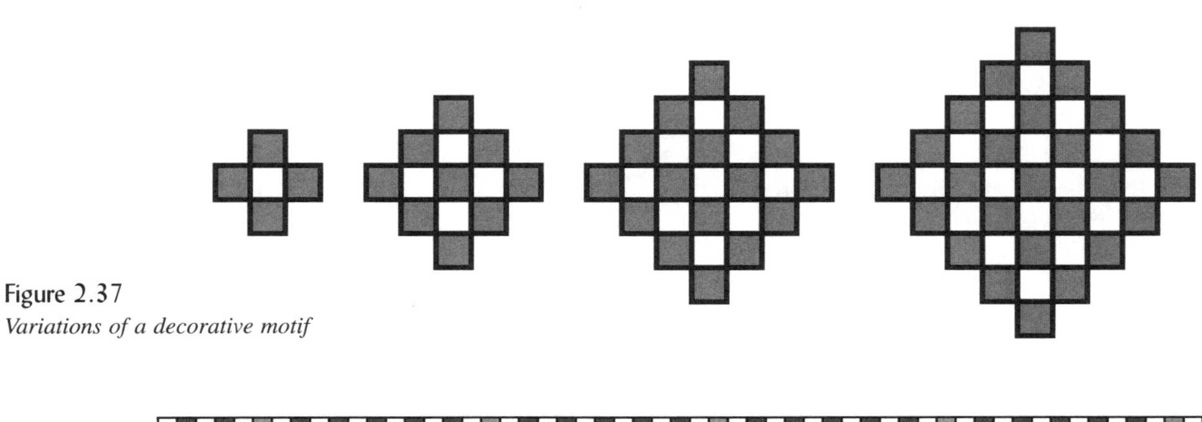

Figure 2.37
Variations of a decorative motif

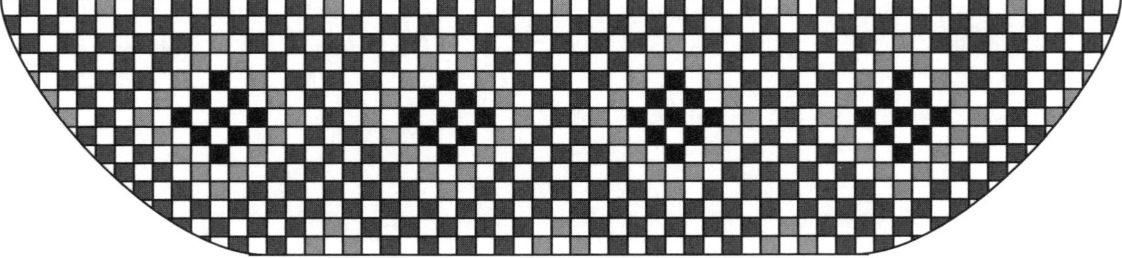

Figure 2.38
Ancient Egyptian painted representation of a basket

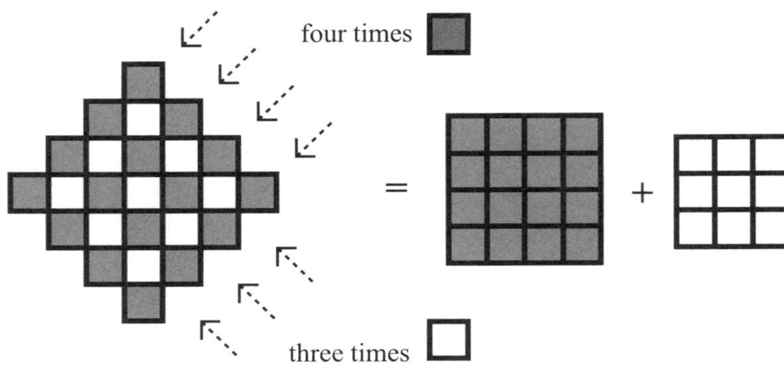

Figure 2.39

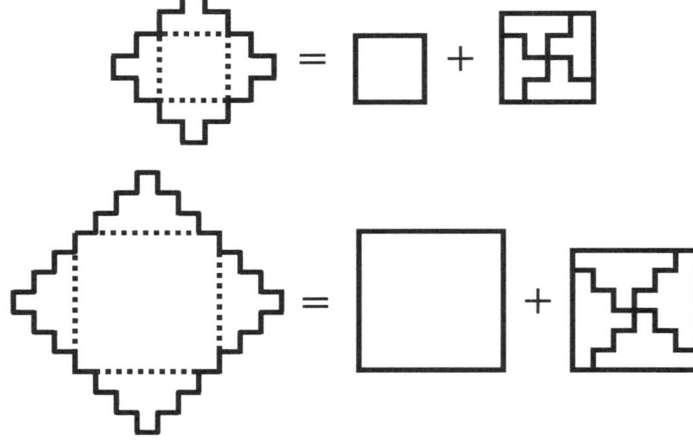

Figure 2.40

Students can also be led in many other ways to draw the conclusion that a 'toothed' square is equal in area to the sum of two real squares. For example, the teacher may ask them to transform a 'toothed' square made of loose tiles of two colors into two unicolored similar figures. Or, we may ask them to cut off the biggest possible square from a 'toothed' square made of paper or cardboard and to analyze which figures can be laid down with the other pieces (see Figure 2.40).

A 'toothed square,' especially one with many teeth, looks almost like a real square. So naturally the following question arises: is it possible to transform a toothed square into a real square of the same area? By experimentation (see Figure 2.41), the students may be led to the conclusion that this is indeed possible.

In Figure 2.39, it was shown that the area of a toothed square (T) is equal to the sum of the areas of two smaller squares (A and B):

$$T = A + B.$$

From Figure 2.41, it may be concluded that the area of a toothed square (T) is equal to the area of a 'real' square

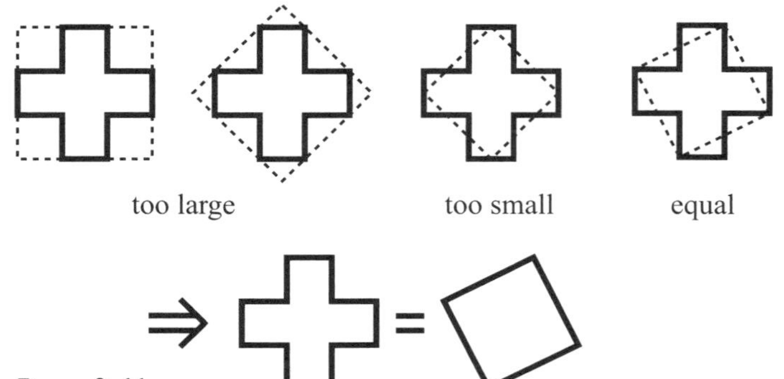

Figure 2.41

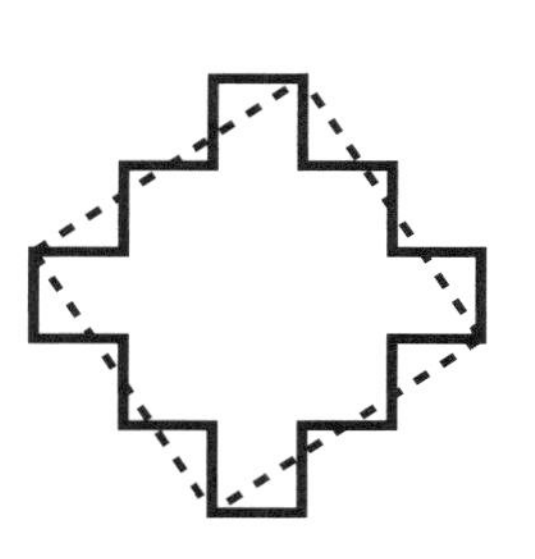
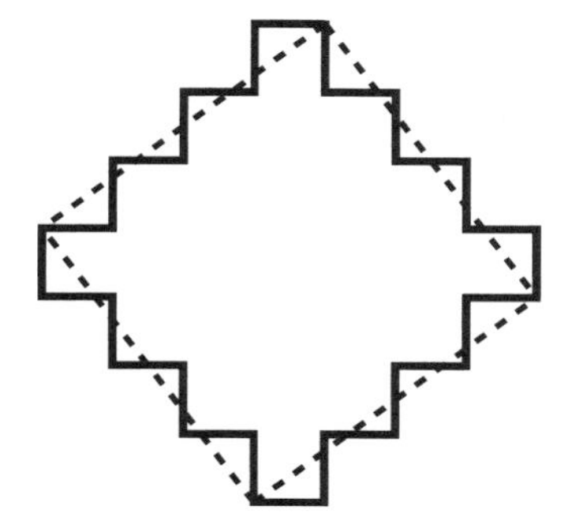

Figure 2.42

(C). Since $C = T$, we may conclude that

$$A + B = C.$$

Is the same transformation of a 'toothed square' in a 'real' square also possible when we increase the number of teeth on its side? (see Figure 2.42).

When we now draw the oblique square (area C) (for instance, on square grid paper) together with the squares with areas A and B, in such a way that the squares become 'neighbors,' situations like the one in Figure 2.43 emerge. This figure illustrates the Pythagorean Proposition for the special case of the (3, 4, 5) right triangle.

On the basis of these experiences, students may be led to conjecture the Pythagorean Theorem in general. In other words, toothed squares may assume a heuristic value for the discovery of this important proposition. Does the same discovery process also suggest any (new) *demonstrations* for the Pythagorean Theorem? What happens when we reverse the process? When we begin with two arbitrary squares and use them to generate a toothed square?

A first proof

Let A' and B' be two arbitrary squares. We look at Figure 2.39 for inspiration: dissect A' into 9 little congruent squares, and B' into 16 congruent squares, and join the 25 pieces together as in Figure 2.44. The toothed square obtained T' is equal in area (T) to the sum of the real squares A' and B':

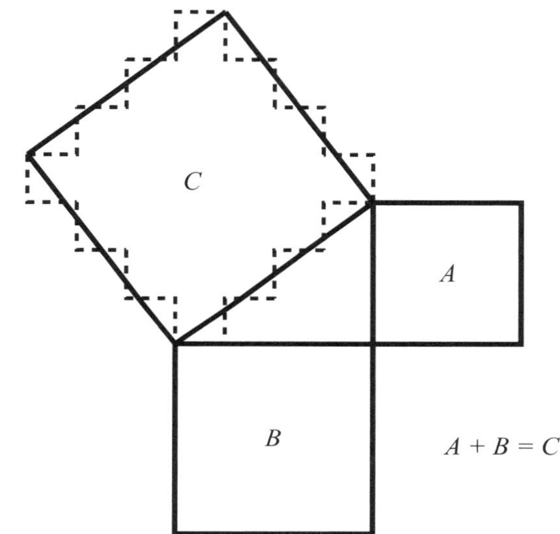

Figure 2.43

$$T = A + B.$$

As, once again, the toothed square is easily transformed into a real square C' of the same area (see Figure 2.45), we arrive at

$$A + B = C,$$

i.e., the Pythagorean Theorem in all its generality.

An infinity of proofs

Instead of dissecting A' and B' into 9 and 16 subsquares, it is possible to dissect them into n^2 and $(n + 1)^2$ congruent subsquares for each (positive) integer value of n. Figure

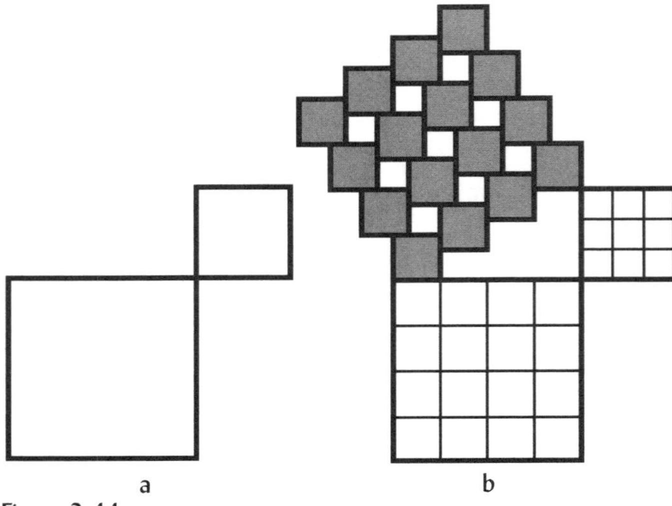

Figure 2.44

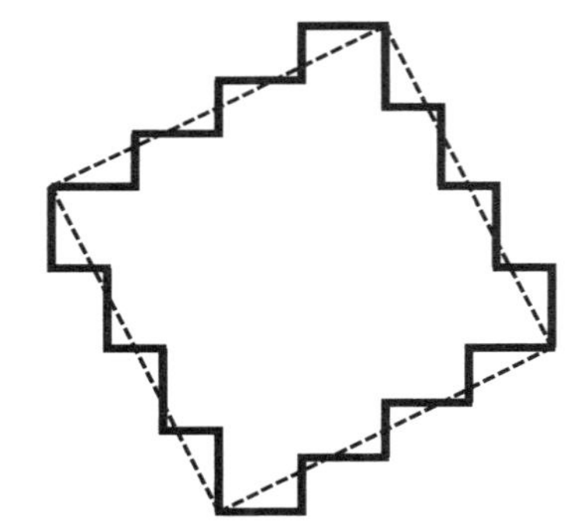
Figure 2.45

2.46 illustrates the case n = 14. To each value of n there corresponds a proof of the Pythagorean Proposition. In other words, an infinite number of demonstrations of the famous theorem exist.

For relatively high values of n, the truth of the Pythagorean Proposition is almost immediately visible. When we take the limit $n \to \infty$, we arrive at one more demonstration of the theorem. For $n = 1$, we obtain a very short, easily understandable proof, as illustrates Figure 2.47.

Pappus' theorem

Analogously, Pappus' generalization of the Pythagorean Proposition for parallelograms can be proved in an infinite number of ways (Figure 2.48 illustrates the case $n = 2$).

Loomis' well known study *The Pythagorean Proposition* gives "... 370 different proofs, each calling for its specific figure" (Loomis, p. 269) and its author invites his audience to "Read and take your choice; or better, find a new, a different proof..." (p. 13). My reflections in this chapter on a widespread decorative motif led not only to an alternative, active way of introducing the Pythagorean Proposition in the classroom, but also to the generation of an infinite number of proofs of the same theorem.

Other educational explorations of 'toothed' squares (cf. Gerdes, 1995, 1997, ch. 11)

The 'toothed squares' may also be explored educationally in other ways, as the following examples illustrate.

Example 1. Let us consider a 'toothed' square and count the number of brown unit squares in each of the horizontal rows (see Figure 2.49). We already know that the total number of brown unit squares is a square number, in this case 4^2. Comparing the two enumerations, we find:

$$1+2+3+4+3+2+1=4^2,$$

or, taking into consideration the dotted divisionary line:

$$(1+2+3+4)+(3+2+1)=4^2.$$

It follows that

$$(1+2+3)+(3+2+1)=4^2-4,$$

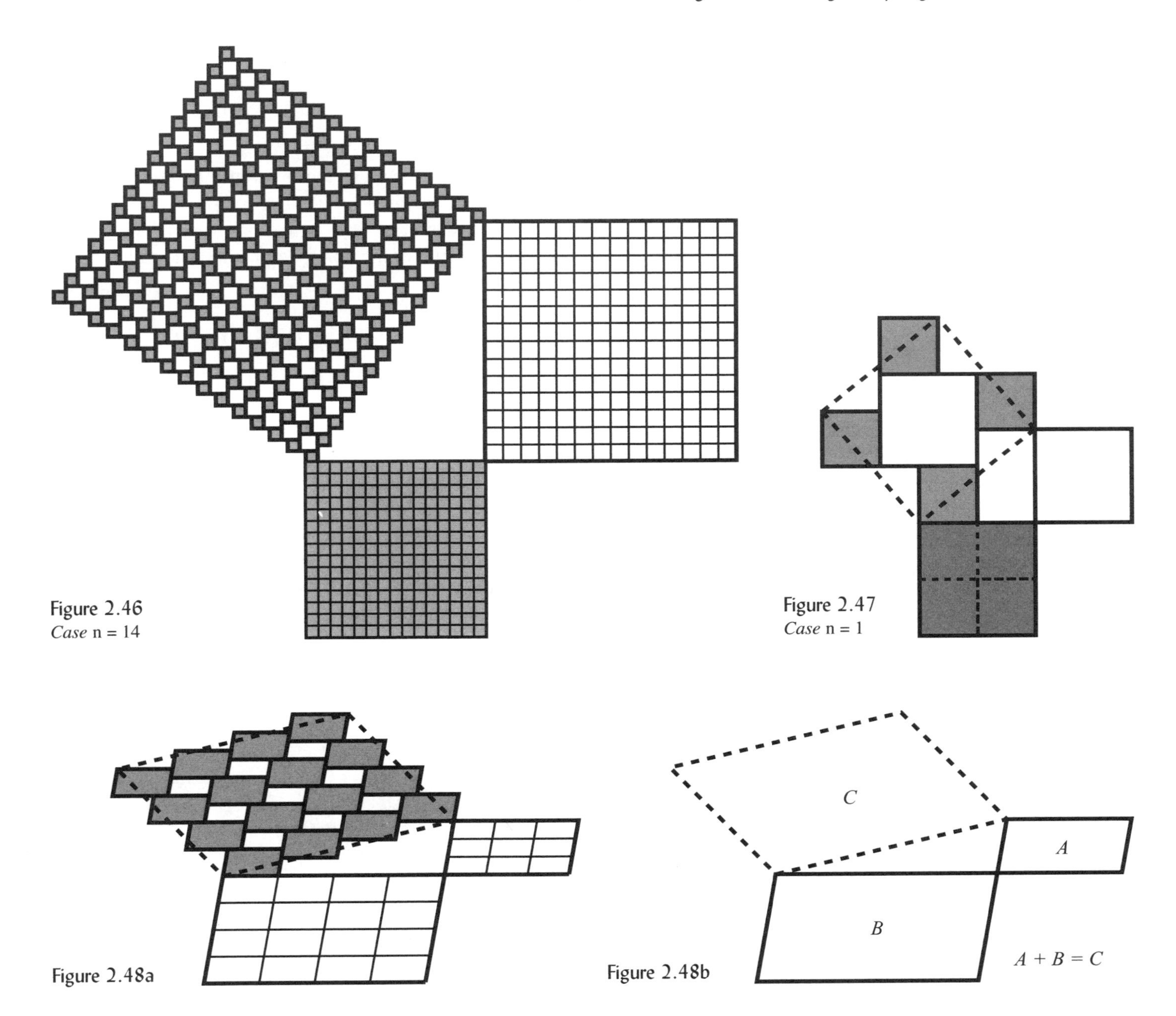

Figure 2.46
Case n = 14

Figure 2.47
Case n = 1

Figure 2.48a

Figure 2.48b

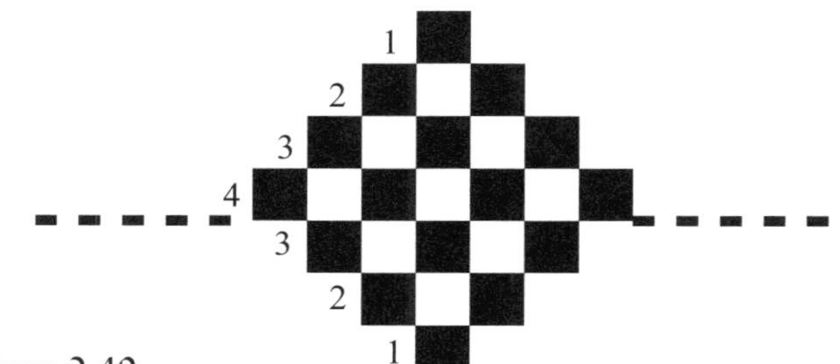

Figure 2.49

Figure 2.50

and

$$1+2+3=\frac{4^2-4}{2}.$$

Experimentation with other 'toothed' squares, varying the number of teeth on the side, may lead to the extrapolation

$$1+2+3+\cdots+(n-1)=\frac{n^2-n}{2},$$

where n denotes a natural number bigger than 1.

Example 2. Let us observe the upper part of Figure 2.49 and count both the brown and the white unit squares row by row. For the brown unit squares, we find $1 + 2 + 3 + 4$. For the white ones (see the right side of Figure 2.50), we find $1 + 2 + 3$. The total sum of unit squares of the "toothed triangle" is thus equal to $(1 + 2 + 3 + 4) + (1 + 2 + 3)$, or $(1+2+3+4) + (3+2+1)$, which was the sum of all brown unit squares of the original 'toothed' square, that is 4^2.

By counting all the unit squares of the 'toothed' triangle, row by row, irrespective of their color, we find $1+3+5+7$. Comparing the two enumerations we may conclude

$$1+3+5+7=4^2.$$

In other words, the sum of the first four odd numbers is 4^2. Experimentation with other "toothed triangles" and comparison of the results may lead to the discovery of the general result that the sum of the first n odd numbers is equal to n^2.

4. From mat weaving patterns to Pythagoras, and Latin and magic squares

Let us consider the following mat weaving problem, supposing we weave with layers of brown (vertical) strands and white (horizontal) strands (Figure 2.51). How can we weave them over-and-under to obtain an overall white mat covered by a square grid of brown dots (Figure 2.52)?

Figure 2.53a shows a solution found by Chokwe mat makers from northeastern Angola: each brown strand goes over one white strand and then under four white strands; two successive brown (from the left to the right) differ two in phase (see Figure 2.53b).

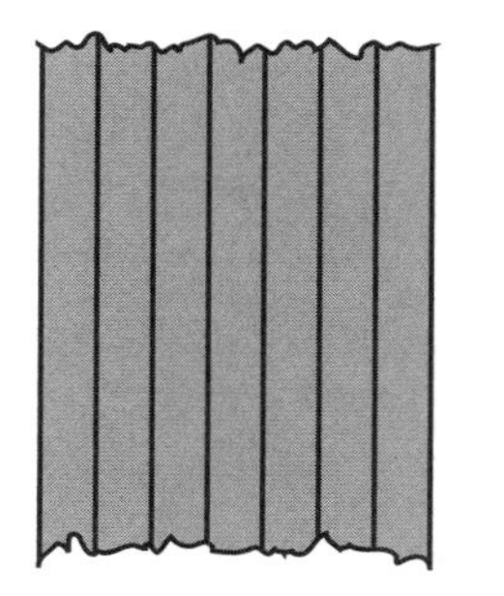

Figure 2.51

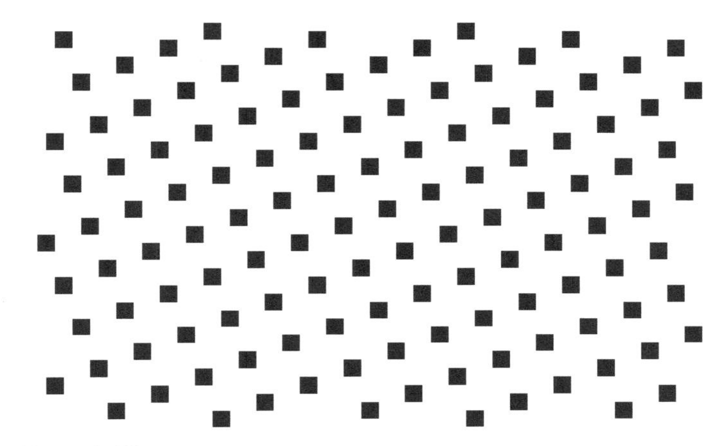

Figure 2.52

Weaving patterns are usually represented by brown-and-white diagrams on squared paper. Figure 2.54 illustrates a diagram corresponding to the Chokwe fabric. Each white square corresponds to a place where the horizontal (white) strand passes over a vertical (brown) strand, and, inversely, the square is colored brown if the vertical (brown) strand passes over the horizontal (white) strand.

From a brown dot on this diagram to the nearest dot on the next column on the right side, we move one unit to the right and two units down. We may call this the (1, –2)-solution of the problem posed at the beginning. In this way, the upper part of the detail of a Kuba mat (Congo/Zaire), shown in Figure 2.55, displays the (1, 2)-solution.

Figure 2.53a
'Toothed' square detail of a Chokwe mat

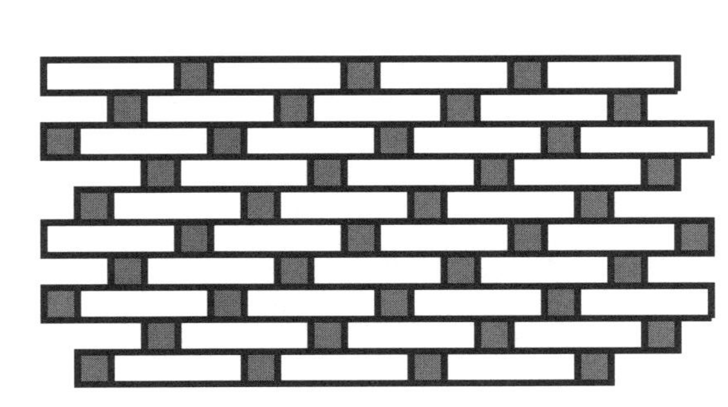

Figure 2.53b

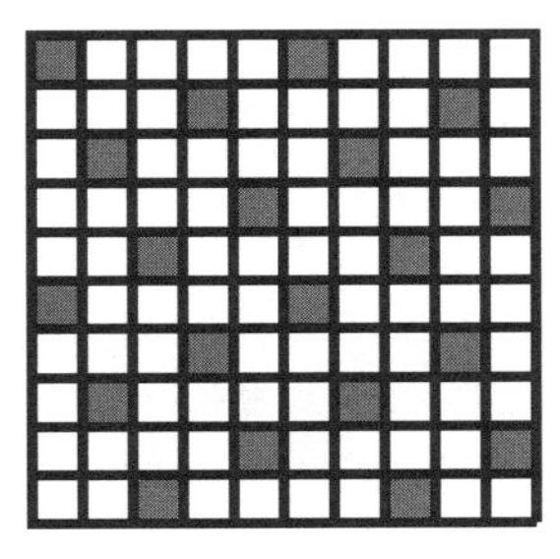

Figure 2.54

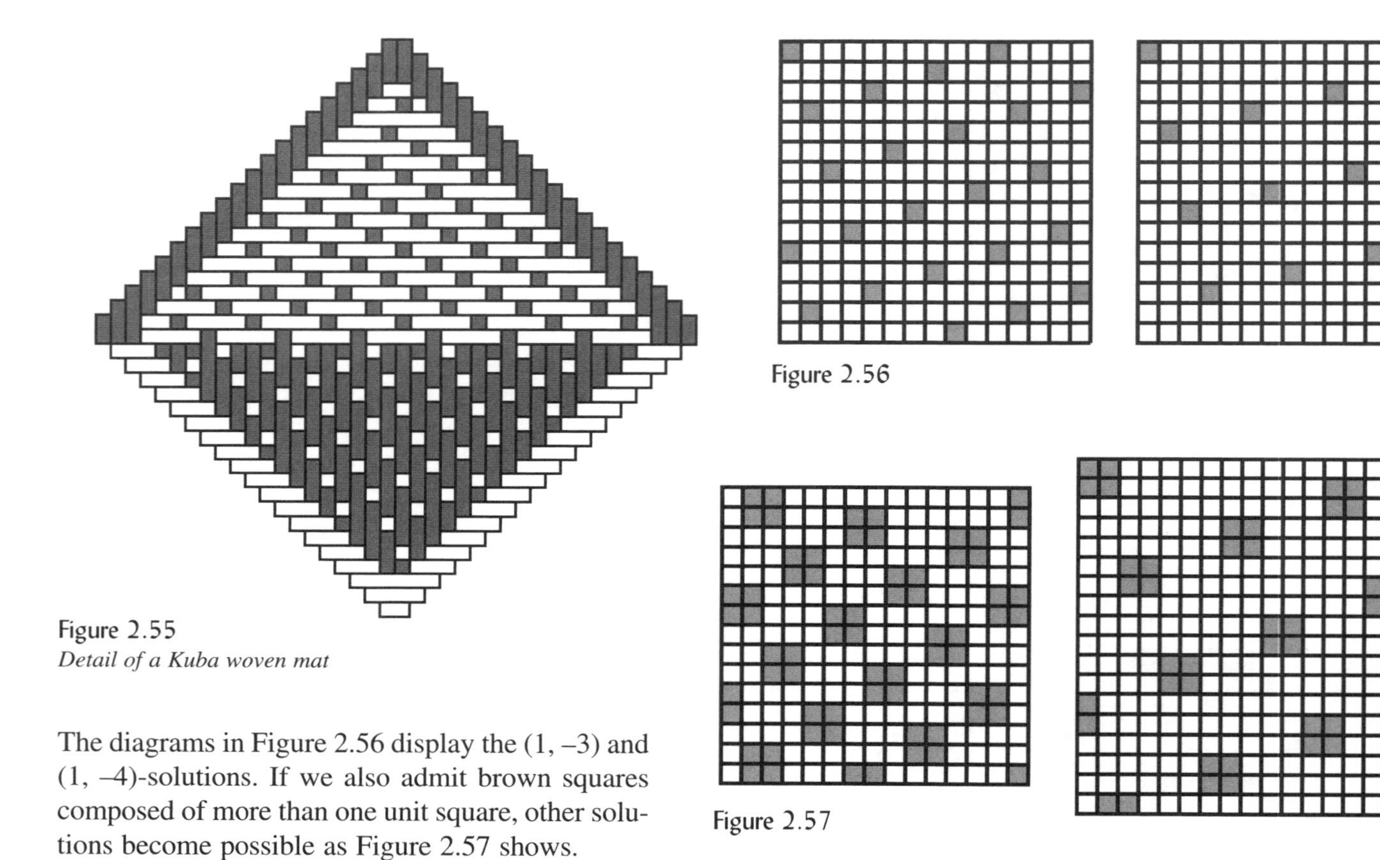

Figure 2.55
Detail of a Kuba woven mat

Figure 2.56

Figure 2.57

The diagrams in Figure 2.56 display the (1, –3) and (1, –4)-solutions. If we also admit brown squares composed of more than one unit square, other solutions become possible as Figure 2.57 shows.

Discovering a theorem

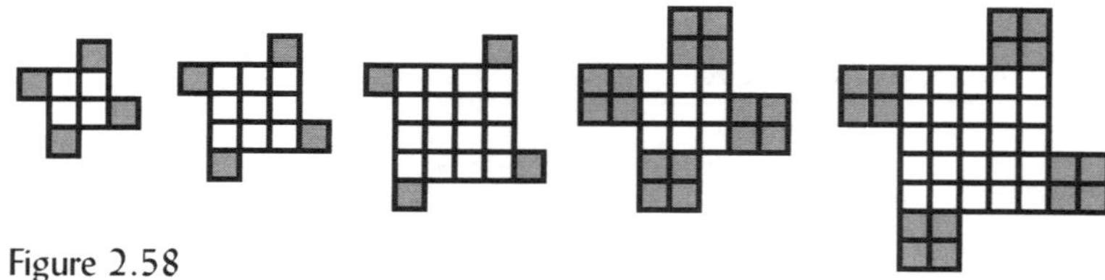

Figure 2.58

All these solutions generate tesselations of the plane. They have in common that four neighboring brown dots 'embrace' a white square (see Figure 2.58). When we draw straight lines through the centers of these neighboring white squares, we obtain a new square grid, this time an oblique one. As each white square is divided by the lines of the new grid into four congruent quadrilaterals and, at the same time, each square of the oblique grid is composed of a brown square at its center, surrounded by four such congruent quadrilaterals (see Figure 2.59), it follows that the area of an oblique square (C)

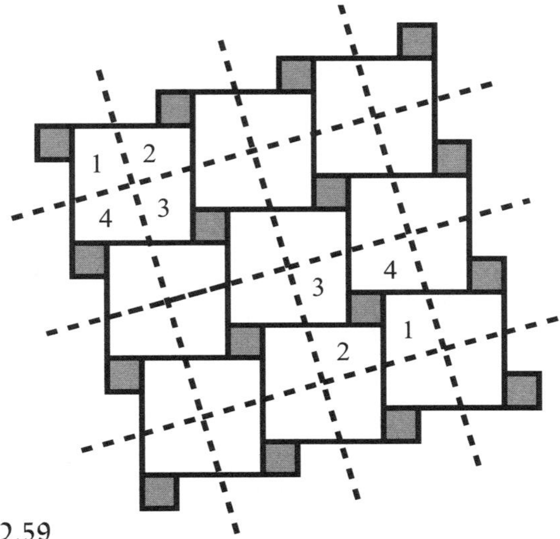

Figure 2.59

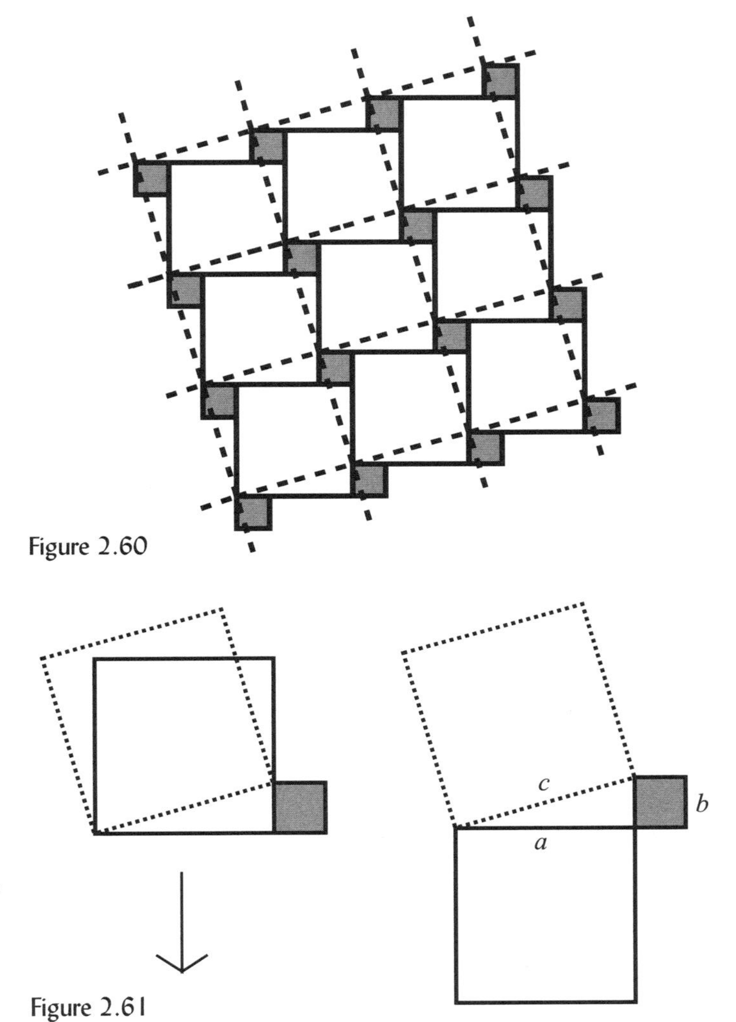

Figure 2.60

Figure 2.61

is equal to the sum of the areas of one brown (A) and one white (B) square: $A + B = C$. Which possible interpretations exist for the equality obtained?

When we translate the oblique grid in such a way that its vertices coincide with vertices of the brown squares, as e.g., in Figure 2.60, we arrive at:

(2) $$a^2 + b^2 = c^2,$$

i.e., at the Theorem of Pythagoras, where a, b and c denote the lengths of the sides of the squares of area A, B and C (cf. Figure 2.61).

A variant

The design formed by four neighbouring brown dots and the 'embraced' white square (Figure 2.62a) displays a rotational symmetry of 90°. This implies that any four corresponding points are the vertices of a square. Figure 2.62b gives an interesting example, as the area of the new square obviously has the same area as the white square and the four brown squares together. When we now rearrange the white and brown squares as in Figure 2.63, we arrive once more at the 'Theorem of Pythagoras.'

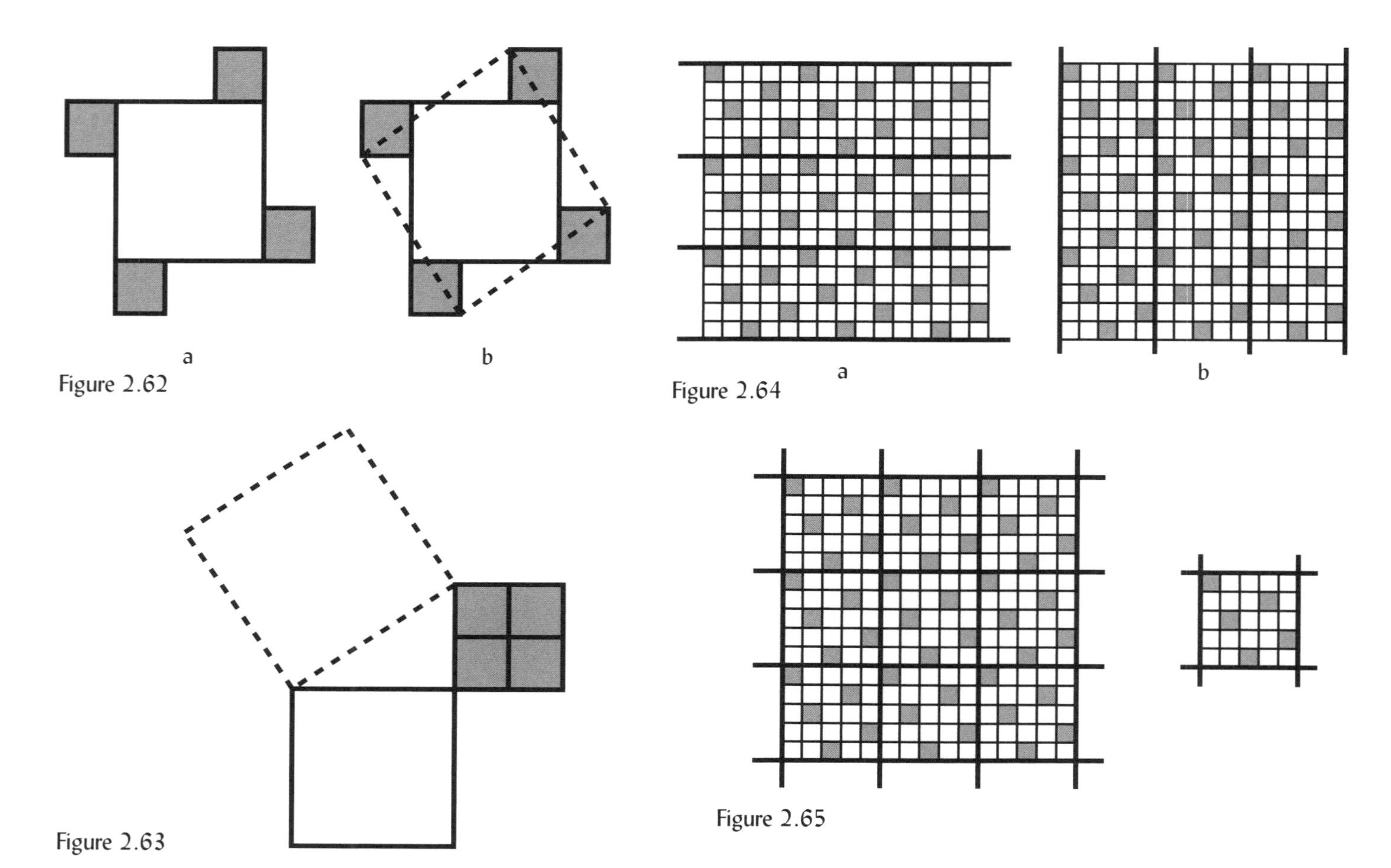

Figure 2.62

Figure 2.64

Figure 2.63

Figure 2.65

Periodicity and fundamental blocks

In the example of the Chokwe mat, the vertical brown strands pass always over one horizontal white strand and then under four white strands, the brown-and-white pattern repeats itself in the vertical direction after 5 (horizontal) strands (see Figure 2.64a). In the horizontal direction, each white strand passes under one brown strand and then over 4 brown strands. In the horizontal direction the pattern also repeats itself after 5 strands (see Figure 2.64b). It may be said that the pattern has period 5 in both directions, i.e., may be considered as composed of equally colored 5×5 blocks (see Figure 2.65). This fundamental 5×5 block contains just one little brown square in each row and each column. Many interesting problems involving fundamental blocks may be presented in the mathematics classroom for investigation, as the following examples illustrate.

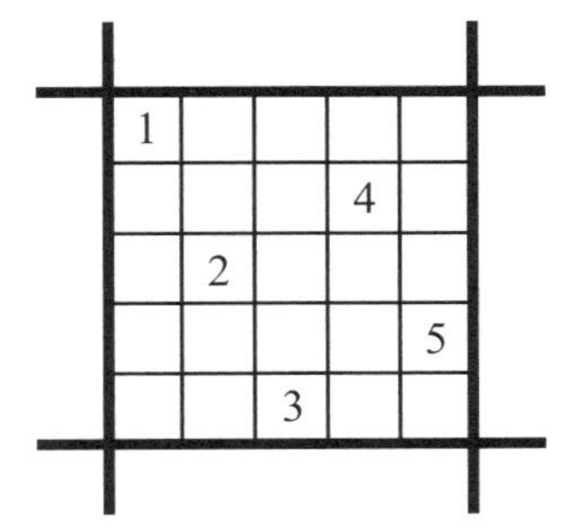

Figure 2.66

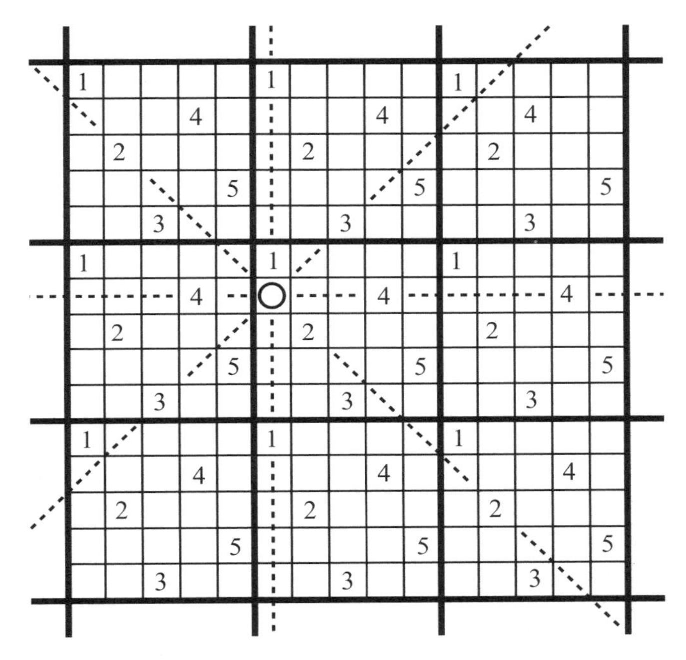

Figure 2.67

Enumerating the unit squares

Let us enumerate, from the left to the right, 1, 2, 3, 4, 5 the successive (brown) squares of a fundamental block, as illustrated in Figure 2.66. Is it possible to attribute, to each small square of a specific 5 × 5 block, a number 1, 2, 3, 4 or 5, in such a way that it is different from the numbers that appear in the same horizontal, vertical or diagonal line? For the unit square marked by a circle in Figure 2.67, we see that there is only one possibility: 5.

A Latin square

Doing the same for the other unit squares (of the same fundamental block), we obtain the 5 × 5 number square shown in Figure 2.68. In each row and each column the numbers 1, 2, 3, 4 and 5 appear precisely once. Such number squares are usually called Latin squares.

The Latin square in Figure 2.68 has interesting properties. Students may observe, for instance, that the numbers 1, 2, 3, 4, and 5 appear in each column downwards in the sequence 5, 4, 3, 2, 1. The position of the 1's (see Figure 2.69a) is equal to the fundamental brown-and-white block (Figure 2.69b) after a reflection about its main diagonal as an axis (see Figure 2.69c).

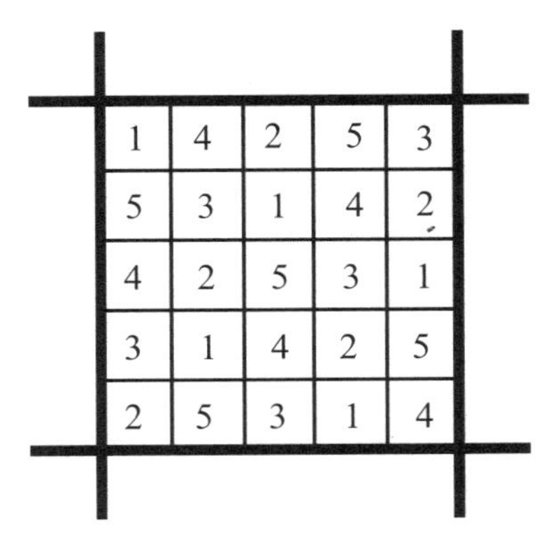

Figure 2.68

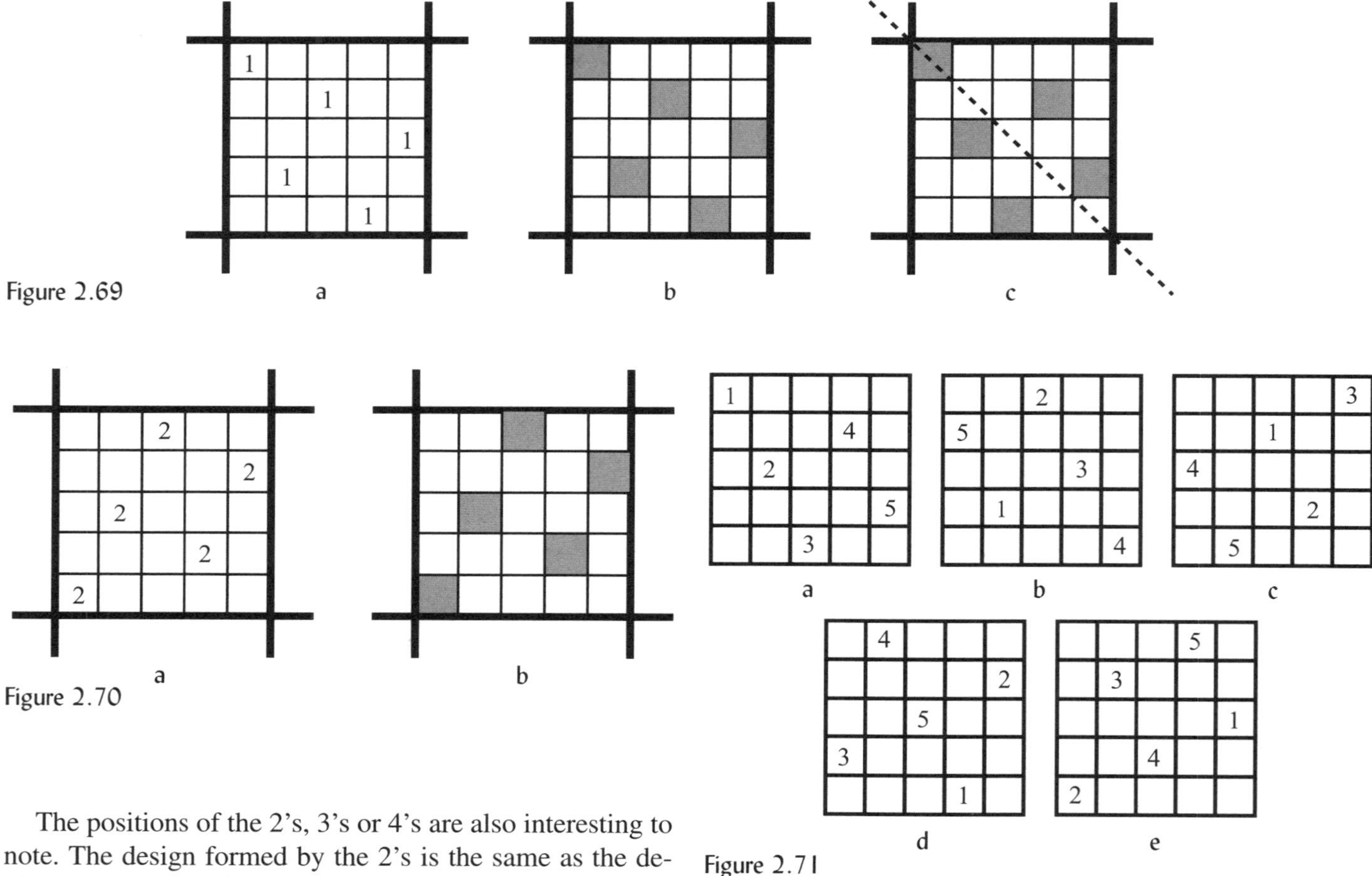

Figure 2.69

Figure 2.70

Figure 2.71

The positions of the 2's, 3's or 4's are also interesting to note. The design formed by the 2's is the same as the design made up of the 1's after having been turned 90° to the left (see Figure 2.70). The 3's constitute the same design after the design of the 1's has been turned 90° to the right. The 4's form the design of the 1's after a rotation of 180°. From a 1 to a 2 in the next column to the right, we always go two squares down. The successive 1, 2, 3, 4, and 5 once more constitute the initial design (see Figure 2.70b), turned 90° or 180°, with the case where the 5 is placed in the center of the 5x5 number square as the only exception (see Figure 2.71).

Magic squares

Our Latin square may be used to construct magic squares. Usually a magic square is defined as a set of integers in serial order, beginning with 1, arranged in square formation so that the sum total of each row, column and main diagonal is the same.

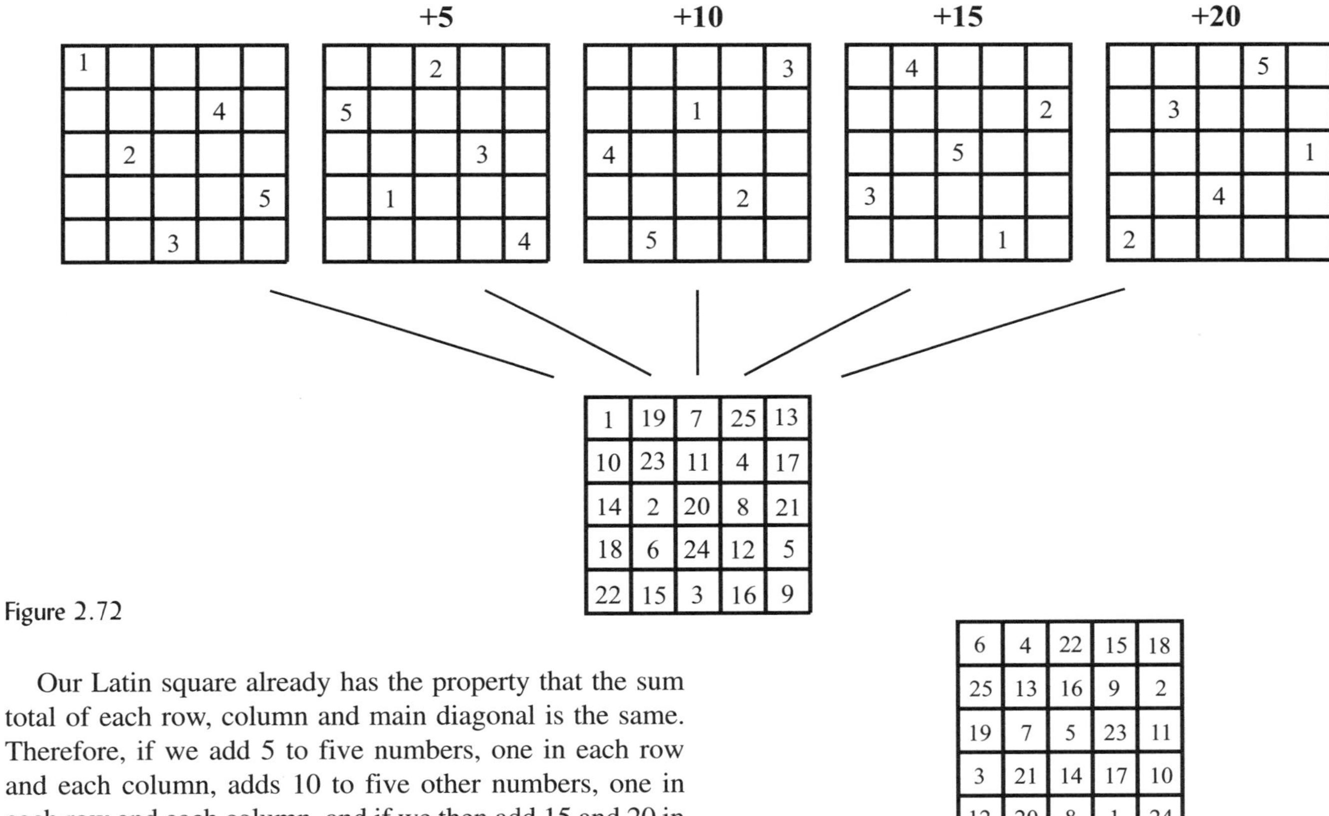

Figure 2.72

6	4	22	15	18
25	13	16	9	2
19	7	5	23	11
3	21	14	17	10
12	20	8	1	24

Figure 2.73

Our Latin square already has the property that the sum total of each row, column and main diagonal is the same. Therefore, if we add 5 to five numbers, one in each row and each column, adds 10 to five other numbers, one in each row and each column, and if we then add 15 and 20 in the same way, we obtain a magic square. Figure 2.72 gives an example. To the numbers of the block displayed in Figure 2.71b, we added 5, to the numbers of the block in Figure 2.71c we added 10, etc. The sum total of the rows, columns and main diagonal is 65.

The magic square given in Figure 2.73 may be obtained by adding 5 to the numbers in block a, 10 to the numbers in block e, 15 to the numbers in block c and 20 to the numbers in block b of Figure 2.71.

These magic squares have further properties. They are also 'diabolic' or 'pandiagonal,' that is the sum totals of the so-called 'broken diagonals' are equal to the same constant 65. 'Broken diagonals' are those that belong partly to a number square and partly to its equal neighboring number squares (see the examples in Figure 2.74). When we join several equal magic squares of this type, as in Figure 2.75a, and then displaces the 5×5 square as in the example in Figure 2.75b, we obtain new magic squares (see Figure 2.75c).

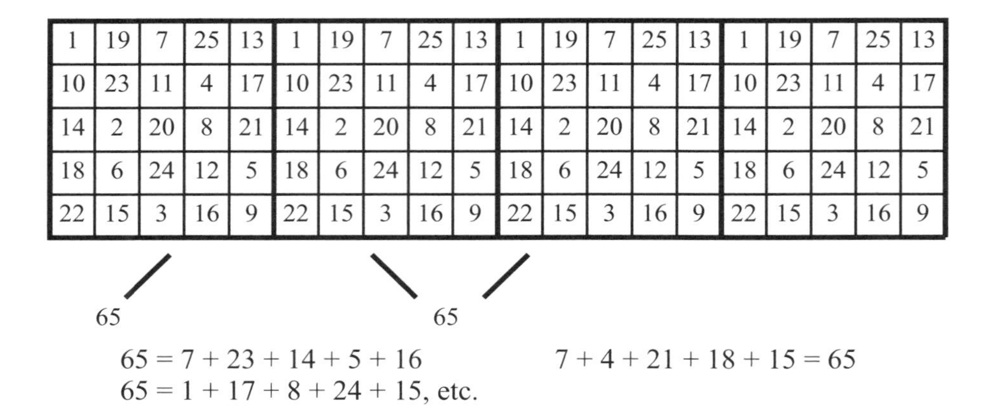

Figure 2.74

The magic square in Figure 2.75 has the additional property that each number on one of the four axes of the square, added to the number symmetrically opposite the square's center, yields the same: $12 + 18 = 14 + 16 = 7 + 23 = \cdots = 30$ (see Figure 2.76).

When we reflect the magic square in Figure 2.72 about its vertical axis, we obtain one of the magic squares (Figure 2.77) presented by Muhammad ibn Muhammad al Fullani, an early eighteenth century astronomer and mathematician from Katsina (now northern Nigeria), in his work *A treatise on the magical use of the letters of the alphabet*, written in Arabic (Cf. Zaslavsky, pp. 138–151; see also Sesiano).

The Kuba design illustrated in Figure 2.55 has the 5×5 block shown in Figure 2.78a as the fundamental block. The alternative solutions in Figure 2.56 have period 10 and 17 respectively. Figures 2.78b and c display their fundamental blocks. As before, by enumerating the successive small squares in the case of the 17×17 block (Figure 2.79a), we may construct Latin and magic squares. Figures 2.79b and 2.79c give an example. For which integers n, can we construct $n \times n$ Latin and magic squares in the described way? How many of them exist?

Arithmetic modulo n

The context of constructing Latin squares starting with African designs may also be further explored for the introduction of arithmetic modulo n. For instance, consider the Latin square in Figure 2.62. In order to go one place to the right in this Latin square, we always add 3 modulo 5. In order to go one place up, we have to add 1 modulo 5 (see Figure 2.80). When we go one place diagonally down to the right, what will be the end effect? First we may go one place to the right (+3) and then one place vertically downwards (–1); the result is +2. When we move one place diagonally downwards to the left, we go, e.g., first one place vertically downwards (–1) and then one place left (–3); the end result is –4, but –4 = +1 modulo 5.

Going from 2 to 5 in Figure 2.81, we add 3. If we pass through 4, we add first 1 and then two times 3, and from 4 to 5, we add first 3 and then subtract two times 1. Comparing the two ways, we see:

$$(1 + 2 \times 3) + (3 + 2 \times [-1]) = 3 \text{ (modulo 5)}$$

or,

$$7 + 1 = 3 \text{ (modulo 5)}.$$

6	4	22	15	18	6	4	22	15	18
25	13	16	9	2	25	13	16	9	2
19	7	5	23	11	19	7	5	23	11
3	21	14	17	10	3	21	14	17	10
12	20	8	1	24	12	20	8	1	24
6	4	22	15	18	6	4	22	15	18
25	13	16	9	2	25	13	16	9	2
19	7	5	23	11	19	7	5	23	11
3	21	14	17	10	3	21	14	17	10
12	20	8	1	24	12	20	8	1	24

a

6	4	22	15	18	6	4	22	15	18
25	13	16	9	2	25	13	16	9	2
19	7	5	23	11	19	7	5	23	11
3	21	14	17	10	3	21	14	17	10
12	20	8	1	24	12	20	8	1	24
6	4	22	15	18	6	4	22	15	18
25	13	16	9	2	25	13	16	9	2
19	7	5	23	11	19	7	5	23	11
3	21	14	17	10	3	21	14	17	10
12	20	8	1	24	12	20	8	1	24

b

8	1	24	12	20
22	15	18	6	4
16	9	2	25	13
5	23	11	19	7
14	17	10	3	21

c

Figure 2.75

6				18
	13	16	9	
19	7		23	11
	21	14	17	
12		8		24

Figure 2.76

13	25	7	19	1
17	4	11	23	10
21	8	20	2	14
5	12	24	6	18
9	16	3	15	22

Figure 2.77

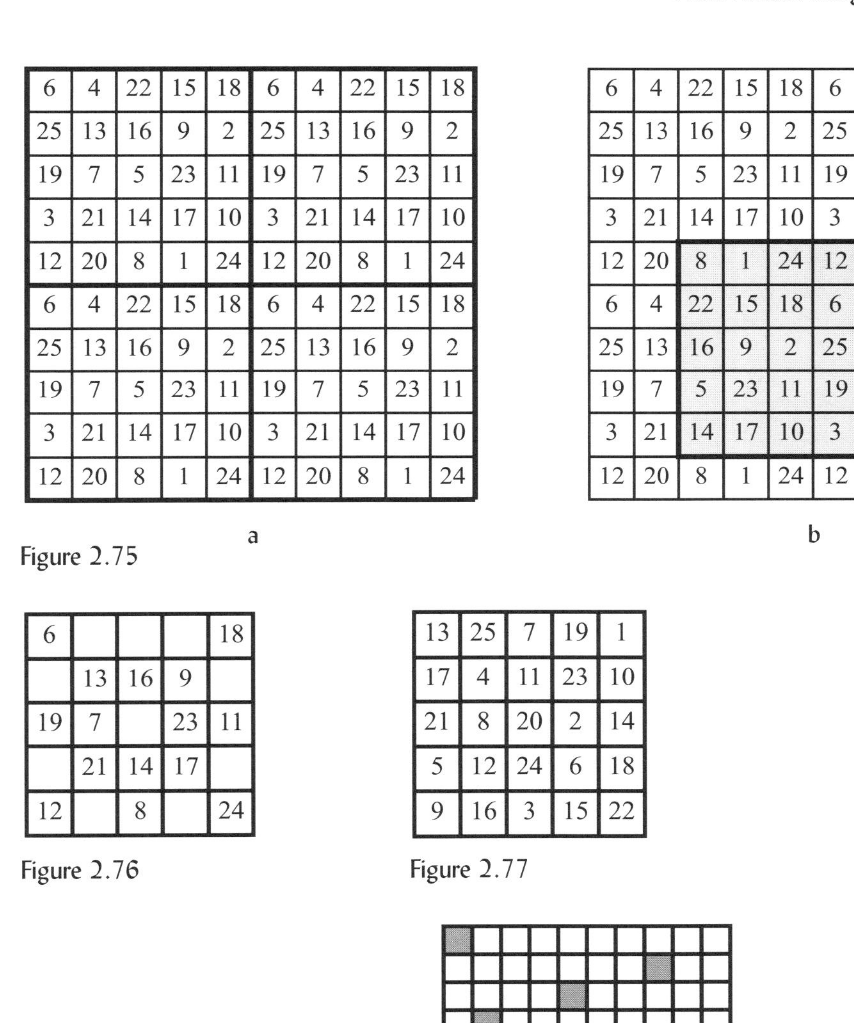

a

b

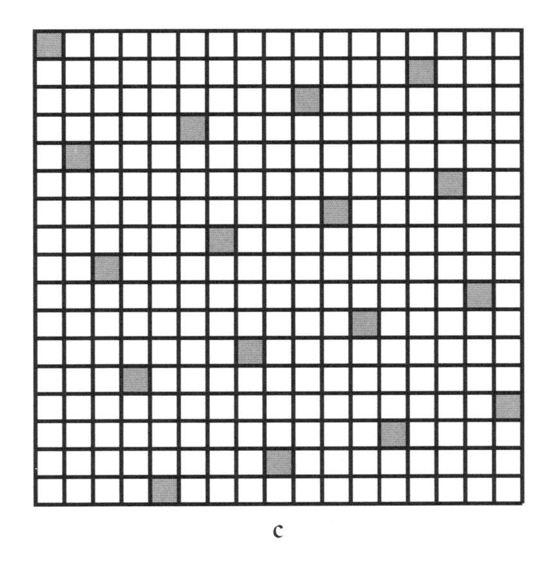

c

Figure 2.78

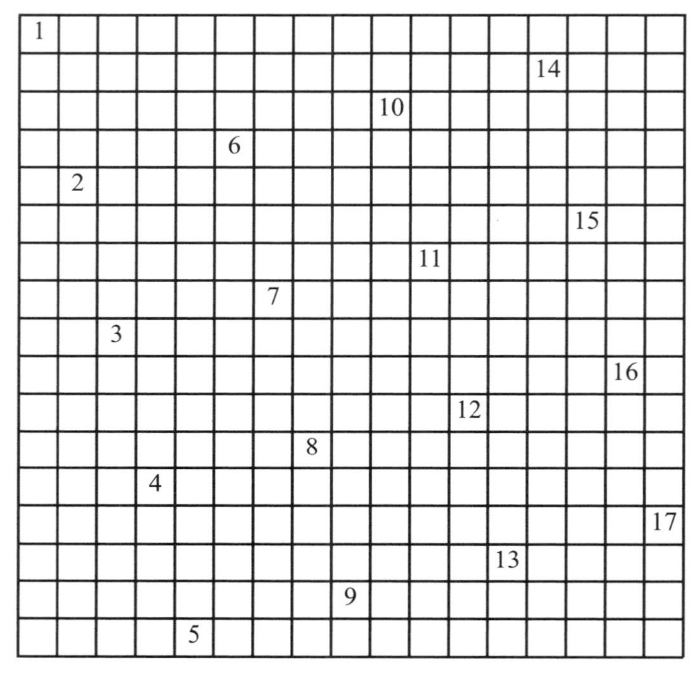

Figure 2.79a

1	4	7	10	13	16	2	5	8	11	14	17	3	6	9	12	15
9	12	15	1	4	7	10	13	16	2	5	8	11	14	17	3	6
17	3	6	9	12	15	1	4	7	10	13	16	2	5	8	11	14
8	11	14	17	3	6	9	12	15	1	4	7	10	13	16	2	5
16	2	5	8	11	14	17	3	6	9	12	15	1	4	7	10	13
7	10	13	16	2	5	8	11	14	17	3	6	9	12	15	1	4
15	1	4	7	10	13	16	2	5	8	11	14	17	3	6	9	12
6	9	12	15	1	4	7	10	13	16	2	5	8	11	14	17	3
14	17	3	6	9	12	15	1	4	7	10	13	16	2	5	8	11
5	8	11	14	17	3	6	9	12	15	1	4	7	10	13	16	2
13	16	2	5	8	11	14	17	3	6	9	12	15	1	4	7	10
4	7	10	13	16	2	5	8	11	14	17	3	6	9	12	15	1
12	15	1	4	7	10	13	16	2	5	8	11	14	17	3	6	9
3	6	9	12	15	1	4	7	10	13	16	2	5	8	11	14	17
11	14	17	3	6	9	12	15	1	4	7	10	13	16	2	5	8
2	5	8	11	14	17	3	6	9	12	15	1	4	7	10	13	16
10	13	16	2	5	8	11	14	17	3	6	9	12	15	1	4	7

Figure 2.79b
Latin square of dimensions 17×17

Bibliography

Anta-Diop, Cheik (1980), *Civilisation ou Barbarie, Anthropologie sans Complaisance*, Présence Africaine, Paris.

Bastin, M.-L. (1961), *Art décoratif Tshokwe*, Publicações Culturais da Companhia de Diamantes de Angola, Lisbon.

Baumann, Hermann (1929), Afrikanische Kunstgewerbe, in: H.Bossert (Ed.), *Geschichte des Kunstgewerbes aller Zeiten und Völker*, Berlin, Vol.2, 51–148.

Cole, Herbert and Aniakor, Chike (1984), *Igbo Arts: Community and Cosmos*, Museum of Cultural History, Los Angeles.

Denyer, Susan (1978), *African traditional architecture*, Heinemann, London/Ibadan/Nairobi.

Gerdes, Paulus (1988), A widespread decorative motif and the Theorem of Pythagoras, *For the Learning of Mathematics*, Montreal, Vol. 8, No. 1, 1988, 35–39.

—— (1992), *Pitágoras Africano: Um estudo em Cultura e Educação Matemática*, Universidade Pedagógica, Maputo.

—— (1994), *African Pythagoras: A study in Culture and Mathematics Education*, Universidade Pedagógica, Maputo.

—— (1995), *Women and Geometry in Southern Africa*, Universidade Pedagógica, Maputo.

—— (1997), *Women, Art and Geometry in Southern Africa*, Africa World Press, Trenton NJ and Asmara, Eritrea.

1	38	75	112	149	186	206	243	280	28	65	102	122	159	196	233	270
145	182	219	239	276	24	61	98	135	155	192	229	266	14	51	71	108
289	20	57	94	131	168	188	225	262	10	47	84	104	141	178	215	252
127	164	201	238	258	6	43	80	117	137	174	211	248	285	33	53	90
271	2	39	76	113	150	187	207	244	281	29	66	86	123	160	197	234
109	146	183	220	240	277	25	62	99	136	156	193	230	267	15	35	72
253	273	21	58	95	132	169	189	226	263	11	48	85	105	142	179	216
91	128	165	202	222	259	7	44	81	118	138	175	212	249	286	34	54
235	272	3	40	77	114	151	171	208	245	282	30	67	87	124	161	198
73	110	147	184	221	241	278	26	63	100	120	157	194	231	268	16	36
217	254	274	22	59	96	133	170	190	227	264	12	49	69	106	143	180
55	92	129	166	203	223	260	8	45	82	119	139	176	213	250	287	18
199	236	256	4	41	78	115	152	172	209	246	283	31	68	88	125	162
37	74	111	148	185	205	242	279	27	64	101	121	158	195	232	269	17
181	218	255	275	23	60	97	134	154	191	228	265	13	50	70	107	144
19	56	93	130	167	204	224	261	9	46	83	103	140	177	214	251	288
163	200	237	257	5	42	79	116	153	173	210	247	284	32	52	89	126

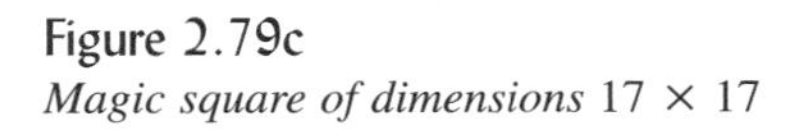

Figure 2.79c
Magic square of dimensions 17 × 17

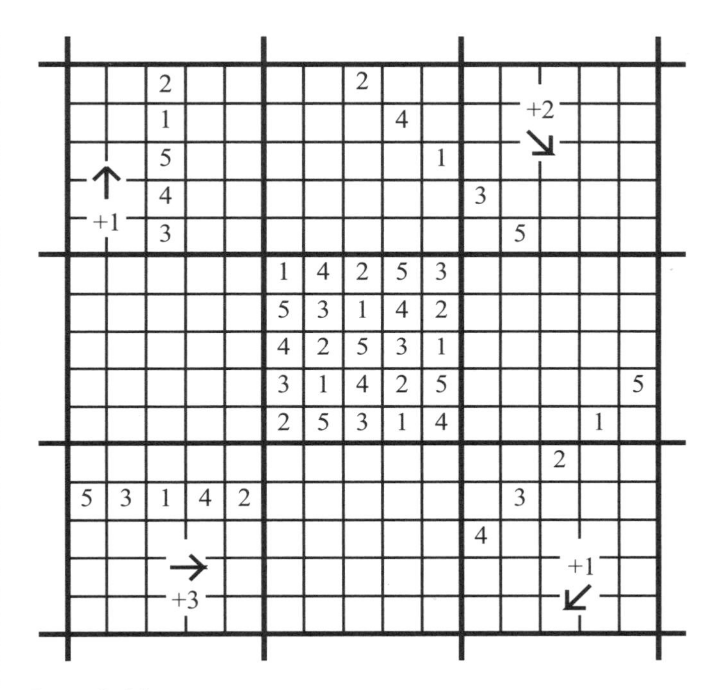

Figure 2.80

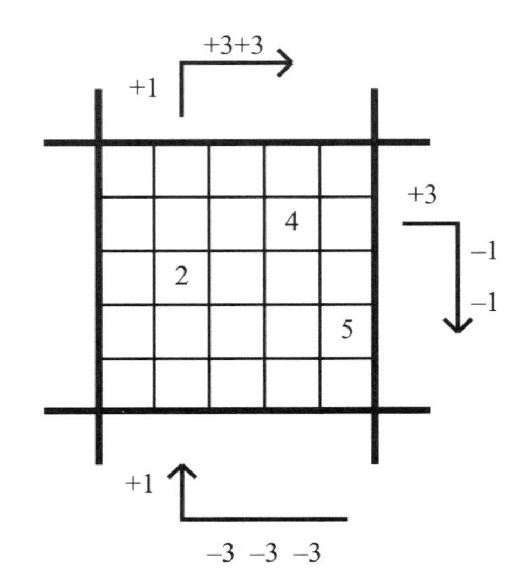

Figure 2.81

Hauenstein, Alfred (1988), *Examen de motifs décoratifs chez les Ovimbundu et Tchokwe d'Angola*, Universidade de Coimbra, Coimbra.

Loomis, E.S. ([1940] 1972), *The Pythagorean Proposition*, NCTM, Reston.

Martins, M. (1986), *Moçambique: Aspectos da cultura material*, Instituto de Antropologia, Coimbra.

Meurant, Georges (1987), *Abstractions aux royaumes des Kuba*, Éditions Dapper, Paris.

Mveng, R.Engelbert (1980), *L'art et l'artisanat africains*, Éditions Clé, Yaoundé.

Sesiano, Jacques (1994), Quelques méthodes arabes de construction des carrés magiques impairs, *Bulletin de la Société Vaudoise des Sciences Naturelles*, Vol. 83.1, 51–76.

Williams, Geoffrey (1971), *African designs from traditional sources*, Dover, New York.

Wilson, Eva (1986), *Ancient Egyptian designs*, British Museum, London.

Zaslavsky, Claudia ([1973] 1990), *Africa Counts : Number and Pattern in African culture,* Lawrence Hill Books, Brooklyn, NY.

"Sipatsi" bags woven by Tonga women (Mozambique. From the author's collection.
Photograph by Marcos Cherinda, Ethnomathematics Research Project, Universidade Pedagógica, Maputo, Mozambique.

3

Geometrical ideas in crafts and possibilities for their educational exploration

In the third section, I analyze geometrical ideas inherent in various crafts and explore possibilities for their educational use. Chapters deal with topics such as symmetrical wall decoration in Lesotho and South Africa; house building in Mozambique and Liberia; weaving pyramidal baskets in Congo/Zaire, Mozambique, and Tanzania; plaited strip patterns from Guinea, Mozambique, Senegal, and Uganda; finite geometrical designs from the Lower Congo region. Exploration of a hexagonal basket weaving technique from Cameroon to Kenya, Congo to Madagascar and Mozambique, will lead to connections between the underlying geometry and chemical models of recently discovered carbon molecules. Pentagrams will be discovered in knots. I develop alternative ways for rectangle constructions and for the determination of areas of circles and volumes of spatial figures, including a twisted decahedron.

1 Wall decoration and symmetries (cf. Gerdes, 1995, 1996, 1998a, ch. 9)

In Lesotho and neighboring zones of South Africa, Sotho women developed a tradition of decorating the walls of their houses with designs. The walls are first neatly plastered with a mixture of mud and dung, and often colored with natural dyes. While the last coat of plastered mud is still wet, the women engrave the walls, using their forefinger. Their art is seasonal: The sun dries it and cracks it, and the rain washes it away. An entire village is redecorated before special occasions such as engagement parties, weddings, and important religious celebrations.

The Sotho women call their geometric patterns *litema* (singular: *tema*). The books *The African Mural* by Changuion et al. (1989) and *African Painted Houses* by Wyk (1998) contain beautiful collections of photographs of *litema*. The National Teacher Training College of Lesotho published a collection *litema* patterns collected by its mathematics students (NTTC, 1976). Symmetry is a basic feature of the *litema* patterns. Figure 3.1 presents part of a *litema* pattern.

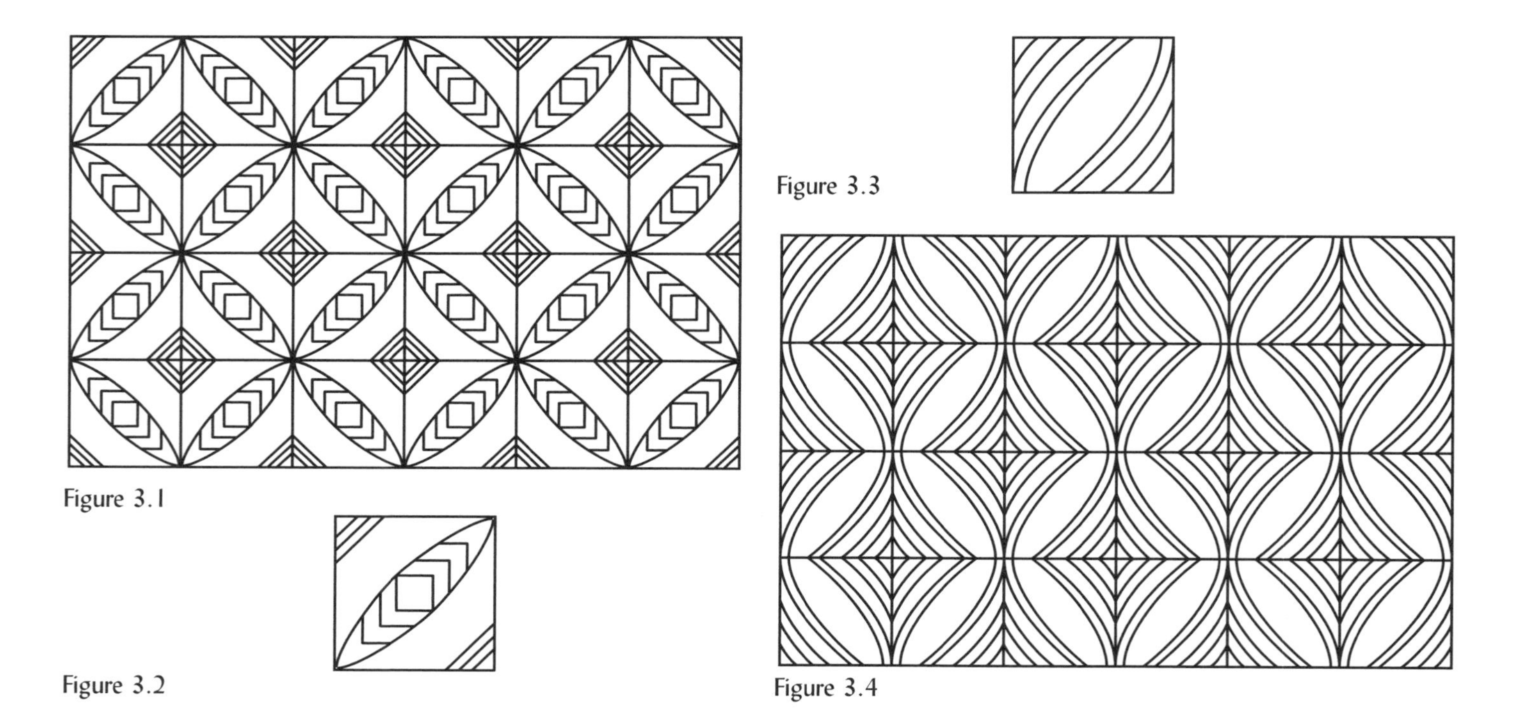

Figure 3.1

Figure 3.2

Figure 3.3

Figure 3.4

As is often the case, this *tema* pattern is built up from a basic square that constitutes the (unit) cell of the pattern. Figure 3.2 displays the cell for the *tema* in Figure 3.1.

The Sotho women lay out a network of squares and then they reproduce the basic design in each square. The number of reproductions or repetitions of the unit cells depends, in practice, on the available space on the wall to be decorated. As in Figure 3.1, a whole pattern is built up out of repetitions of a 2×2 square, in which the unit cell appears in four positions, obtained by horizontal and vertical reflection about the axes of the 2×2 square. The symmetries of a whole pattern—imagined to be extending infinitely in all directions in the same plane as the drawing—depend on the symmetries of the unit cell. The unit cell in Figure 3.2 has two diagonal axes of symmetry. The unit cell (see Figure 3.3) of the *litema* pattern in Figure 3.4 has no axial symmetry however it is invariant under a half turn. The unit cell (see Figure 3.5) of the *litema* pattern in Figure 3.6 has one axis of symmetry.

Students can be asked to invent their own *litema* patterns. Furthermore, they could analyze which symmetries are possible in constructing *litema* patterns.

Brown-and-white patterns

Several painted *litema*, and those for which changes in the relief of the dun surface of the plaster suggest two distinct colors, may be represented on paper as brown-and-white

Figure 3.5

Figure 3.6

Figure 3.7

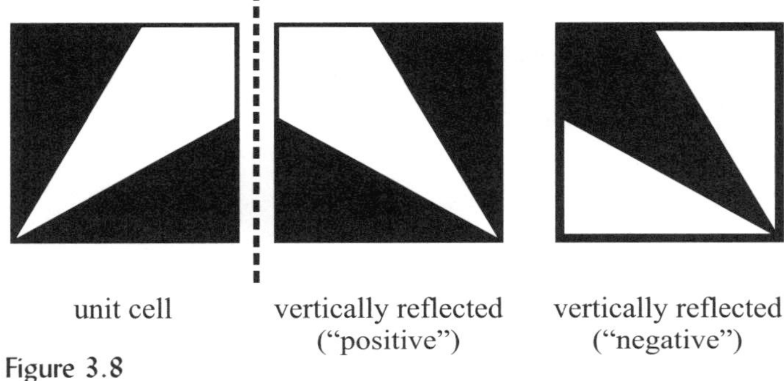

Figure 3.8

Figure 3.9

patterns. Some, as the one in Figure 3.7, are built up in the same way as the earlier patterns considered.

Others are two-color patterns where in each horizontal or vertical reflection of the unit cell the colors are reversed. The image of a unit cell is the 'negative' (in photographic terms) of the reflected cell (see the example in Figure 3.8, leading to the litema pattern in Figure 3.9).

Once more, the symmetries of a whole pattern depend on the symmetries of the unit cell, but in this context we

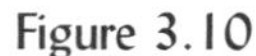
Figure 3.10

Figure 3.11

Figure 3.12

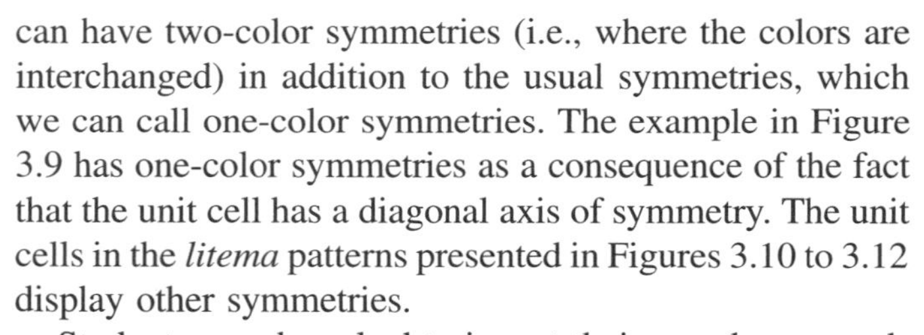
can have two-color symmetries (i.e., where the colors are interchanged) in addition to the usual symmetries, which we can call one-color symmetries. The example in Figure 3.9 has one-color symmetries as a consequence of the fact that the unit cell has a diagonal axis of symmetry. The unit cells in the *litema* patterns presented in Figures 3.10 to 3.12 display other symmetries.

Students may be asked to invent their own brown-and-white patterns of the considered type. Furthermore they could analyze which symmetries are possible for this type of *litema* pattern.

2 Rolling up mats

In Mozambique and several other parts of Africa, a woven rectangular, sitting or sleeping mat may be rolled up when not in use. It then takes approximately the form of a cylinder (see Figure 3.13). Rolling up the mat does not change its weight and volume. May this cultural experience be explored for students to obtain an idea of how to calculate the volume of a cylinder? How has the (approximately) circular base of the cylinder been produced?

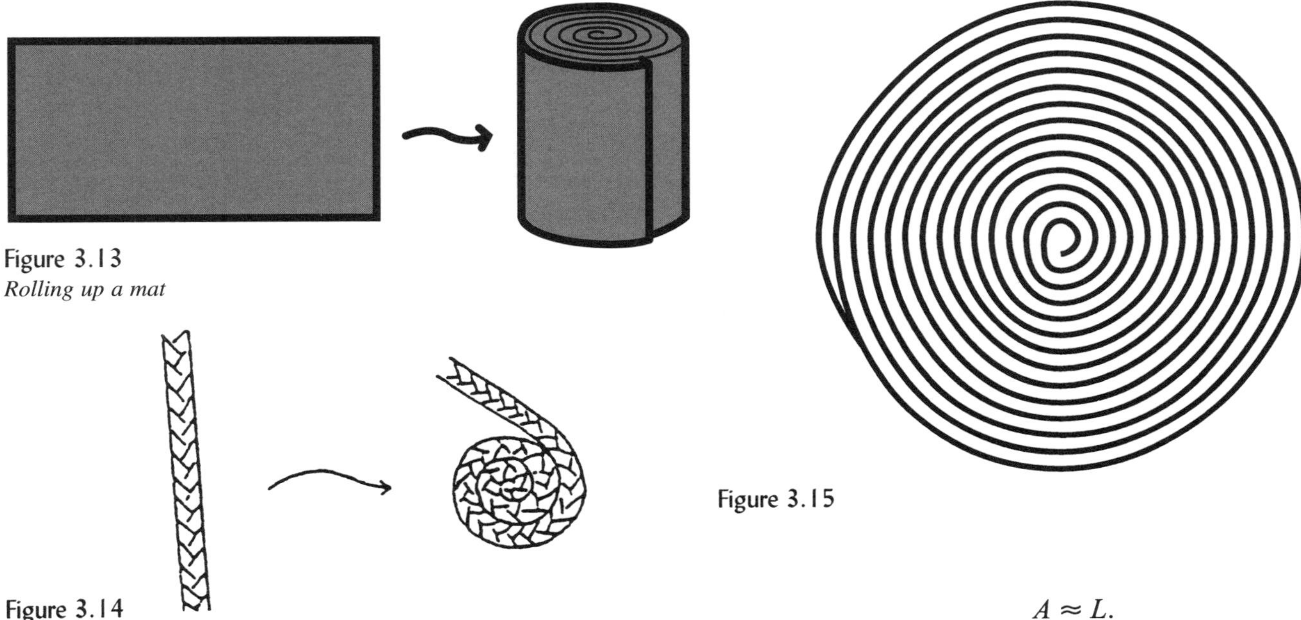

Figure 3.13
Rolling up a mat

Figure 3.14

Figure 3.15

Area of a circle (cf. Gerdes 1985b)

Makhuwa artisans in the north of Mozambique also produce circular mats. They start by braiding a long strand of sisal, and then rolling it into a coil, sewing together the successive circuits of the coil (Figure 3.14); and finally cutting the end off a little bit to give the impression of a perfect circle (see Figure 3.15).

Looking from above, the initial strand of sisal may be considered as a rectangle. It may be assumed that rolling up the strand does not significantly affect its (top) area: on the outer side the sisal braid is stretched, while on the inner side it is slightly pressed together. Taking the width of the sisal strand as the unit of linear measurement and L the number of such units in the strand's length, the area (A) of a coiled circle (before cutting off its end), is given by

$$A \approx L.$$

The diameter (d) of the coiled circle can be counted from the endpoint of the coil to the opposite side of the mat. Students may be asked to experiment with rolling up a strand and trying to discover a relationship between L and d. For instance, they may find that $d = 9$ when $L = 64$. This implies, that

$$A \approx \left(\frac{8d}{9}\right)^2,$$

which is equivalent to the way the area of a circle was approximated in ancient Egypt (cf. Gerdes, 1985b). This approximation corresponds to $\pi \approx 4(8/9)^2 \approx 3.1605$. Higher values of L lead to improved approximations of A and, by consequence, of π. This context may be used as an introduction to the notion of π and the discovery of the formula $A = \pi r^2$. Furthermore, the context may serve as an introduction to the limit concept.

Volume of a cylinder (cf. Gerdes, 1985a)

Now the volume and surface area of a cylinder may be determined analogously. Taking the thickness of the initial mat as unit of linear measurement, we have that the volume of the mat is equal to length (L) times width (W). After rolling up, we have the top area A equal to L. Therefore the volume of a cylinder is approximately given by:

$$V \approx L \times W \approx A \times W = A \times h,$$

where h represents the height of the cylinder (which is the same as the width W of the mat). Advancing with the limit, the volume of a cylinder is found to equal $\pi r^2 h$.

Surface area of objects with rotational symmetry

Baskets with a vertical axis of rotational symmetry are widely produced in Africa, using the coiled weaving technique. They may display various outlines. Figure 3.16 shows typical outlines of such baskets from Swaziland.

The length of the strand that forms the coil corresponds to the surface area of the basket. How may the production technique be explored to determine approximately the surface area of an object with rotational symmetry? Secondary school students can experiment with strands or bands of varying sizes to obtain a reasonably approximation for the surface of a hemisphere. Calculus students could translate their approximation into a precise method.

Figure 3.16
Outlines of coiled baskets

3 Exploring rectangle constructions used in traditional house building

Most African peoples south of the Sahara traditionally build houses with circular or rectangular bases. Among the Mozambican peasantry, two methods for the construction of the rectangular bases are common. Readers from a different cultural environment may be surprised to see that rectangles can be constructed without starting by constructing the right angles one by one.

In this chapter, I will first present two construction methods and then suggest several possibilities for their educational exploration.

Rectangle constructions (cf. Gerdes, 1988)

Figure 3.17 illustrates the first method (M1):

M1 The house builders start by laying down on the floor two long bamboo sticks of equal length (a). Then these first two sticks are combined with two other sticks also of equal length, but normally shorter than the first ones (b). Now the sticks are moved to form the closure of a quadrilateral (c). The figure is further adjusted until the diagonals—measured with a rope—become equal (d). Then, from where the sticks are now lying on the floor, lines are drawn and the house builders can start.

Figure 3.18 illustrates the second method (M2).

M2 The house builders start with two ropes of equal length, that are tied together at their midpoints (a). A bamboo stick, whose length is equal to that of the desired width of the house, is laid down on the floor and at its endpoints pins are hit into the ground. An

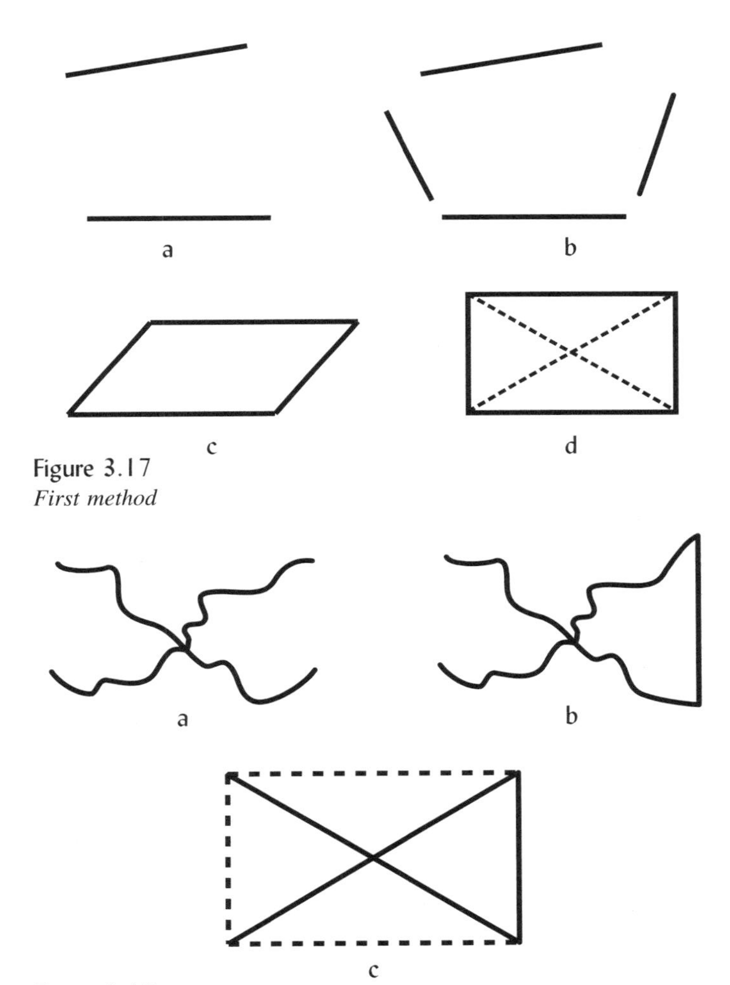

Figure 3.17
First method

Figure 3.18
Second method

endpoint of each of the ropes is tied to one of the pins (b). Then the ropes are stretched and at the remaining two endpoints of the ropes, new pins are hit into the ground. These four pins determine the four vertices of the house to be built (c).

The knowledge behind these rectangle constructions is equivalent to theorems in Euclidean geometry:

T1 A parallelogram with congruent diagonals is a rectangle.

T2 A quadrilateral with congruent diagonals that intersect at their midpoints is a rectangle.

Rectangle axioms

Looking for possibly interesting didactic alternatives of axiomatic constructions for Euclidean geometry in secondary education or in mathematics teachers' education, the 'rectangle axiom' proposed by Alexandrov of the former Soviet Union may be used as a substitute for Euclid's famous fifth postulate. Alexandrov's 'rectangle axiom' (RA) says the following (see Figure 3.19):

RA If in a quadrilateral $ABCD$, $AD = BC$ and $\angle A$ and $\angle B$ are right angles, then $AB = DC$ and $\angle C$ and $\angle D$ are also right angles.

The knowledge underlying the traditional Mozambican house building techniques might be used to formulate alternative 'rectangle axioms'.

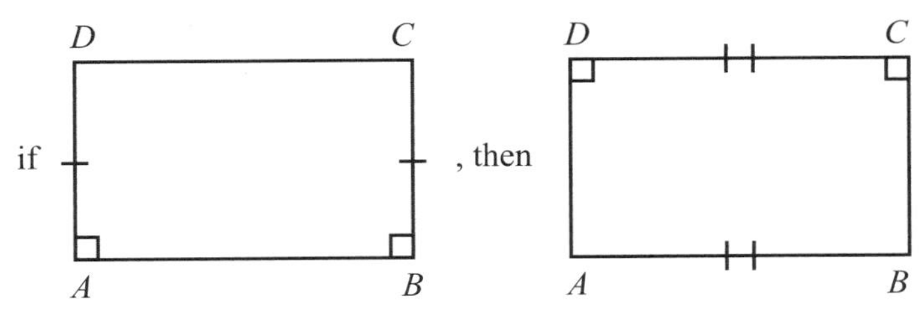

Figure 3.19
Alexandrov's "rectangle axiom"

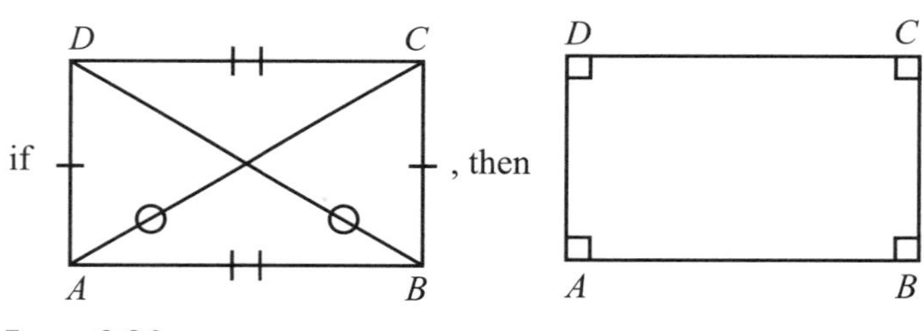

Figure 3.20

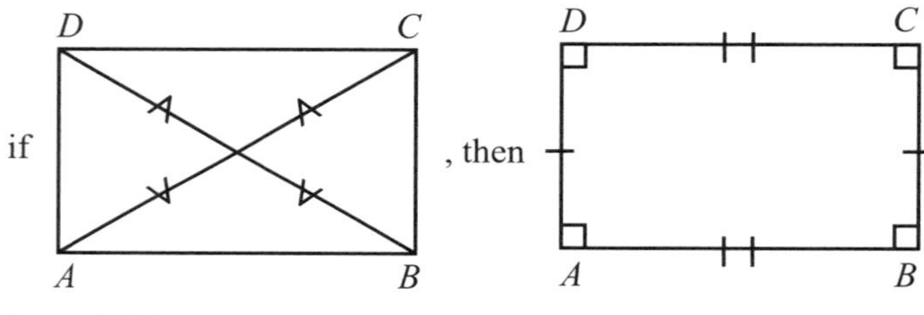

Figure 3.21

RA1 If in a quadrilateral $ABCD$, $AD = BC$, $AB = DC$ and $AC = BD$, then $\angle A$, $\angle B$, $\angle C$ and $\angle D$ are right angles (see Figure 3.20);

RA2 If in a quadrilateral $ABCD$, $M = AC \cap BD$ and $AM = BM = CM = DM$, then $\angle A$, $\angle B$, $\angle C$, and $\angle D$ are right angles, $AD = BC$ and $AB = DC$ (see Figure 3.21).

To use these rectangle constructions from traditional house building as source of inspiration to formulate alternative rectangle axioms, is one of several interesting possibilities of exploring them educationally and mathematically.

Generalizations and variations of the first construction method

Another possibility consists of looking for generalizations by weakening the initial conditions. For instance, what

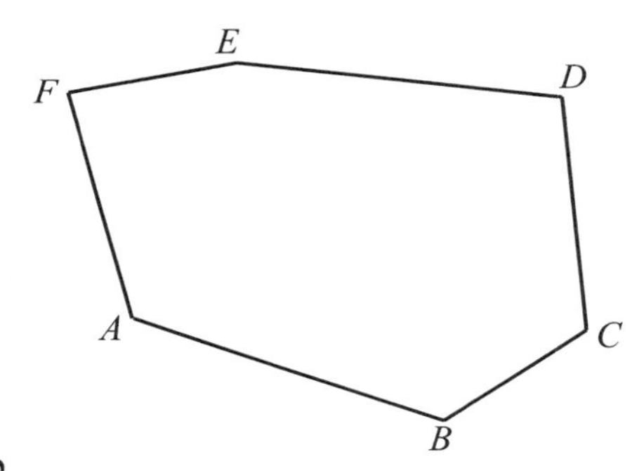

Figure 3.22

happens in the case of M1, if we demand that only one pair of sticks are of equal length. Now, when the figure is adjusted until the diagonals become equally long, its form becomes that of an isosceles trapezoid.

Let us consider another situation: Start with three pairs of sticks, each pair being of equal length, and form a closed hexagon in such a way that the opposite sides are congruent (see the example in Figure 3.22).

What will happen if the new figure is adjusted until the principal diagonals (AD, BE, CF) become equally long? Is this adjustment always possible? Is it unique? When are the principal diagonals concurrent? When are the opposite sides parallel?

What will happen if the initial figure is adjusted until the 2-diagonals (AC, BD, CE, DF, EA) become equally long? Is this always possible? Considering these questions, what if all six sticks have the same length? Figure 3.23 shows that figures other than the regular hexagon are possible that have congruent principal diagonals.

What will happen if we start with two groups of three equally long sticks, and alternate them to lay down a hexagon? Figure 3.24 shows an example where it is possible to adjust the principal diagonals in such a way that they become equally long. Is such a solution unique?

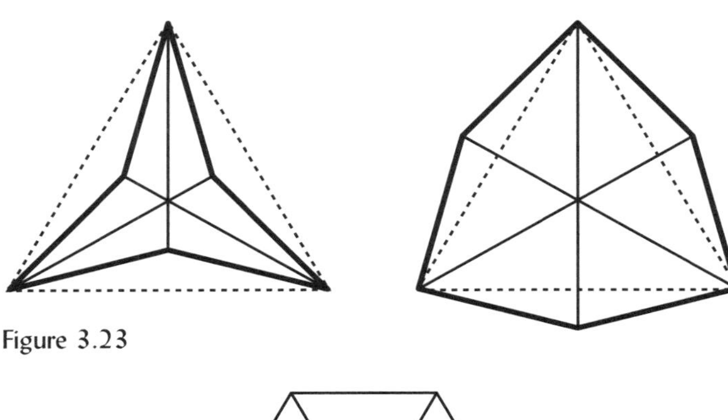

Figure 3.23

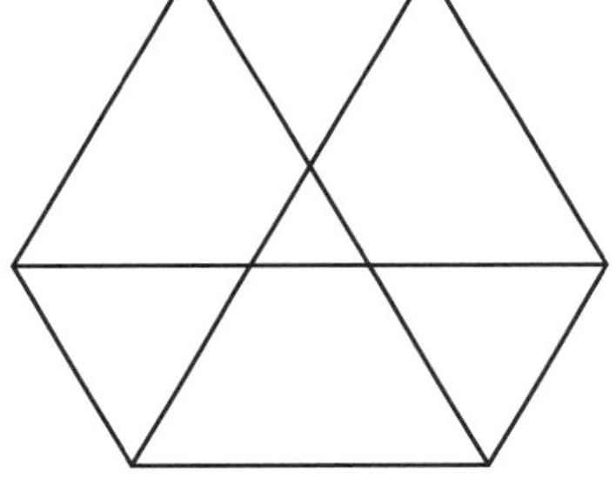

Figure 3.24

Imagine taking any diagonal and sliding it around in such a way that it continues to intersect the other two diagonals. Now the number of solutions is infinite. With *Geometer's SketchPad* or *Cabri* one could make animated, computer-generated films to present the solutions.

And what if we start with two groups of four equally long sticks and alternate them to lay down a octagon?

Hexahedra and octahedra

Analogous to the traditional house builders' transformation in the plane of a parallelogram into a rectangle, is the transformation in space of a parallelopiped into a rectangular block by adjusting its four diagonals until they become equally long. Would it be sufficient to adjust only three diagonals?

Similarly, what will happen to an octahedral figure whose opposite sides are congruent, when its three diagonals are adjusted to become equally long. What can be said about the resulting octahedron?

Generalizations and variations of the second construction method

In the case of the second rectangle construction method (M2), we may change the initial conditions and analyze what will happen. For example, what form will be obtained if we start with two ropes of equal length, that are tied together at corresponding points (see Figure 3.25).

What if two unequal ropes are tied together at their midpoints (see Figure 3.26)? What if two unequal ropes are

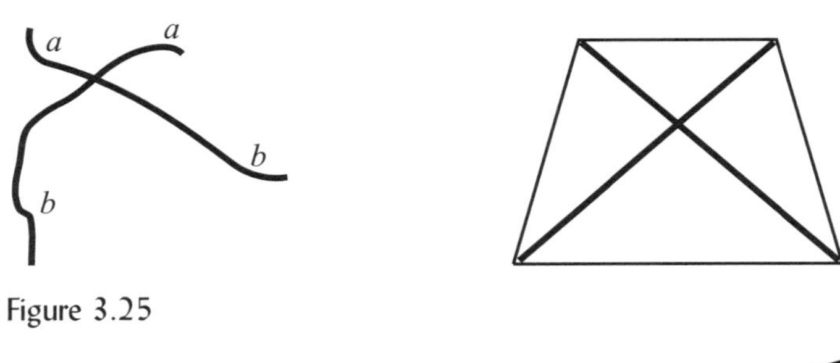

Figure 3.25

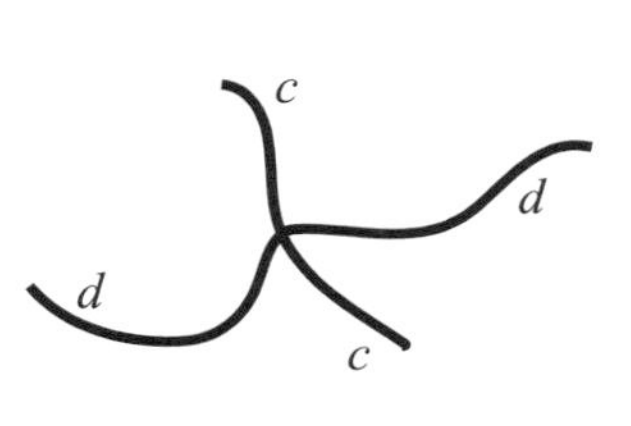

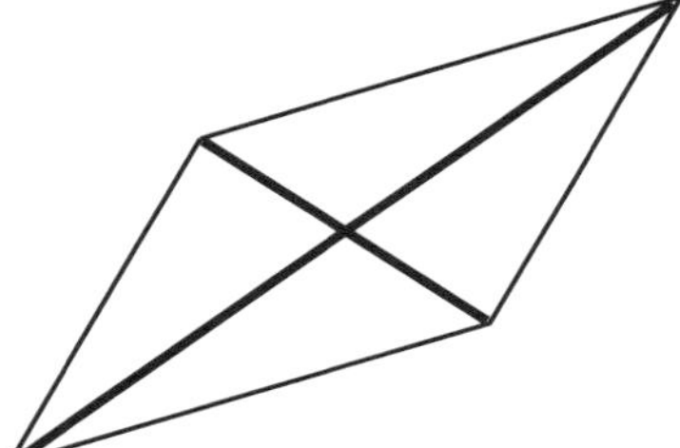

Figure 3.26

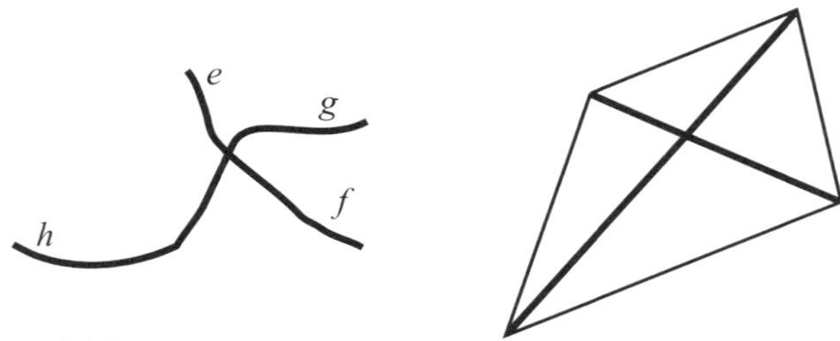

Figure 3.27

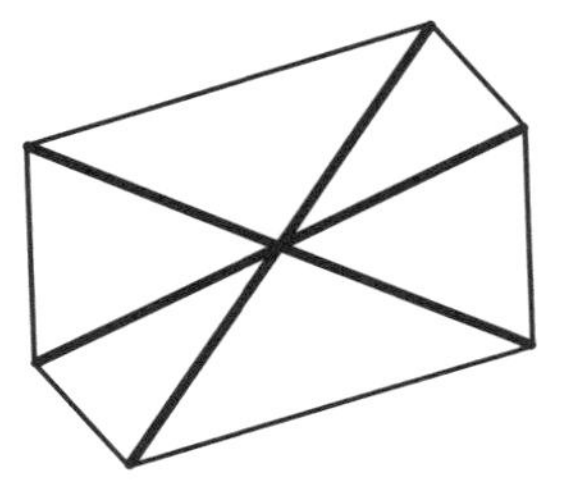
Figure 3.29

tied together at proportionally corresponding points ($e{:}f = g{:}h$; see Figure 3.27)?

What will happen if we only demand that the ropes are of equal length? Obviously, we are led to form an equidiagonal quadrilateral, but does such a quadrilateral have any particular characteristics? If we draw through its vertices lines that are parallel to the diagonals, these constitute the sides of a rhombus (see Figure 3.28), and students may try to figure out why this happens.

If instead of starting with two ropes, one starts with three ropes of equal length, that are tied together in their midpoints, what type of figures will be formed? If the ropes are in the same plane, hexagons are formed with several interesting properties: for example, their opposite sides are congruent and parallel; they are invariant under a half turn around the center (see the example in Figure 3.29). When the ropes are not in the same plane, then octahedra are formed, whose properties may be investigated by the students.

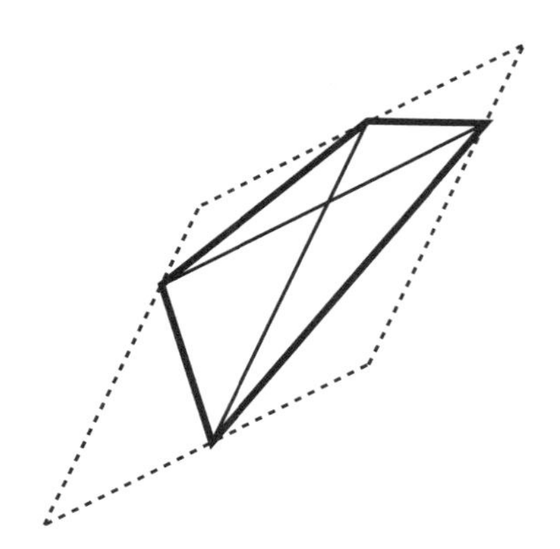
Figure 3.28

Link with vector geometry

The knowledge involved in the two methods for the construction of the rectangular bases of traditional houses may also be translated into the language of vector geometry and linear algebra, giving a sufficient and necessary condition for the perpendicularity of two vectors. In the case of the first construction method (M1), we have:

P1 $|p + q| = |p - q| \Leftrightarrow p \perp q$ (see Figure 3.30), where p and q represent vectors.

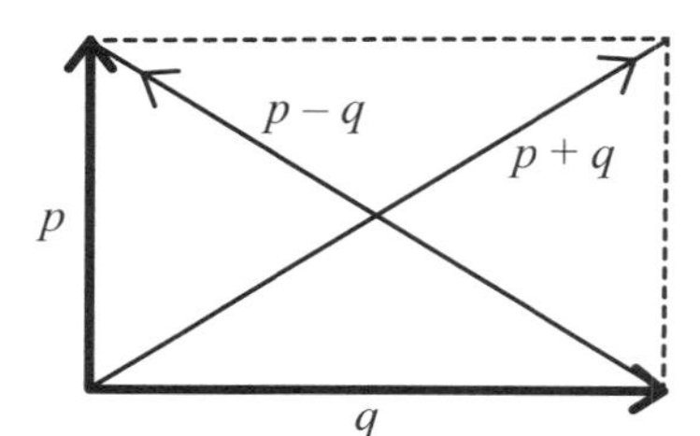

Figure 3.30

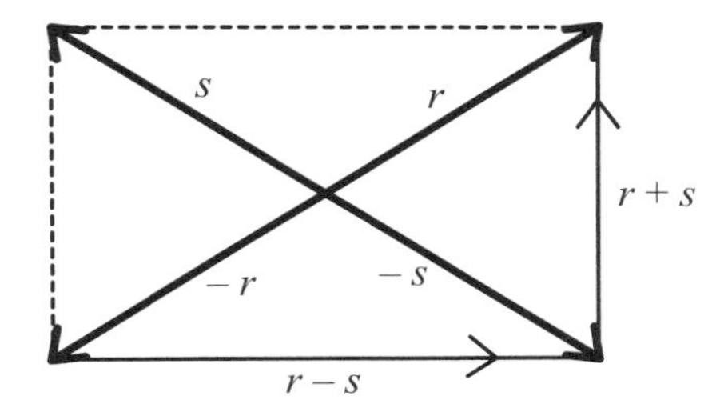

Figure 3.31

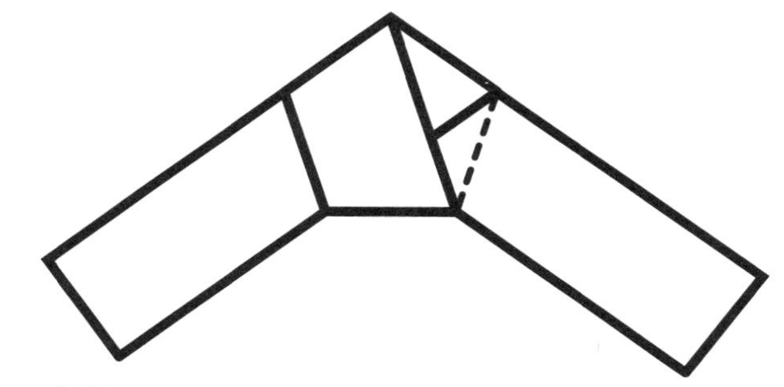

Figure 3.32
Pentagonal knot

In the second case, we have:

P2 $|r| = |s| \Leftrightarrow (r + s) \perp (r - s)$ (see Figure 3.31), where r and s represent vectors.

Looking back

In this section we explored several possibilities for an educational use of two methods for the construction of the rectangular bases used in house building among the Mozambican peasantry. The suggestions varied from the formulation of alternative rectangle axioms, a link with vector geometry to investigations of possible generalizations and variations both in the plane and space.

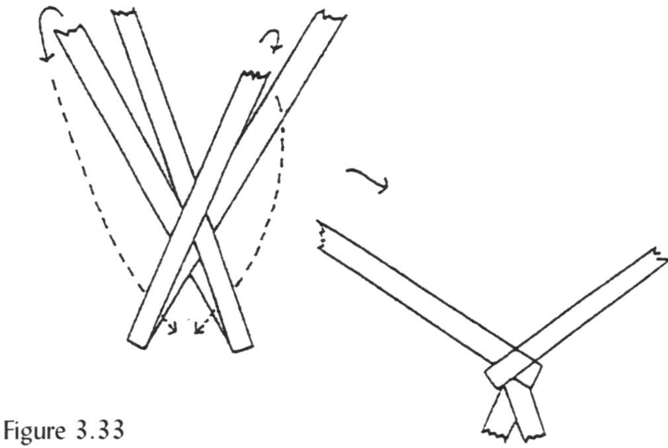

Figure 3.33
Mozambican pentagonal knot

4 A woven knot as a starting point

(cf. Gerdes, 1990b, 1999a; Gerdes and Bufalo, 1994)

When one makes a simple knot in strip of paper and flattens the knot carefully, the knot takes on the form of a regular pentagon (see Figure 3.32). Held up to the light, four of its sides and four of its diagonals are visible. It is an interesting exercise to ask students to figure out why this pentagon is regular.

Artisans in southern Mozambique start the weaving of their hand and shoulder bags by tying together two palm strands of equal width—often, two pairs of two strands each—in a flexible pentagonal knot (see Figure 3.33). Then a series of knots are woven together to constitute the wall of a bag (see Figure 3.34).

Let us make a knot with two strips of thin paper. Holding it up, after flattening, to the light, several sides and diagonals of a seemingly regular pentagon are visible (see Figure 3.35). Why should its form be regular?

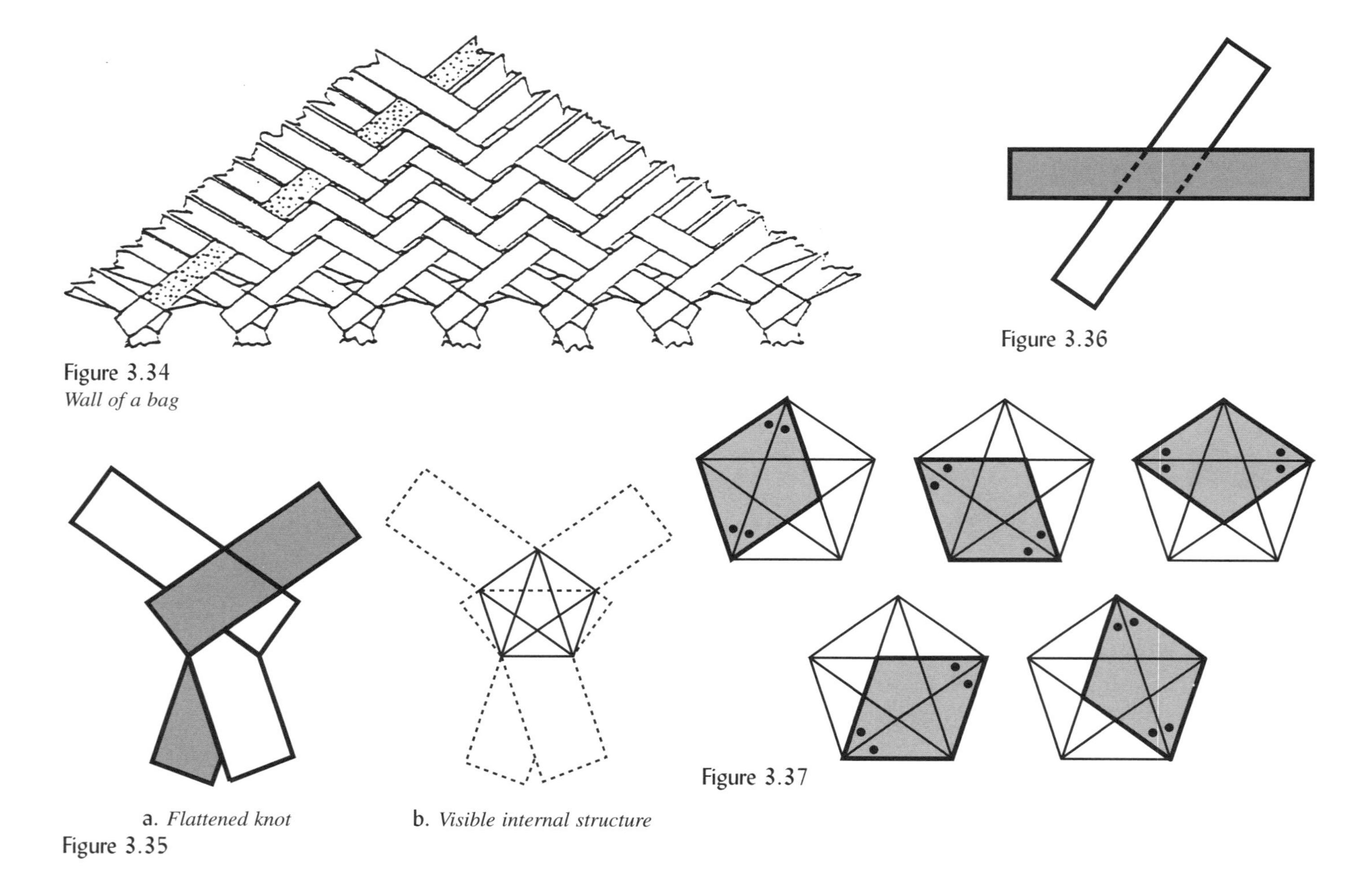

Figure 3.34
Wall of a bag

Figure 3.36

Figure 3.37

a. *Flattened knot* b. *Visible internal structure*
Figure 3.35

An attractive way—in the view of my student teachers—to prove the regularity of the pentagonal knot is based on the property that two crossing strips of equal width determine a rhombus (see Figure 3.36).

In a rhombus the diagonals are bisectors. Exploring this property for the various rhombi that appear in the pentagonal knot (Figure 3.37), it follows that all the angles marked with a dot are congruent and all sides marked with a stripe are also congruent (Figure 3.38). This implies the regularity of both the pentagon and the pentagram.

Students may be invited to investigate if, analogously, it is possible to construct knots with more than two strips of the same width, which when held up to the light reveal regular polygons of higher order.

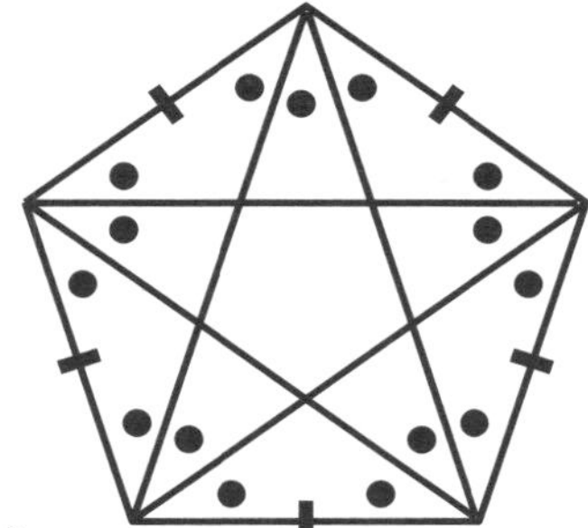

Figure 3.38

Figure 3.39
Eheleo-*funnel*

5 Exploring a woven pyramid

(cf. Gerdes, 1988, 1990b, 1999a)

In the North of Mozambique, in the South of Tanzania, in the Congo/Zaire region, and in Senegal, a pyramidal basket is woven (cf. Weule, p.10 and Table 19; Torday and Joyce, p. 337; Lestrange and Gessain, p. 118). In Mozambique and Tanzania it is used as a funnel in the production of salt. As Figure 3.39 shows, the funnel is hung on a skeleton and salt containing earth is put in it. A bowl is placed beneath the funnel, hot water is poured on the earth in the funnel, and saltwater is caught in the bowl. After evaporation, salt remains in the bowl.

The basket has the form of a triangular pyramid: its base (i.e., the top of the funnel) is an equilateral triangle and its other three faces are isosceles right triangles. In the following chapter, I will present some ideas of how both the form and the production process of the funnel—called *eheleo* in the Makhuwa language in the North of Mozambique—can be explored in the context of mathematics education.

Volumes of pyramids and regular polyhedra

Figure 3.40 shows an *eheleo*-pyramid as part of a cube. Let s be the side length of the cube. It is not necessary to know the formula for the volume of a pyramid to determine the volume of a *eheleo*-pyramid. Its volume is equal to one half of the pyramid $ABCDH$ with the same height (DH) and with the cube's base as base (see Figure 3.41), as may be assumed by comparing all corresponding horizontal layers of both pyramids.

As the last pyramid fits three times into the cube ($ABCDH$, $ABFEH$, and $BCGFH$), the volume of the *eheleo*-

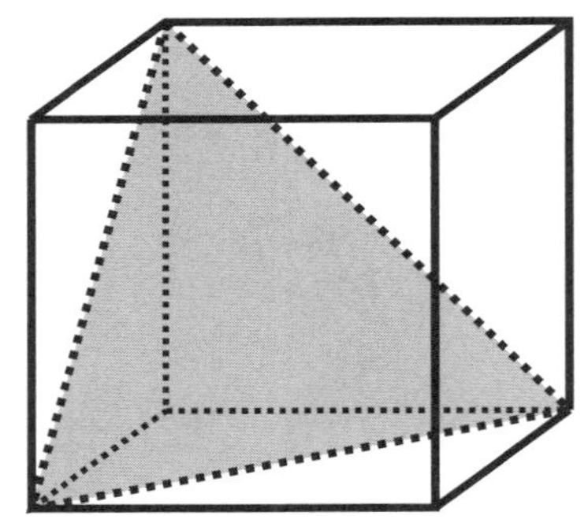

Figure 3.40
Eheleo-*pyramid as part of a cube*

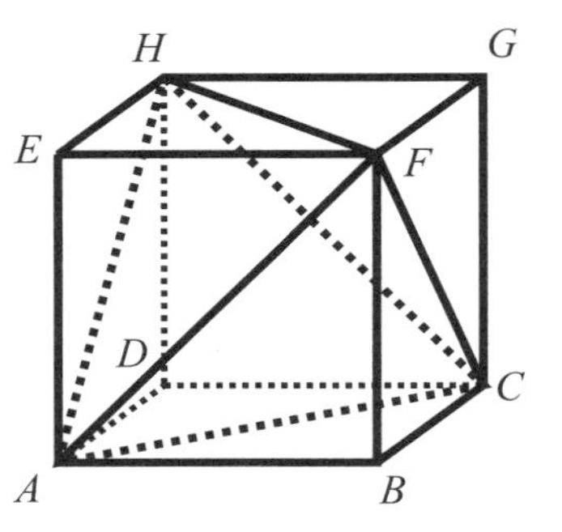

Figure 3.42

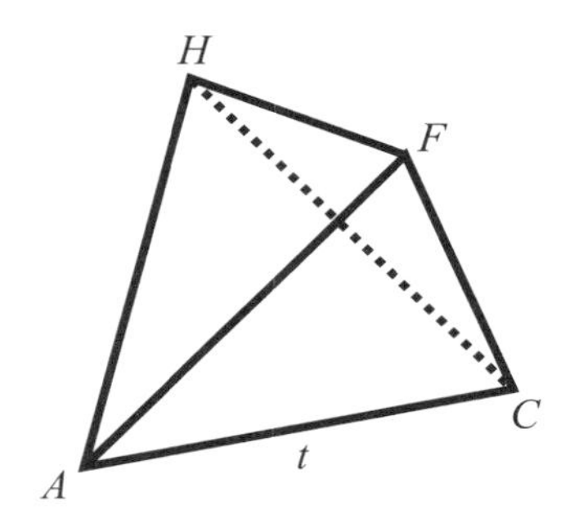

Figure 3.43

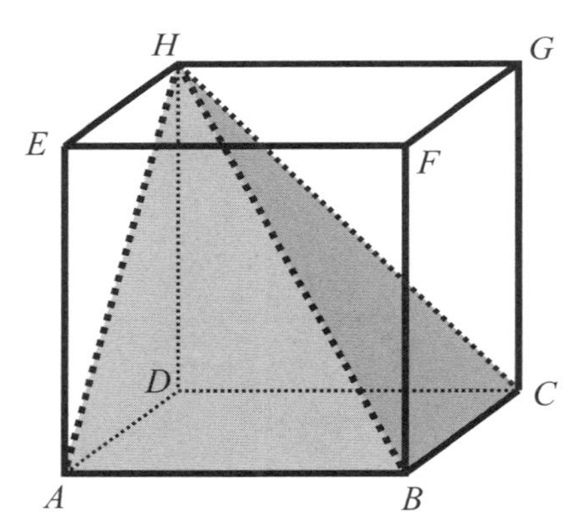

Figure 3.41

pyramid (V_E) is one sixth of the volume of the cube:

$$V_E = \frac{s^3}{6}. \tag{1}$$

Trying to fill a cube with *eheleo*-pyramids, one sees that six do not fit into the cub, but the four *eheleo*-pyramids *ADCH, ABCF, CGHF,* and *HEAF* surround the regular tetrahedron ACFH and together these five pyramids fill the cube (see Figure 3.42). Also, the volume (V_T) of the regular tetrahedron *ACFH* (Figure 3.43) is twice the volume of the *eheleo*-pyramid or one third of the volume of the cube. In other words:

$$V_T = \frac{s^3}{3} = \frac{t^3\sqrt{2}}{12}, \tag{2}$$

where t represents the side length of the regular tetrahedron. Joining eight *eheleo*-pyramids as in Figure 3.44, one obtains a regular octahedron. Its volume (V_O) is eight times the volume of the *eheleo*-pyramid, and we find that

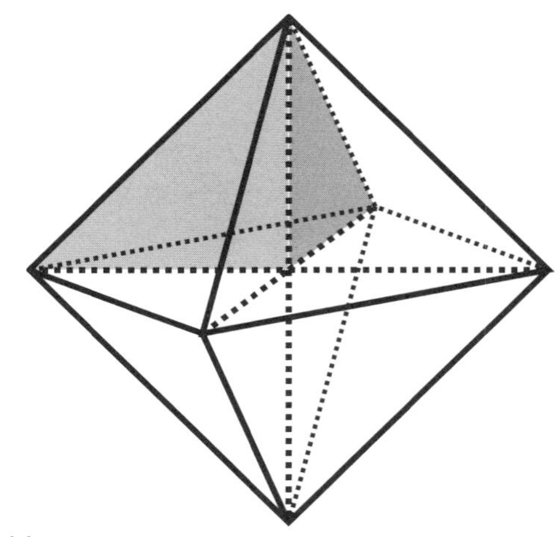

Figure 3.44

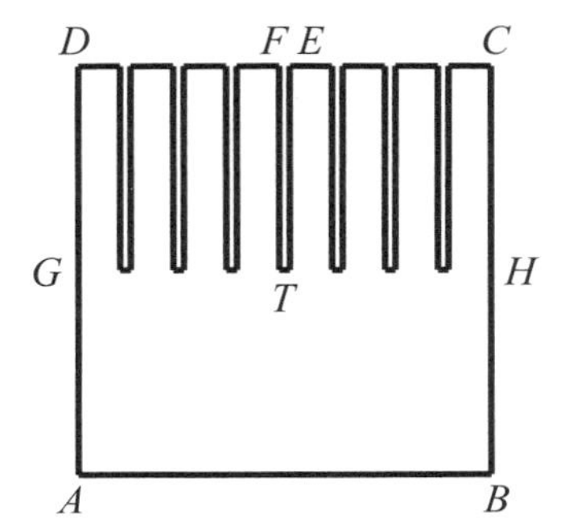

Figure 3.45

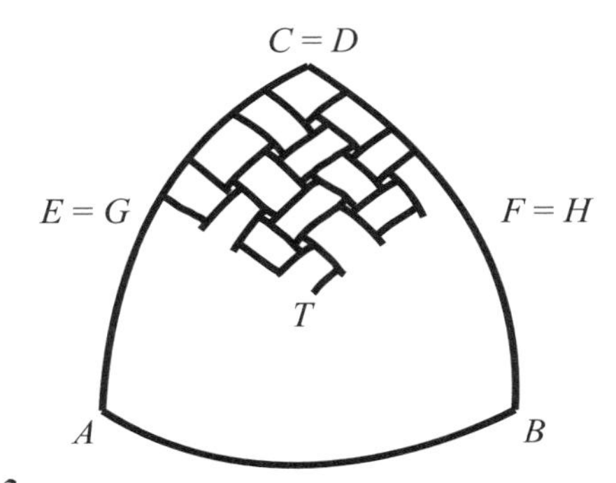

Figure 3.46

(3) $$V_O = \frac{8s^3}{6} = \frac{t^3\sqrt{2}}{3},$$

where t represents the side length of the regular octahedron. In other words, the volume of a regular octahedron is four times the volume of a regular tetrahedron of the same side length.

An alternate construction of regular polygons

Artisans weave a *eheleo*-funnel in the following way. They start by making a square mat *ABCD*, but do not finish it: with the strands in one direction (horizontal in our figure), the artisan weaves only until the middle (*GH*, see Figure 3.45, where we represented only 8 strands in the vertical direction—in reality there are some tens).

Then, instead of introducing more horizontal strips, he interweaves the vertical strands on the right (between *C* and *E*) with those on the left (between *F* and *D*). In this way, the mat does not remain flat but is transformed into a basket (see Figure 3.46). The center *T* goes downwards and becomes the vertex of the *eheleo*-funnel. To guarantee a stable rim, the artisan fortifies the edges *AB*, *BC*, and *AC* with little branches (see Figure 3.47). Finally, the funnel has the form of a triangular pyramid.

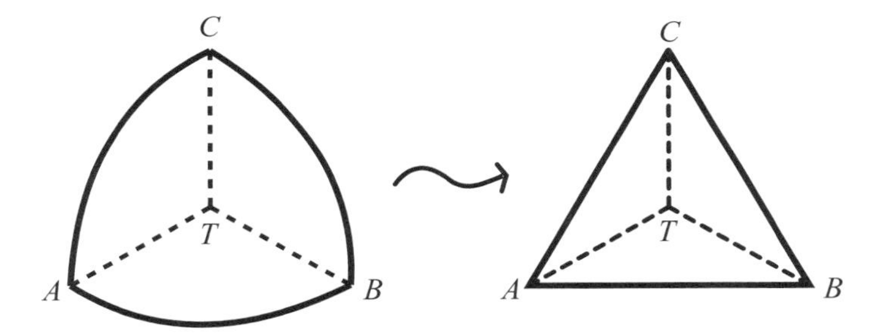

Figure 3.47

The artisan transforms the square mat *ABCD* into a pyramidal basket *ABCT*, whose base *ABC* is an equilateral triangle. Does this production process suggest a method to construct an equilateral triangle? The objective of the artisan was not to produce an equilateral triangle, but a useful funnel. Can we simplify the artisans' method if we only want to construct an equilateral triangle?

By simplifying the *eheleo*-funnel production process, the following diagrams (Figure 3.48 to Figure 3.51) show how to produce a pyramid with an equilateral triangle as base, out of a square of cardboard or paper.

Step 1) Folding the diagonals AC and BD (Figure 3.48).

Step 2) Folding the apothem FT (see Figure 3.49).

Step 3) Joining the triangles DFT and CFT until C and D coincide; F goes up, T goes down (see Figure 3.50).

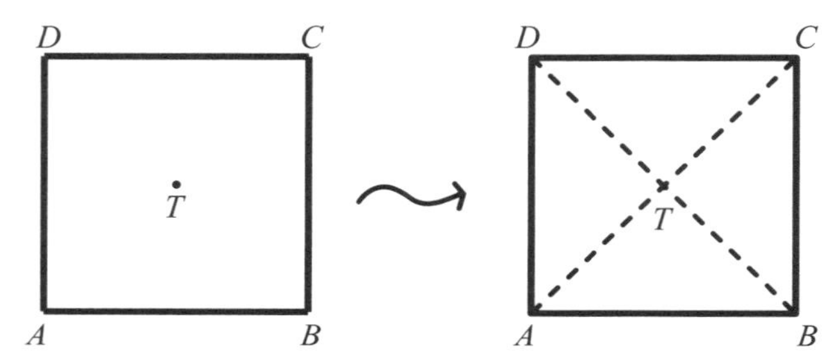

Figure 3.48

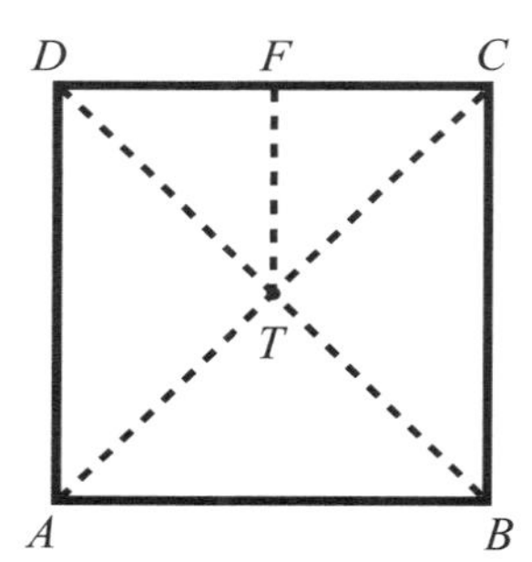

Figure 3.49

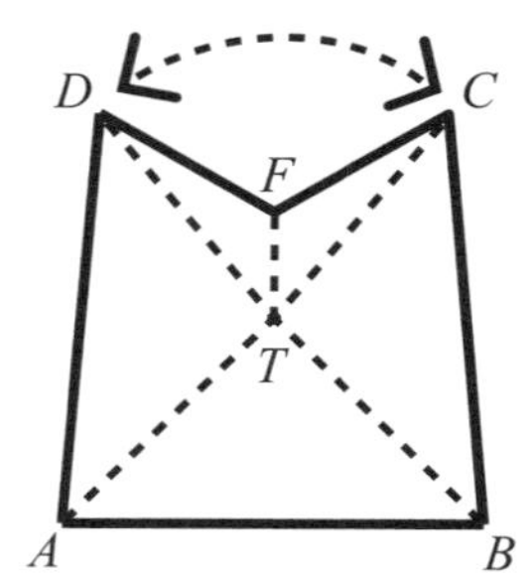

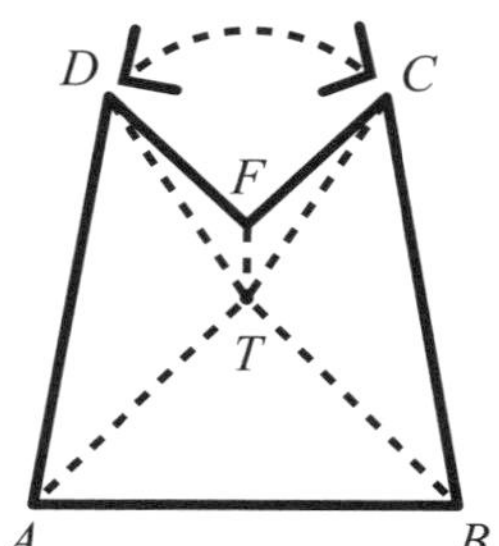

Figure 3.50

Step 4) Fixing the 'double triangle' DFT to the face ATC, for example, with a paper clip (see Figure 3.51).

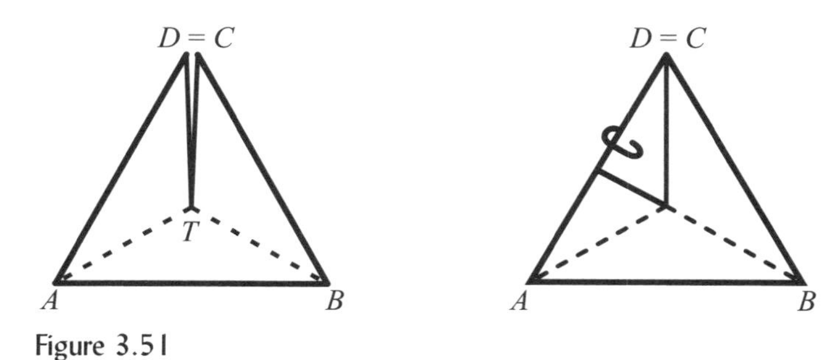

Figure 3.51

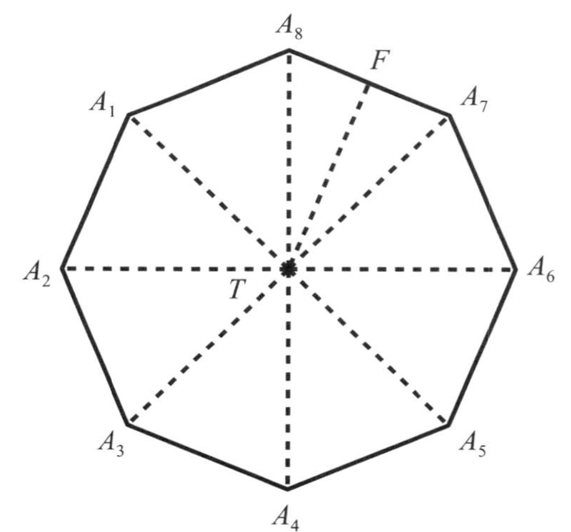

Figure 3.52

Students may be asked if this simplified method can be generalized. Starting with a regular octagon, is it possible to transform it analogously into a regular heptagonal pyramid? Figure 3.52 shows the first two steps of the transformation of the regular octagon: folding the principal diagonals and bisector FT. Next, one joins the triangles A_7FT

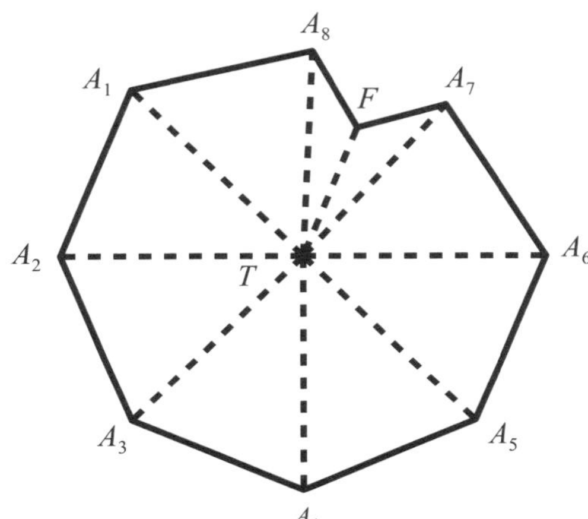

Figure 3.53

and A_8FT until A_7 and A_8 coincide; F goes up, T goes down (see Figure 3.53).

Now one has to fix the 'double triangle' A_8FT to the face A_1TA_8 (see Figure 3.54). In this way, the regular octagon was easily transformed into a pyramid that has a regular

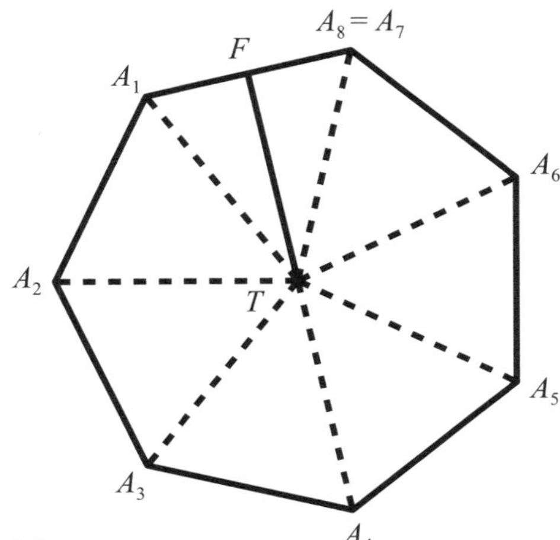

Figure 3.54

heptagon as base. Repeating the folding of an apothem, the regular heptagonal pyramid, may be transformed quickly into a regular hexagonal pyramid. Once more, folding another apothem, one can transform the regular hexagonal pyramid into a regular pentagonal pyramid.

Reflecting on this experience, students can be asked to construct a regular 15-gon and 11-gon.

As a (2^n)-gon is easy to fold, starting by doubling the central diagonals of a (2^{n-1})-gon and each time that the simplified *eheleo*-method is applied, the number of sides of a regular polygon (or of the regular polygonal base of a pyramid) decreases by 1, it can be concluded that all regular polygons can be constructed in this way. Once arrived at this point, it is possible to look back and ask the students: "Did we learn something from the artisans who weave the *eheleo*-funnels?" "Is it possible to construct a regular heptagon using only a ruler and a compass?" "And with the *eheleo*-method?"

Let D represent the operation by which an m-gon is transformed into a $(2m)$-gon, and C the operation by which a n-gon is transformed into a $(n - 1)$-gon, m and n being natural numbers (m>2, n>3). In this way D^4 followed by C^5, or abreviately D^4C^5, applied to a square, corresponds to the transformation of the square into a regular 64-gon followed by the transformation of the regular 64-gon into a pyramid with a regular 59-gon as a base. With D^3C^8 we may produce a regular 24-gon starting with a square ; are there other ways to do so? With D^3C^4, D^2C^2D and DCD^2 we are able to produce a regular 28-gon starting with a square; which is the quickest way? And so on.

Looking back

In this section, I presented some ideas of how both the form and the production process of the *eheleo*-funnel may be explored in the context of mathematics education. The ideas ranged from a study of volumes of pyramids and regular tetra- and octahedra to an alternative construction of regular polygons.

6 Exploring square mats and circular basket bowls

In eastern and southern Africa (cf. Stuhlman, 1910, p.43) a type of basket bowl is common that is used as a sieve or as a dish for food (see the example in Figure 3.55). Among the Makonde in north Mozambique this basket bowl is called *chelo*.

A Makonde weaver makes the bottom of a *chelo* by plaiting a square mat. To produce the border of the basket, the weaver bends a wooden board and fastens its two ends one to another. Now the artisan fastens the sides of the mat at their respective midpoints to the border (see Figure 3.56).

Figure 3.55
A bowl basket seen from above

Figure 3.56
Fastening a mat to the circular border

Then the weaver wets the mat to turn it more flexible and with a foot presses the mat uniformly inwards. To conclude, he cuts off the outstanding parts of the mat and fastens the rest of the bottom to the border. The reader is invited to reflect on what would happen to the basket bowl if the initial mat was not square.

To guarantee that the initial mat is really a square and to know easily where are the mid points of its sides, the artisan weaves the mat with a design like the one in Figure 3.57 that shows him all the information at a glance: both middle lines and the center of the square are immediately visible. These middle lines become perpendicular diameters of the basket bowl (see Figure 3.58).

Discovering properties of circles

A flat basket bowl may be explored in the primary or middle school classroom to discover properties of circles. For example, in an experimental text for grade 8 in Mozambique (Gerdes, 1990a), I used the flat basket bowl heuristically to find the unknown center of a circle and to construct tangent lines.

Consider a strand. Let A and B be its end points at the circular border. AB is parallel to one of the visible diameters (PQ), and therefore perpendicular to the other diameter (RS)

Figure 3.57

(see Figure 3.59). Turning the basket bowl around its axis *RS*, one interchanges the positions of *A* and *B*. Based on this symmetry, we have $AM = MB$, where *M* denotes the place where the strands *AB* and *RS* cross. This observation may lead students to discover a circle property: its center lies on each of the perpendicular bisectors of its chords. From here it is not difficult to find a method to complete a circle, given an arc.

If one places a smaller circular border on a square mat in such a way that their centers coincide, it may be seen that some strands have two 'points' below the border, whereas others have no 'point' at all (Figure 3.60).

Going from the left strand in Figure 3.60 to the right strand, one passes through one very particular strand: the one where the strand has only one point below the circular border (See Figure 3.61). Abstracting from this experience (Figure 3.62), the students may discover the basic property of a tangent line to a circle: it is perpendicular to its corresponding diameter.

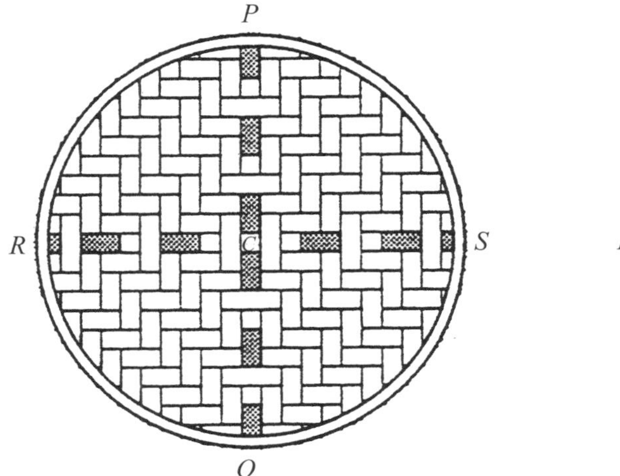

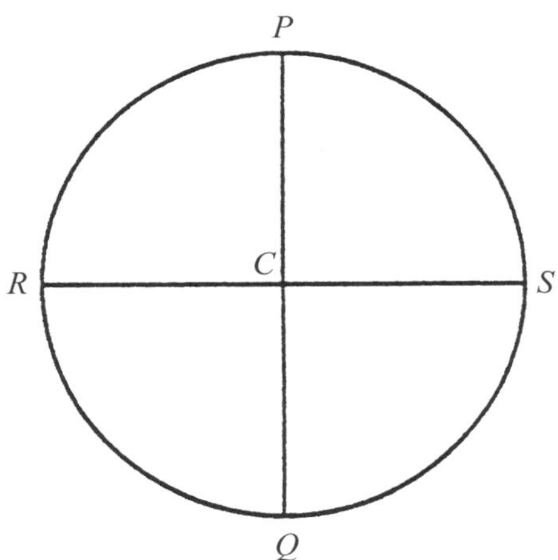

Figure 3.58

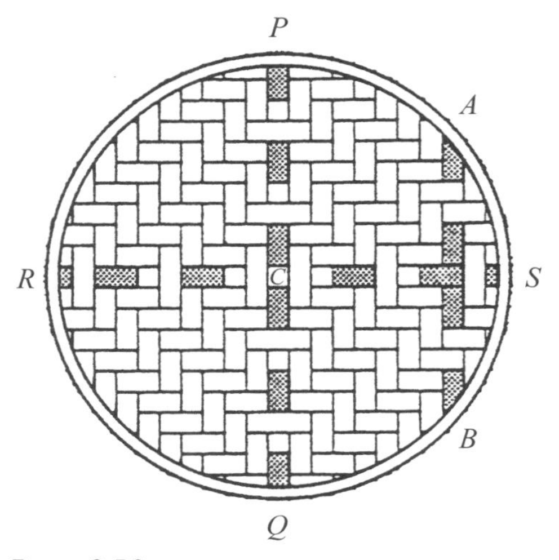

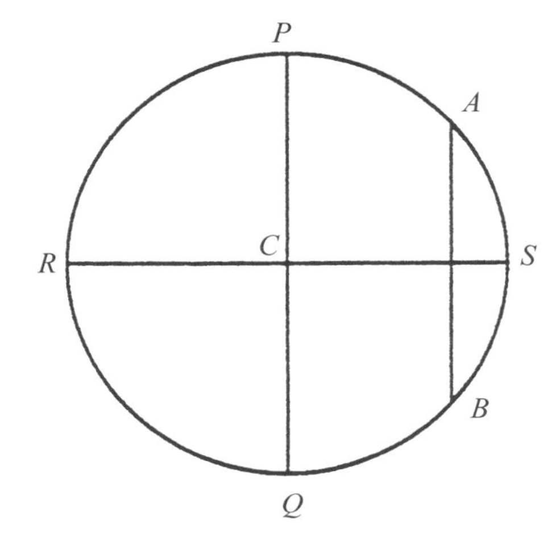

Figure 3.59

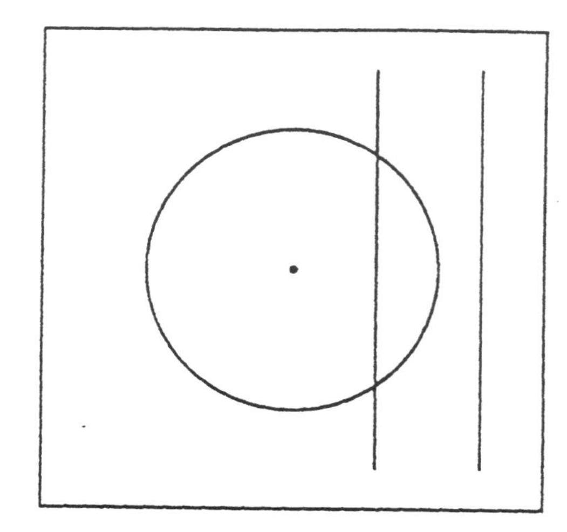

Figure 3.60

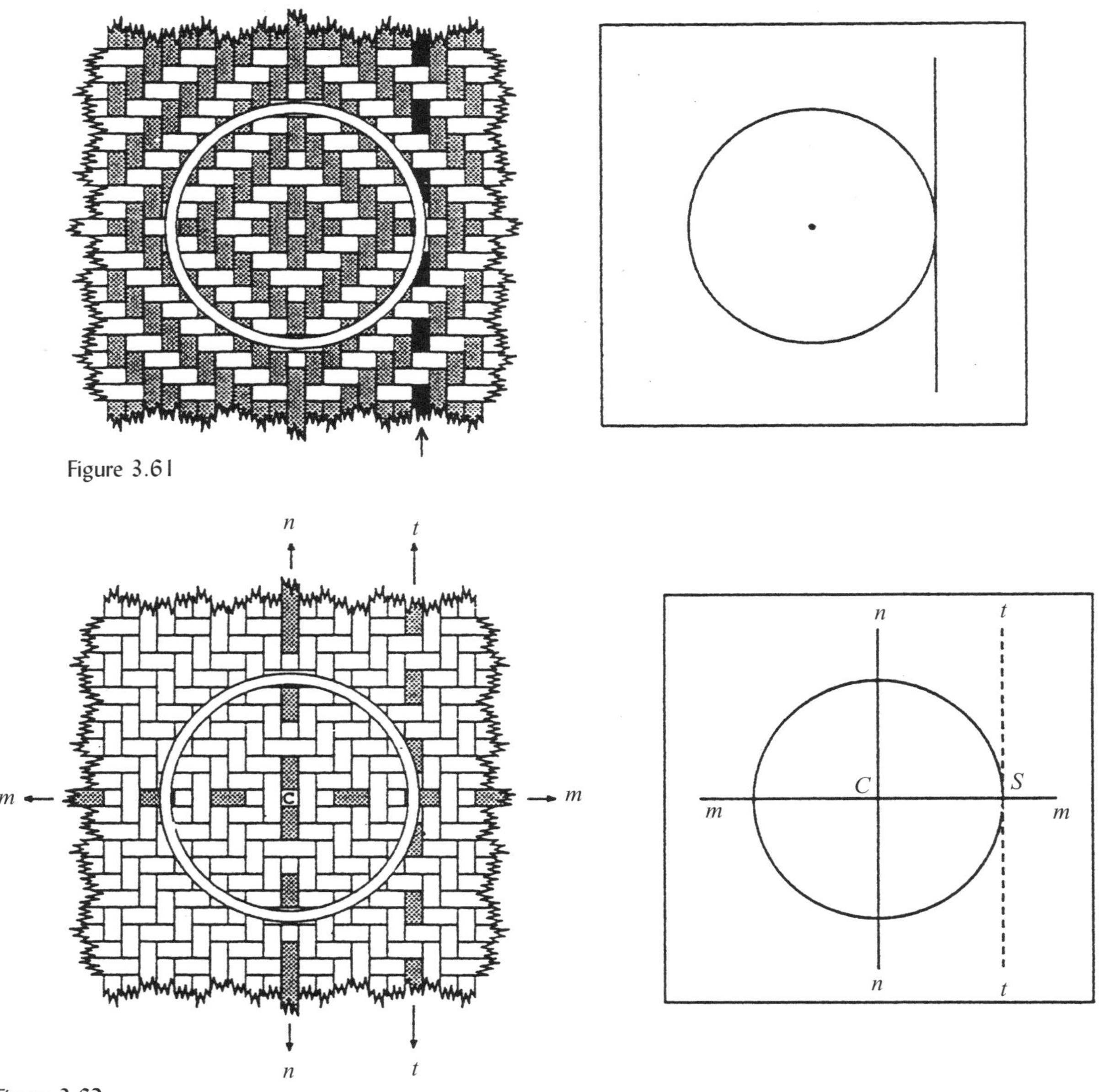

Figure 3.61

Figure 3.62

Discovering an approximation of the area of a circle

The experience with the designs of basket bowls may be explored to discover an interesting approximation formula for the area of a circle. Students may be invited to construct circles in a network of equidistant concentric squares. Let this be done in such a way that the circles have the same centers as the squares. When does there appear a circle that seems to have an area almost equal to one of the squares?

When the circle radius (r) measures four times the half diagonal of the first square, the circle area (A_c) seems more or less to be equal to the area of the square (A_s) of which the half diagonal measures five times the half diagonal of the smallest square (See Figure 3.63). In other words:

$$A_c \approx A_s.$$

As A_s is twice the square of its semidiagonal and this semidiagonal is five fourth of the circle radius, it follows that:

$$A_c \approx A_s = 2\left(\frac{5}{4}r\right)^2 = \frac{25}{8}r^2.$$

This good approximation corresponds to $\pi \approx 25/8 = 3\ 1/8 =$ 3.125 (relative error is about half a percent), and is equivalent to the ancient Babylonian formula for the more precise determination of the area of a circle (cf. Bruins & Rutten, 1961; Gerdes, 1990b, 260–263; 1999a).

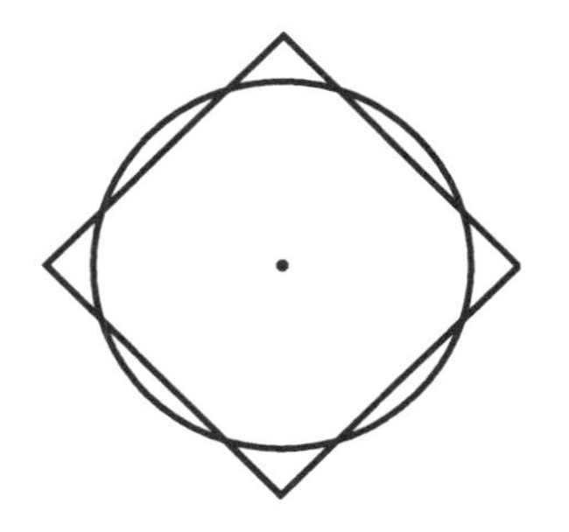

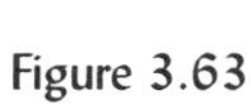

Figure 3.63

7 Exploring hexagonal weaving: Part I

In Mozambique, artisans make hats (see Figure 3.64) by knitting together successive circuits of a spiral. Figure 3.65 illustrates the idea. The spiral consists of a woven zigzag band.

The zigzag band is made out of two long palm strips of the same width. Artisans also use this technique to construct handbags. The same type of zigzag band is also known in other parts of Africa, e.g., in Nigeria it is similarly used for

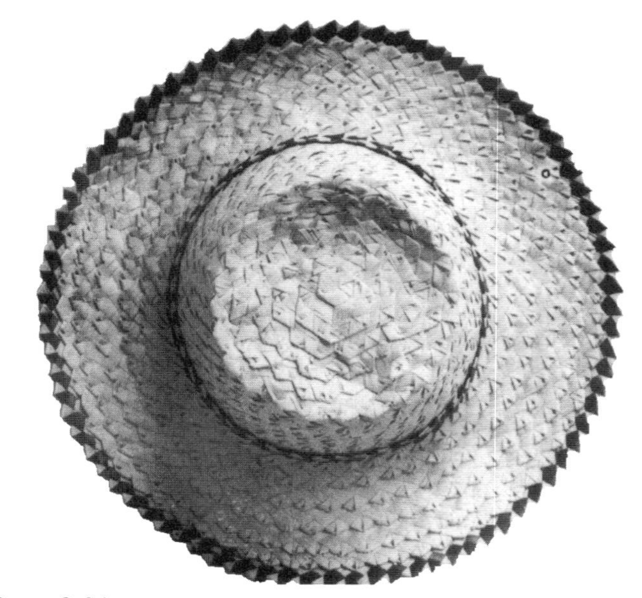

Figure 3.64
Hexagonally woven hat seen from above

Figure 3.65

Figure 3.67

Figure 3.68

making hats and in Kenya it appears as a decoration of circular mats*.

Constructing a zigzag band

A zigzag band may be produced in the following way: One strip (in the schools cardboard strips can be used) is wrapped around another of the same width in such a way that the angle between both is 60° (see Figure 3.66a and b).

Now the horizontal strip is wrapped around the first strip, as in Figure 3.67. In this way, the two strips lie parallel next to each other after wrapping; this makes further weaving possible, as will be seen. If the initial angle were different from 60°, then further weaving would be impossible, as the reader may verify.

Wrapping the left strip around the other strip and weaving it one-over-one-under the strip parts in the middle, one obtains a stable knot, as in Figure 3.68. Continuing to wrap and weave over-and-under, first from the left side, and then alternately twice on the right side and twice on the left side, one produces a straight zigzag band (see Figure 3.69).

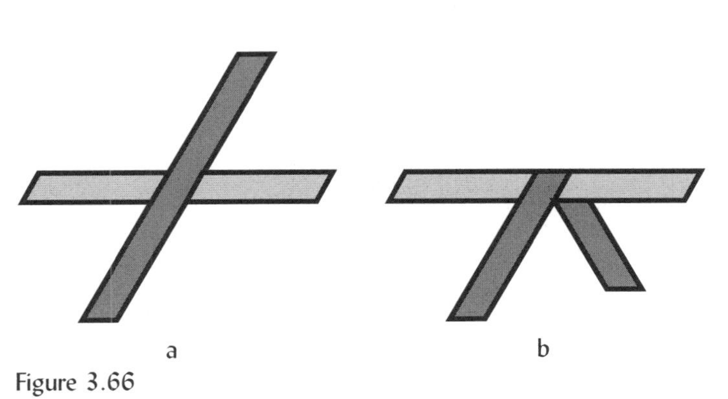

Figure 3.66

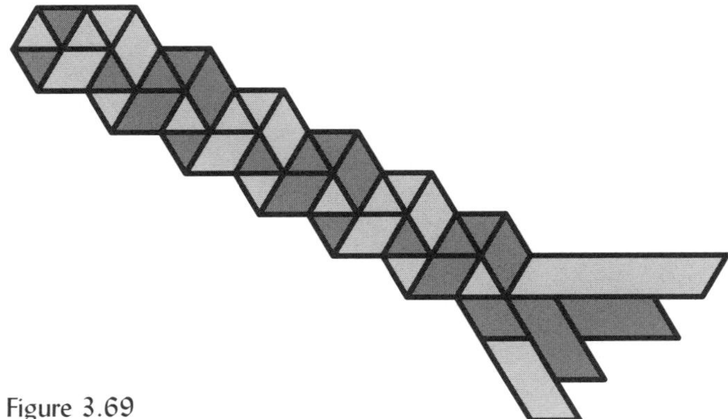

Figure 3.69

* The hexagonal woven strip is also used on Hawai, in Vietnam, China, New Zealand. and among indigenous peoples of Latin America (cf. Gerdes, 1990b, p. 85; 1999a, # 3.3).

Figure 3.70

If during the weaving process the alternation twice-twice is broken, other bands appear. Students can be invited to analyze how a hat weaver might start to obtain a spiral zigzag-band which successive circuits may be knitted together.

Students can be invited to invent their own zigzag bands. They can be asked which letters can be woven (see the example of letter V in Figure 3.70). Furthermore they may be asked to invent other designs that can be built up starting with knots as shown in Figure 3.68. The hexagonal stars presented in Figure 3.71 are two possible designs. How many strips are needed to weave each of them?

Figure 3.71

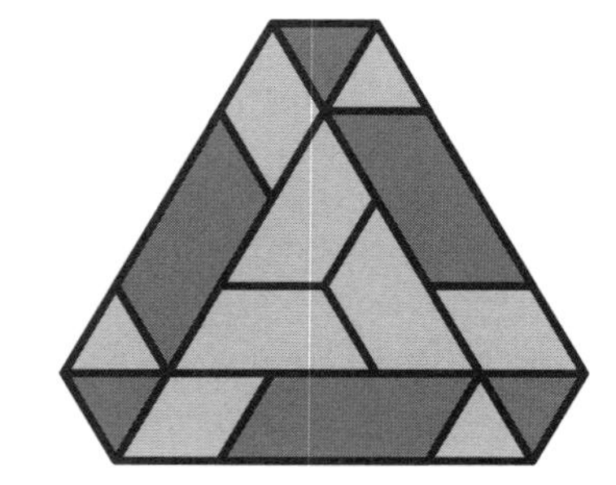

Figure 3.72

Figure 3.73

Figure 3.72 shows two designs each of which can be woven with two strips.

This type of designs may be used to build mosaics, and pose questions like which geometrical figures can be filled up with, for example, only trapezoids of the form illustrated in Figure 3.73.

Students learning to weave and explore the hexagonal zigzag band may be stimulated to develop their creativity: inventing new forms and reflecting on them. They will also feel in their fingers why the angle of 60° is crucial in making the designs, and any deviation from this angle at any moment will provoke bigger distortions later on, and eventually make further weaving impossible.

8 Exploring hexagonal weaving: Part 2

In the north of Mozambique, Makhuwa craftsmen weave their light transportation baskets, called "litenga", and their fish traps "lema" with a pattern of regular hexagonal holes.

Figure 3.74
Three direction weaving pattern

The strands are woven one-over-one-under in three directions leading to a very stable fabric (see Figure 3.74).

The same hexagonal basket weaving technique has been used in several other regions of Africa. It is used in Madagascar to make fish traps and transport baskets. In Kenya the technique is used for making cooking plates, and among the Pygmies (Congo/Zaire) and in Cameroon for weaving carrying baskets, in northeastern Congo/Zaire among the Meje for covering pots and among the Mangbetu for weaving hats (cf. Faublée, pp. 19, 28, 38; Somjee, p.96; Meurant and Thompson, p. 162; Etienne-Nugue, p. 67; Schildkrout and Keim, Fig. 6.16, p. 168). Not only among African but also among several indigenous peoples of the Americas and Asia there also exist traditions of using the hexagonal basket weaving technique*.

There exist many ways to explore this weaving technique in mathematics education, as the examples I present in this chapter will illustrate. The examples range from the study of tilings, finite designs, goniometric functions, and polyhedra. It will also become clear that attractive connections with artistic and science education may be constructed.

Constructing the regular hexagon pattern

The following is one way for students to discover the regular hexagonal weaving pattern. They can be asked how to wrap a flexible cardboard strip around another of the same width. If the angle between the two strips is rather small, like in Figure 3.75, then there remains open space between successive windings. If the angle is rather big, then wrapping is materially impossible.

What is the best angle for which wrapping is possible, that is the angle with which successive windings are adjacent? The correct angle seems to be an angle of 60° (see Figure 3.76). How may this be proven?

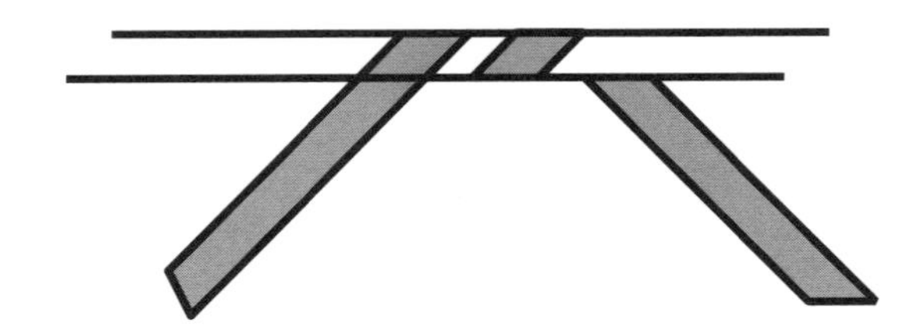

Figure 3.75

* For instance, the Ticuna and Omagua in Brazil, the Huarani in Ecuador and the Yekuana in Venezuela and Guyana use the hexagonal weaving technique to make carrying baskets. The Micmac-Algonkin Indians of East Canada use it for their large snow-shoes, as do Eskimos in Alaska. In Asia, the use of the hexagonal basket weaving technique is well spread, from the Munda in India, the Kha-Ko in Laos, to Malaysia, Indonesia, China, Japan, and the Philippines (cf. Gerdes, 1990b, 1999a).

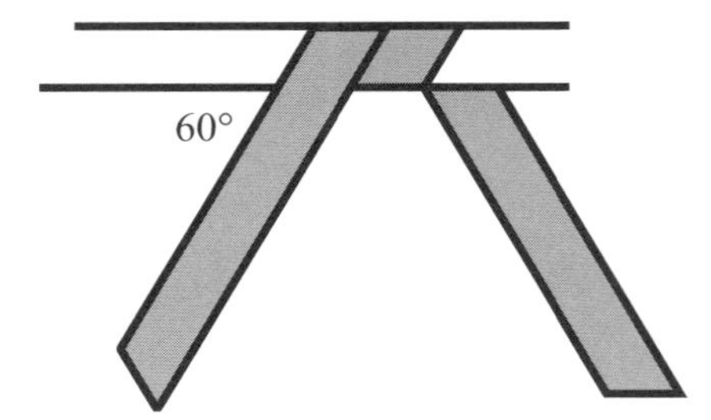

Figure 3.76

By wrapping similarly more strips around the initial horizontal strip (see Figure 3.77) and then introducing more horizontal strips (Figure 3.78), one gets the weaving pattern of regular hexagonal holes with a top border.

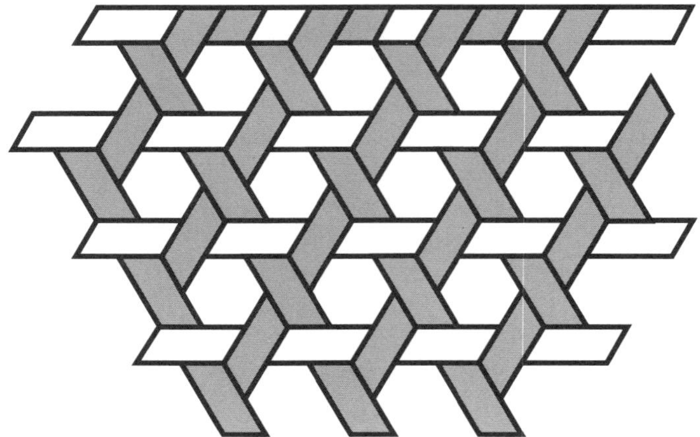

Figure 3.78

Tilings (cf. Gerdes, 1988)

Once students construct a piece of the regular hexagonal weaving pattern, they can discover several tilings of the plane. First they may be asked to place the piece on the top of a sheet of paper and then draw the sides of the hexagonal holes. Continuing only with the sheet of paper, they may link certain vertices by segments and discover the regular hexagonal tiling, as Figure 3.79 illustrates.

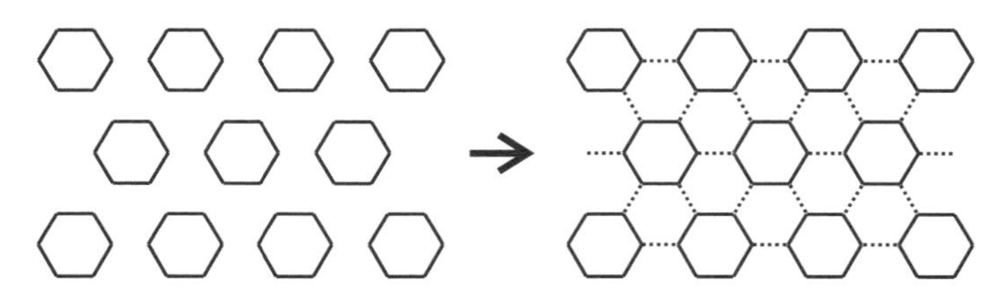

Figure 3.79

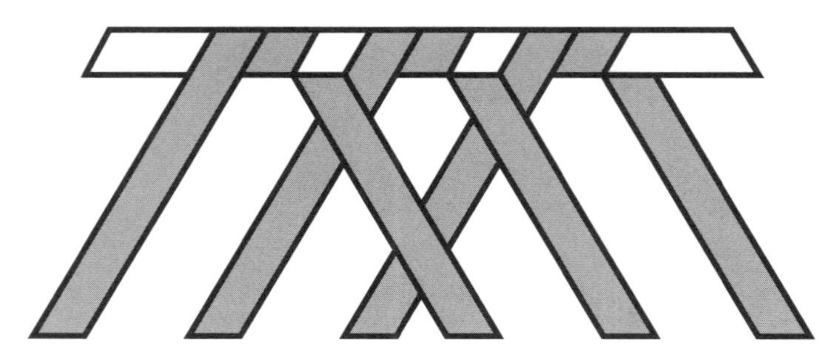

Figure 3.77

Once they have found the regular hexagonal tiling, students can discover other tilings by subdivision, as the examples in Figure 3.80 show. And once students find the tiling with equilateral triangles, they can construct other polygons. By considering these figures, students may formulate conjectures:

- the sum of the measures of the internal angles of a n-gon is equal to $3(n-2) \times 60°$ (see Figure 3.81);
- areas of similar figures are proportional to the squares of their sides, the sum of the first n odd numbers is n^2 (see Figure 3.82).
- the sum of the first n odd numbers is n^2 (see Figure 3.82).

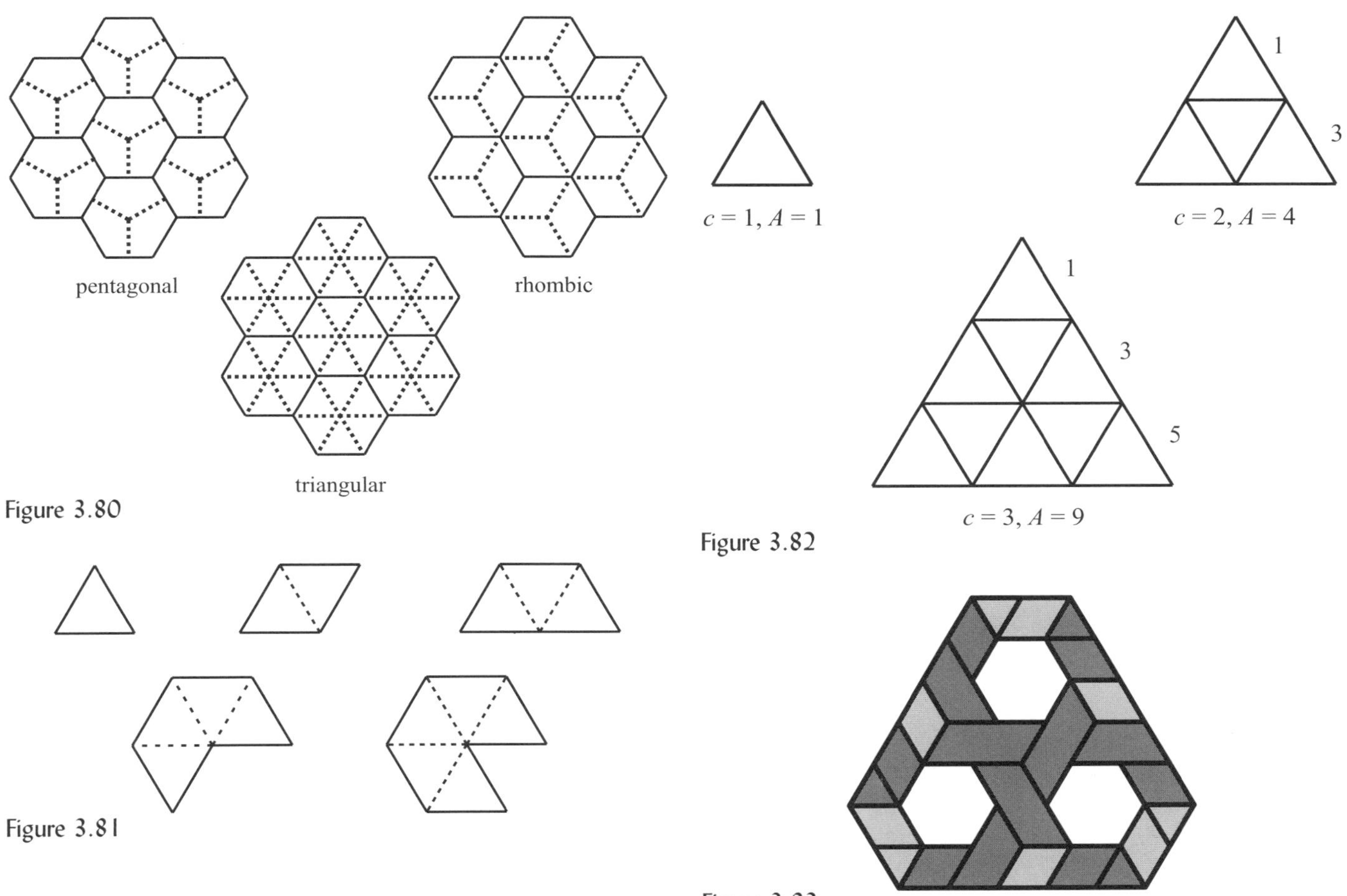

Figure 3.80

Figure 3.81

Figure 3.82

Figure 3.83
Emblem of Mozambique's Ethnomathematics Research Project

Once these conjectures are formulated, there arises the question of how to prove them.

Weaving designs

Returning to the weaving pattern with a top border shown in Figure 3.78, it is interesting to explore with students ways to complete such a woven design with additional borders.

Figure 3.83 displays an example of a *closed hexagonally woven design*. It may be woven out of two strips. This design is the emblem of Mozambique's Ethnomathematics Research Project, as it symbolizes the development of new mathematical ideas and forms inspired by traditional cultural knowledge and know-how.

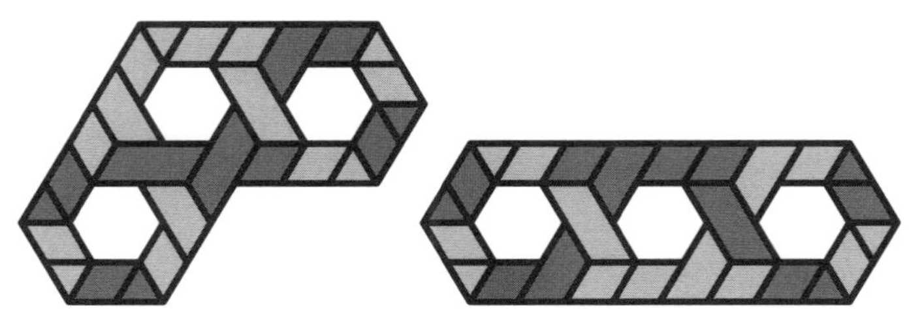

Figure 3.84

This emblem is not the only closed hexagonally woven design with three holes in it. Figure 3.84 presents the other two possibilities. Both may also be woven with only two strips. The three designs display different symmetries. Students can be asked to invent other closed hexagonally woven designs with given properties. For example, how many are there with four holes, with five holes, etc.* Do there exist closed hexagonally woven designs that may be woven out of only one strip?

Figure 3.85 presents two closed hexagonally woven designs with four and six holes respectively, each of which can be woven out of only one strip. How can this be realized?

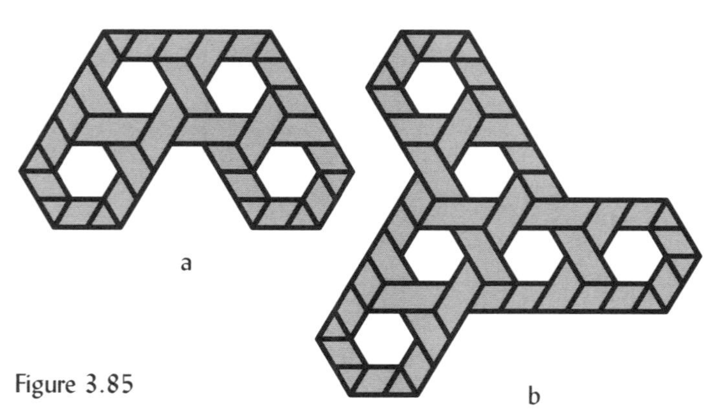

Figure 3.85

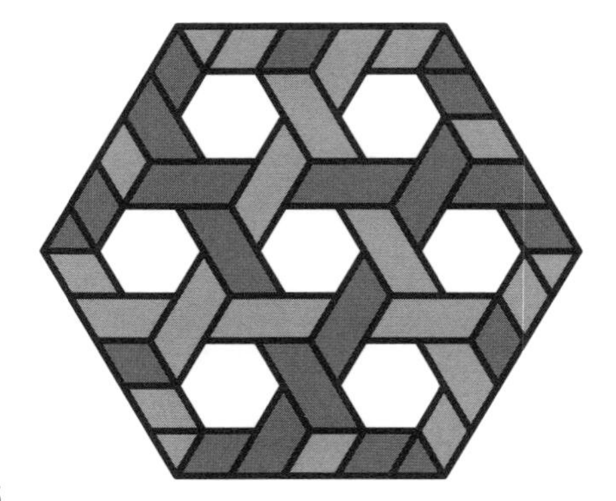

Figure 3.86

Figure 3.86 presents a closed hexagonally woven design, made out of two strips, with sixfold symmetry (if we do not take into account the color of the strips). Is it the smallest possible closed hexagonally woven design with this symmetry?

By introducing a small variation in the way this closed design was woven, another beautiful design can be produced, also made out of two strips, with an hexagonal star-like hole in its middle (see Figure 3.87). Is it the smallest closed hexagonally woven design that can be adapted in such a way that a star appears in its center?

A trigonometric function

In all the exploratory weaving with strips so far, the angle of 60° (and its supplementary 120°) plays an essential role. This angle was materially enforced (see once more Figure 3.76) when using two strips of the same width. We can ask what will happen when we start with strips of different

* This problem is equivalent to finding the number of n-hexes. Polyhexes or hexagonal animals are connected combinations of regular hexagons (cf. Golomb, pp. 90–93, 123–126; see figures).

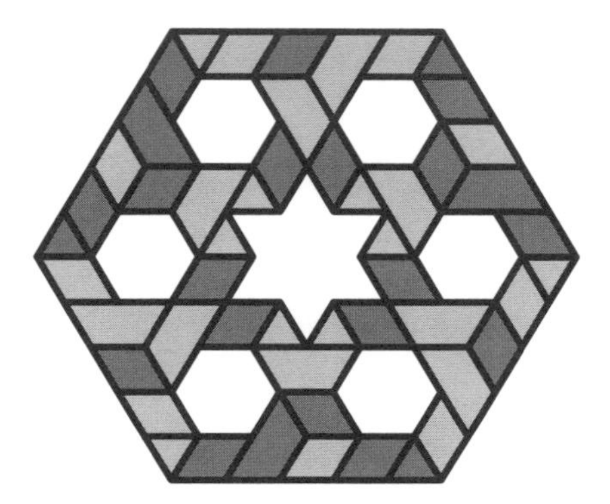
Figure 3.87

widths: how will that special angle depend on them? Let us take the width of the strip to be wrapped around the 'horizontal' strip as unit of measurement, and let the 'horizontal' strip have width a. (see Figure 3.88).

How does the wrapping angle α depend on a?

$$\alpha = \text{hex}(a)$$

How does α vary? Students may measure both α and a and construct a graph (Figure 3.89). How does this trigonometric function relate to the well known trigonometric functions? The following relationship can be easily established

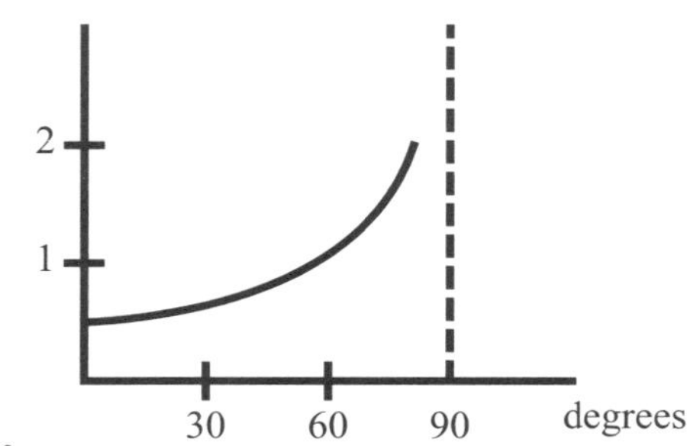

Figure 3.89

$$a = \text{hex}^{-1}(\alpha) = \frac{1}{2\cos\alpha}.$$

Another question for students to investigate, is whether it is possible to weave patterns and closed designs with strips of two different widths (1 and a)?

Cylinders

Basket makers who use the hexagonal weaving technique know that it is possible to make cylindrical parts of baskets using the technique. Students may discover that for themselves when trying to curve a hexagonally woven band with upper and lower borders (see in Figure 3.90 a band

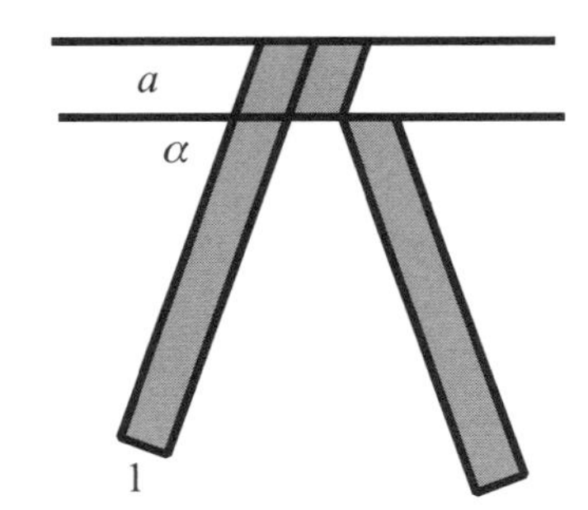

Figure 3.88

Figure 3.90

with three layers of hexagons) and joining the outstanding strips.

Students may now be asked to try to invent other spatial figures that can be woven having only hexagonal holes and borders. For example, they may discover interesting twisted figures, such as the one that may be thought of as twice the design in Figure 3.85b whereby the outer hexagons coincide.

Bowls and domes

Basket makers who use the hexagonal weaving technique also know that when they want to curve a woven surface more than what is done with a cylindrical surface—in other words, when they want a basket with corners—that this is possible if they introduce one or more smaller holes, normally pentagonal in shape (see Figure 3.91).

Students may discover this when we ask them, for example, to reflect on what would happen with the closed hexagonally woven design in Figure 3.86. If they weave it in such a way that the central hexagonal becomes a pentagonal hole, they will find that the design 'comes upwards' with the five border sides, while the pentagonal hole 'stays on the ground.' In other words, the design is curved in space, forming, let us say, a woven bowl. Turning the bowl upside down, we have an interesting dome structure.

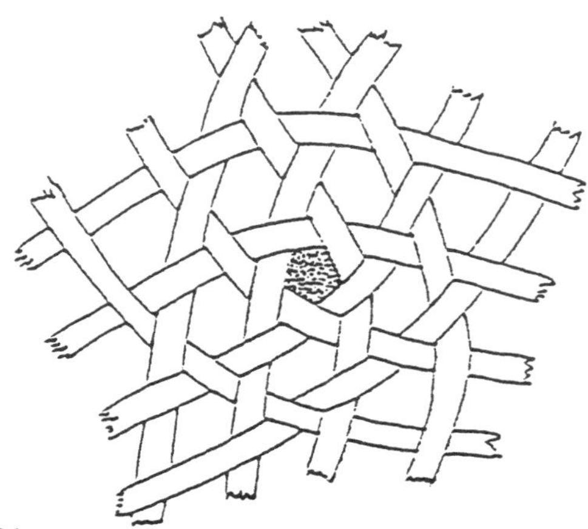

Figure 3.91

Having arrived at this point, students may conjecture and construct various attractive bowl and dome structures. For example,

- How will the dome look if we construct it with a hexagonal hole on the top surrounded by pentagonal holes?;
- How will the dome look if we construct it with a hexagonal hole on the top surrounded alternately by pentagonal and hexagonal holes?;
- How will the dome look if we construct it with a pentagonal hole on the top surrounded by one layer of pentagonal holes ?;
- How will the dome look if we construct it with a pentagonal hole on the top surrounded by one layer of hexagonal and one layer of pentagonal holes?;
- How will the dome look if we construct it with an pentagonal hole on the top surrounded by two layers of pentagonal holes? [dome A]

Students will discover that these bowl and dome structures are very stable, just as the hexagonally woven baskets with pentagonal holes at their corners.

What would happen if some of the woven holes are heptagonal instead of hexagonal?

Students familiar with positive, negative and zero curvature may see relationships between these types of curvature and the weaving process of creating pentagonal, heptagonal, and hexagonal holes. For other students, the building of these domes may serve to introduce them to different types of curvature in space.

a. *woven ball*

b. *soccer ball*

Figure 3.92

Balls and polyhedra

Those who constructed dome A will have discovered that its border base is a pentagonal hole of the same size as the other pentagonal holes. In other words, this dome has a global spherical form. This can be discovered in other ways. For example, students can investigate what types of 'closed' baskets can be woven that display pentagonal holes at all their vertices. Interestingly, the smallest possible 'basket', made out of six strips has a similar structure as the well-known modern soccer ball made out of pentagonal and hexagonal pieces of leather (see Figure 3.92).

The structure of the soccer ball is that of a semi-regular polyhedron, formed by 20 regular hexagons and 12 regular pentagons. By extending, on the one hand, the 20 hexagons, one generates the regular icosahedron; by extending, on the other hand, the 12 pentagons, one generates the regular dodecahedron. That is, the ball's structure is at the same time that of a truncated icosahedron and that of a truncated dodecahedron.

If we increase the curvature and produce instead of pentagonally woven 'vertices', square-hole-vertices or triangular-hole-vertices, will it be possible to generate the other Platonic solids? To facilitate the analysis of this question, it is useful to prepare the strips of thin cardboard, before weaving the polyhedra.

Figure 3.93
Hexastrip

Hexastrips

If we compare the structure of the soccer ball with that of the woven ball, we see that the crossing of any two strips on the woven ball corresponds to a common border of two hexagonal pieces on the soccer ball, and vice versa. If we make these 'crossings' visible on the strip and take out the strip and lay it down, it would look like the strip presented in Figure 3.93.

I call such a strip a *hexastrip*. Hexastrips are cardboard strips in which a series a folds have been introduced in such a way that they facilitate the weaving together of the strips in three-directions. Figure 3.94 shows how the first and

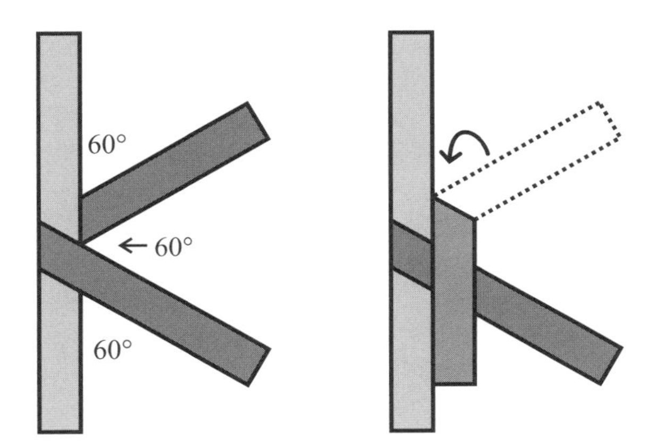

Figure 3.94

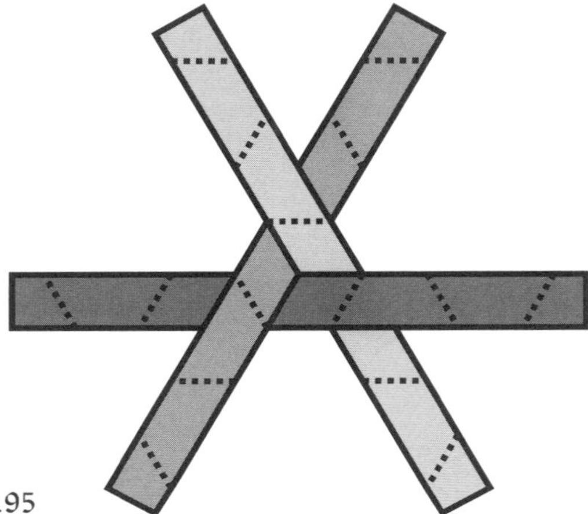

Figure 3.95

successive folds may be produced in a strip to make a hexastrip. First we wrap one strip around the other to obtain the first folding line, using an angle of incidence of 60°. Then the second fold is marked by folding the upper part of the second strip in such a way that it becomes adjacent to the first strip. This process is repeated to produce the various folds.

Figure 3.95 displays how to join three hexastrips—over-and-under. The strips may be held together with paper clips or by gluing their overlapping rhombi.

To weave the truncated icosahedron, one needs six hexastrips each with ten folds. To weave a closed structure with only square holes, one needs four hexastrips each with six folds. To weave a closed structure with only regular triangular holes, one needs three hexastrips each with four folds. The first structure is that of a truncated octahedron, bounded by 6 squares and 8 regular hexagons, that is equal to a truncated cube. The second structure is that of a truncated tetrahedron, bounded by 4 regular hexagons and 4 equilateral triangles. Figure 3.96 displays both structures.

In other words, by constructing the three regular hexastrip polyhedra, students may discover the five regular polyhedra (Platonic solids). Moreover, they may understand the duality of the regular icosahedron and regular dodecahedron; the duality of the regular octahedron and the regular hexahedron or cube; and the self duality of the regular tetrahedron. They can wonder whether these are indeed all regular polyhedra. For example, what will happen if one tries to weave a structure

Figure 3.96

with five hexastrips, each with eight folds? How can one prove with hexastrips that there exist exactly five Platonic solids?

The students may construct each of the three hexastrip polyhedra with cardboard strips of different colors. Observing the final woven polyhedra, what may be said about the number of ways three strips are crossing as in Figure 3.95? What may be said about the number of ways by which three, four or five strips constitute the sides of regular triangles, squares and pentagons? Herein lies a possibility to establish a link with combinatorics.

Students may be invited to construct other hexastrip polyhedra. If the holes are hexagonal and pentagonal, can something be observed about the number of hexagonal or the number of pentagonal holes?

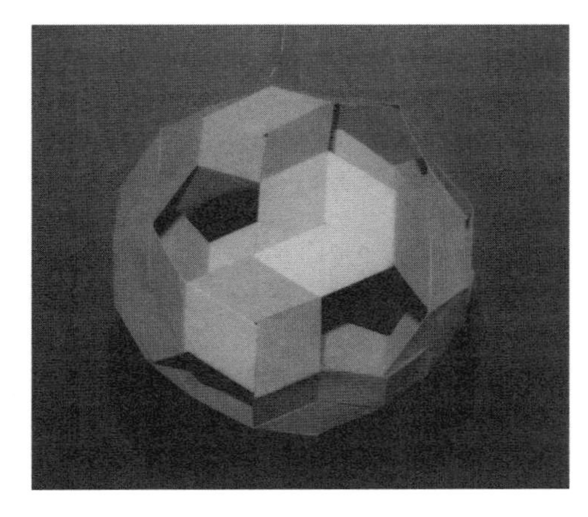

Figure 3.97
Hexastrip molecular model for C_{60}

Links with chemistry (cf. Gerdes, 1998b, 1999b)

In 1996 Curl, Smalley (USA) and Kroto (UK) were awarded that year's Nobel Prize in Chemistry for their 1985 discovery of C_{60} and for their conjecture that it would have the symmetrical structure of a truncated icosahedron (cf. Kroto et al., 1985). They named the molecule buckminsterfullerene (or Bucky ball for short) after R. Buckminster Fuller (1895–1983), the designer of geodesic domes, and indicated soccerene as a possible alternative name. Interestingly, our woven ball (Figure 3.92a) and the corresponding hexastrip structure (see Figure 3.97) constitute a molecular model for C_{60}. Moreover, our woven ball and hexastrip structure are special models of C_{60} that have various advantages that I will discuss.

Prior to the discovery of buckminsterfullerene only two forms of crystalline carbon were known: graphite and diamond. Let us return for a moment to the hexagonally woven plane fabric (Figure 3.74). It constitutes a model for a layer of graphite: Imagine the carbon atoms arranged at the vertices of the hexagonal holes; the edges of these holes represent single bonds between the carbon atoms, and the crossings of two strands between two neighboring vertices of two neighboring hexagonal holes represent the double bonds (see Figure 3.98).

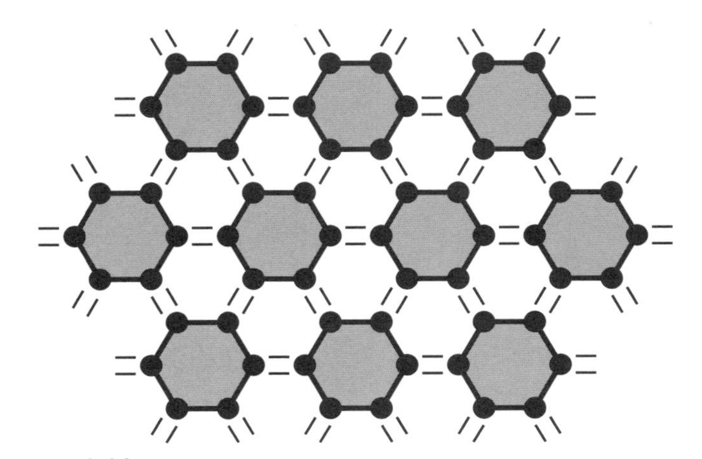

Figure 3.98
Molecular model for a layer of graphite

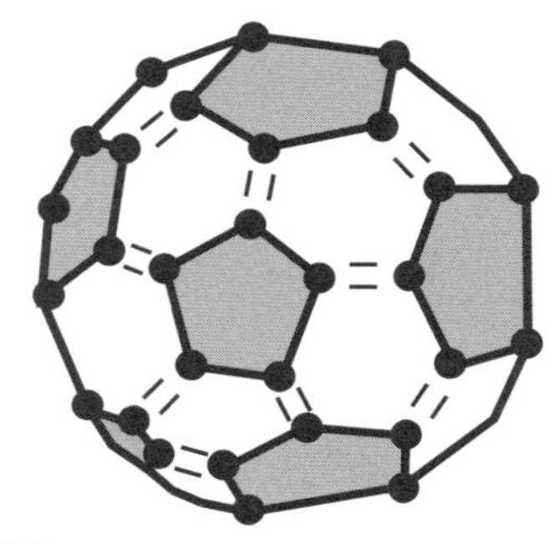

Figure 3.99
Molecular model for C_{60}

Figure 3.100
Hexastrip model of C_{72}

Similarly, the structure of the woven ball and corresponding hexastrip structure (Figure 3.97) constitute a model for buckminsterfullerene C_{60}: Imagine once more the carbon atoms arranged at the vertices of the holes (this time, 12 pentagonal holes, leading to 60 atoms); the edges of these holes represent the single bonds between the carbon atoms, and the crossings of two strands between neighboring vertices of neighboring pentagonal holes represent the double bonds (see Figure 3.99). The hexagonal rings of carbon atoms are held together by alternately single and double bonds. There are 20 hexagonal rings.

Since the 1990 success of Krätschmer (Germany) and Huffman (USA) in synthesizing measurable quantities of C_{60}, many other fullerenes (C_n) and related molecules have been studied. Fullerenes are defined as closed cage molecules comprised entirely of so-called sp^2-hybridized carbons arranged in hexagons and pentagons.

Figure 3.100 shows two views a hexastrip model of C_{72} with two polar hexagonal holes. It is the smallest example of a cylindrical fullerene tube, and has a six-fold rotational axis of symmetry.

Hexastrip models of two isomers of C_{120} are shown in Figure 3.101. The first isomer has a cylinder structure: it is composed of two diametrically opposed hemispherical C_{60} caps, joined by a fivefold cylindrical wall of two rows of hexagonal holes. The second isomer has global tetrahedal symmetry: the twelve pentagonal holes are clustered in four groups of three at the corners of a truncated tetrahedron and are surrounded by single bands of hexagonal holes. The reader is invited to construct hexastrip models of other isomers of C_{120}. How many are possible? What is the smallest fullerene with a tetrahedal symmetry? Is it possible to make a hexastrip model of it?

Photograph 3.102 presents a hexastrip model of the isomer of C_{276} with a globally octahedral form: At the corners of the woven truncated octahedron there are two opposite pentagonal holes surrounded by a layer of six hexagonal holes.

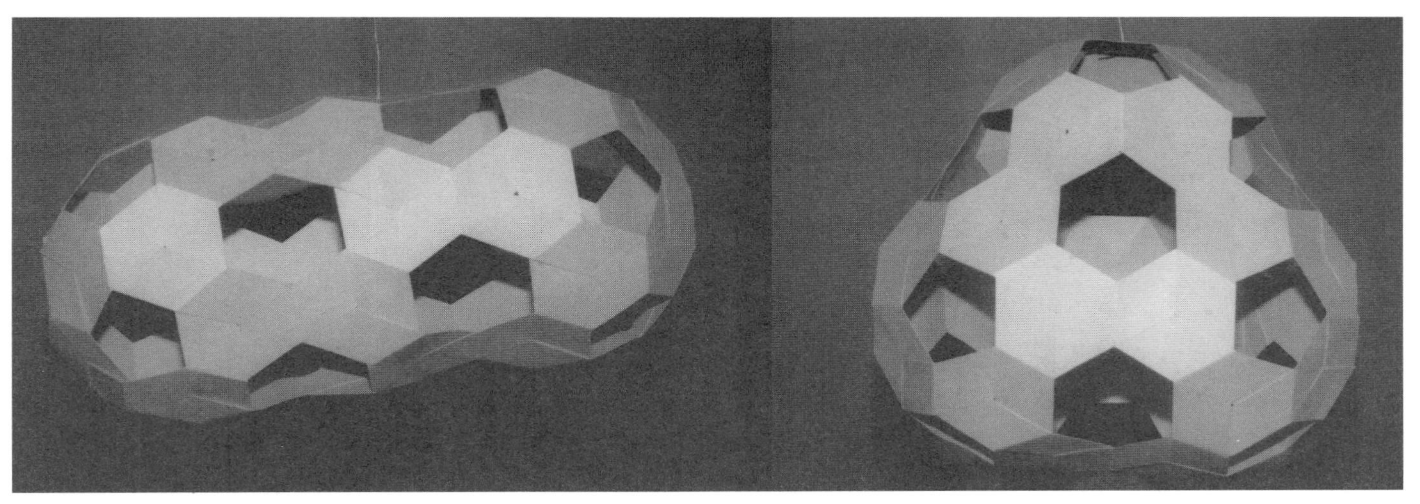

Figure 101. *Hexastrip models of two isomers of* C_{120}

Figure 3.102. *Hexastrip model of an isomer of* C_{276}

Figure 3.103 shows the hexastrip model of an isomer of C_{240} with a truncated regular icosahedral form: between each pair of neighboring pentagonal holes there is one hexagonal hole. If there are always two hexagonal holes in between the neighboring pentagonal holes, we get a hexastrip model of an isomer of C_{540}.

Hexastrip models of fullerenes and related carbon structures are not only beautiful and relatively easy to make from inexpensive materials—and as such are attractive from a didactic viewpoint—but also they give a clear picture of the bonding situation: the edges of the holes represent the single bonds and the folds (that is, where woven hexagons are adjacent) the double bonds. Given a hexastrip model, it is not difficult to determine the number of carbon atoms implied, as there are no problems with double counting of vertices. Conversely, if a number n is equal to $60+6m$, where m is zero or an integer greater than 1, then it is possible to construct a hexastrip model of C_n. The number of theoretically possible isomers of C_n increases quickly as n

Figure 3.103
Hexastrip model of an isomer of C_{240}

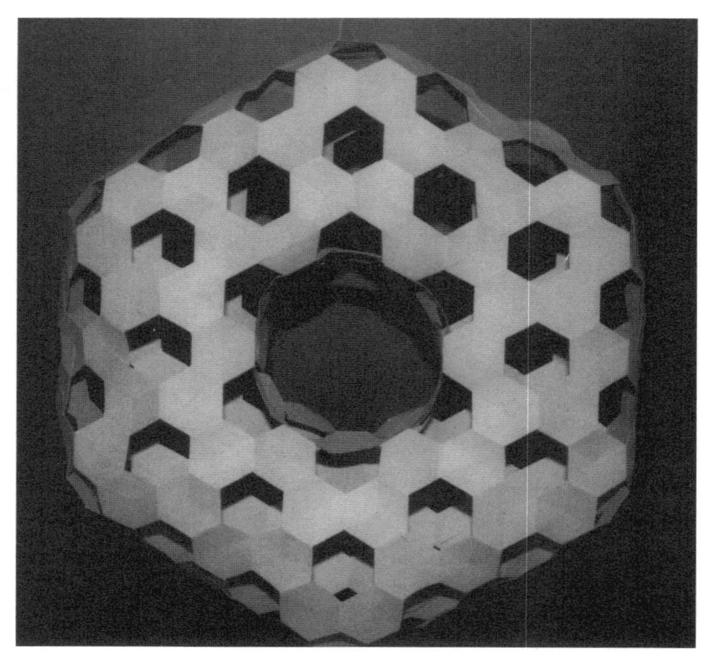

Figure 3.104
Hexastrip model of a quasi-fullerene C_{576}

increases. For example, for $n = 78$, there are over twenty thousand general fullerenes. In the process of discovering real existing or producible fullerenes, it is important to consider the so-called Clar Sextet isomers. These have in common a special sort of Kekulé structure possessed by no other fullerenes. In this structure, each pentagon has five external double bonds and every double bond is seen to take part in two conjugated 6-circuits. As hexastrip models correspond to this particularly stabilizing Kekulé structure that may make them useful in narrowing down the search for possible fullerene isomers (for more detail and references, see Gerdes, 1998b, 1999b). It is interesting to note that I introduced hexastrip models in the early 1980's—before the discovery of fullerenes—during the exploration of possibilities of incorporating the hexagonal basket weaving technique into the teaching of geometry in Mozambique.

The hexastrip structures presented in photograph 3.96 are models of the small quasi-fullerenes C_{24} and C_{12}. Closed carbon cages containing other than 5- and 6-membered rings are known as quasi-fullerenes. Figure 3.104 shows a hexastrip model of a quasi-fullerene C_{576} in the form of a torus; it has 12 pentagonal and 12 heptagonal holes. The heptagonal holes produce the concave regions.

Hexastrip weavable polyhedra and Euler's theorem

In the case of isomers of fullerenes C_n for which it is possible to construct a hexastrip model, it is easy to establish that there are always 12 pentagonal rings, independent of the number of hexagonal rings. Let N_5 be the number of pentagonal holes in the hexastrip model, and N_6 the number of hexagonal holes. For the number of vertices (V), faces (F), and edges (E) of the hexastrip polyhedron, we have:

$$V = 5N_5 + 6N_6;$$

$$F = N_5 + N_6 + \frac{V}{3};$$

$$E = \frac{3V}{2}.$$

This implies that $V + F - E = N_5/6$. As on the other hand, $V + F - E = 2$ in agreement with Euler's polyhedron theorem, it follows that the number of pentagonal holes is equal to 12.

What can be said about the number of pentagonal rings in other fullerenes? What can be said about the number of square rings, triangular rings, heptagonal rings, etc. in quasi-fullerenes?

Figure 3.105
Mangbetu weaving variant

Looking back

In this section, I presented some suggestions for the exploration of the hexagonal basket weaving technique in mathematics education. Connections with art and craft education, with architecture (cf. Gerdes, 1998c), with today's chemical research and teaching were presented, that may increase the students' interest and motivation for geometrical activity.

The hexagonal basket weaving technique explored in this section is not the only hexagonal weave used by African artisans. Figure 3.105 presents a variant used by Mangbetu artisans in northeastern Congo/Zaire: By introducing extra horizontal strips, trapezoidal holes are produced. Figure 3.106 presents two weaves from Uganda. In the first weave, two directions are orthogonal and the third makes an angle of 45° with them. The second weave produces triangular holes.

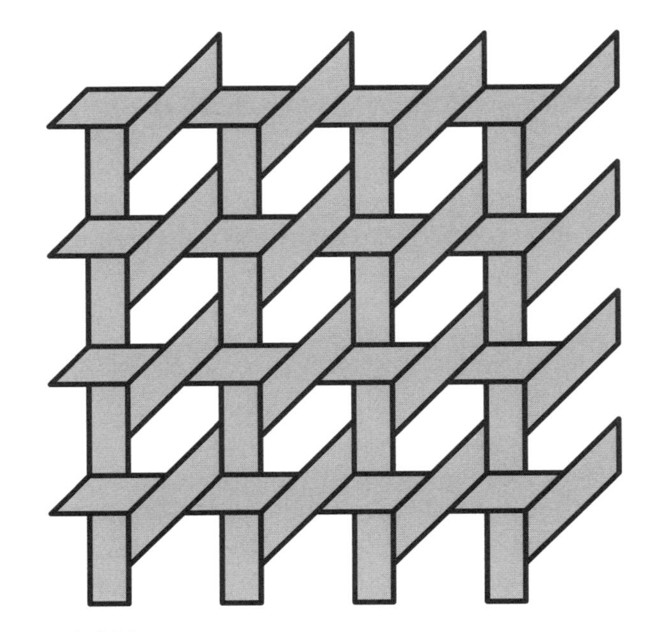

Figure 3.106a
Weave from Uganda

Figure 3.106b
Weave from Uganda

9 Exploring finite geometrical designs on plaited mats

Plaited mats from the Lower Congo river area

(cf. Gerdes, 1999c, d)

Among the ethnographic collections of the Royal Museum of Central Africa (Tervuren, Belgium) is a series of richly decorated, plaited mats, acquired by the museum before 1909, coming from Lower Congo river area in today's southwestern Congo/Zaire (cf. Coart). The photograph in Figure 3.107 gives an example of a mat collected from Cataracts. The mats were woven by Yombe and Sundi women.

The mats display attractive geometrical designs with axial, two-fold or four-fold symmetries. The designs are produced by a well thought-out variation in the interweaving of the natural colored 'white' strands in one direction

Figure 3.107
Mat from the Lower Congo

(vertical in our reproductions) with the artificially colored (mostly black [brown in the figures]) strands in the other direction (horizontal in our reproductions). The designs can be seen in reverse on the other side of the mat. The global form of the designs is that of a 'toothed square'—a square

Figure 3.108
a2. *Dimensions* 27 × 27; b2. *Dimensions* 30 × 29

with adjacent, congruent teeth on its sides—whose sides make angles of 45° with the sides of the rectangular mats. We may call the numbers of strands in both directions used to generate a toothed square its dimensions. In this way, the dimensions of the toothed square design in Figure 3.108a are 27 × 27, whereas those of the design in Figure 3.108b are 30 × 29. In the first case, a brown strand always passes over or under 1, 2, 3, 5, or 7 white strands. For instance, the position of the central horizontal brown strand may be described as: over four white strands, under three, over two, under one, over two, under three, over two, under one, over two, under three, over four. Or shortly: o4-u3-o2-u1-o2-u3-o2-u1-o2-u3-o4. In the second case, a brown strand always passes over or under 2, 4 or 8 white strands. The position of the central horizontal brown strand is described by: o2-u4-o2-u2-02-u2-o2-u2-o2-u2-o2-u4-02.

Odd by odd

Now we will first consider toothed square designs that appear on mats collected from Cataracts and Luvituku. Their dimensions are always odd numbers. Most of them have four axes of symmetry like the examples in Figure 3.109. Some have two

Figure 3.109a
Dimensions 45 × 45

Figure 3.109b
Dimensions 45 × 45

Figure 3.109d
Dimensions 63 × 63

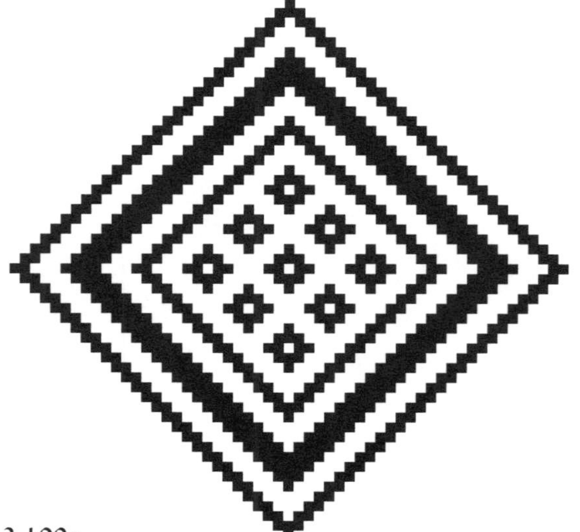

Figure 3.109c
Dimensions 55 × 55

Figure 3.109e
Dimensions 65 × 65

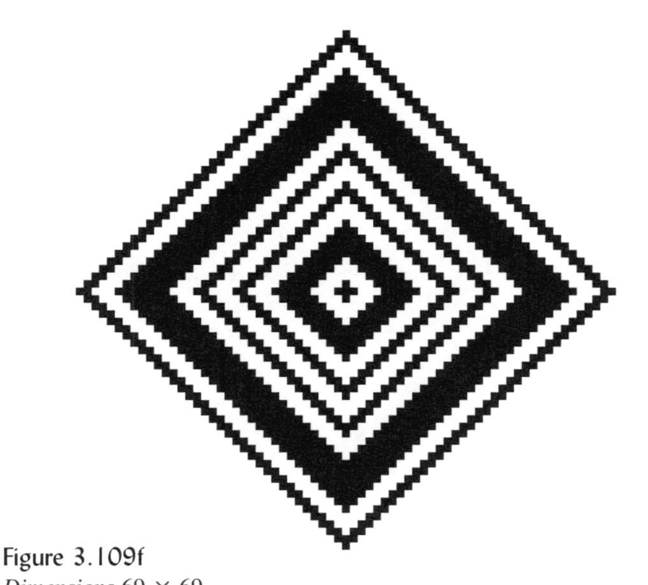

Figure 3.109f
Dimensions 69 × 69

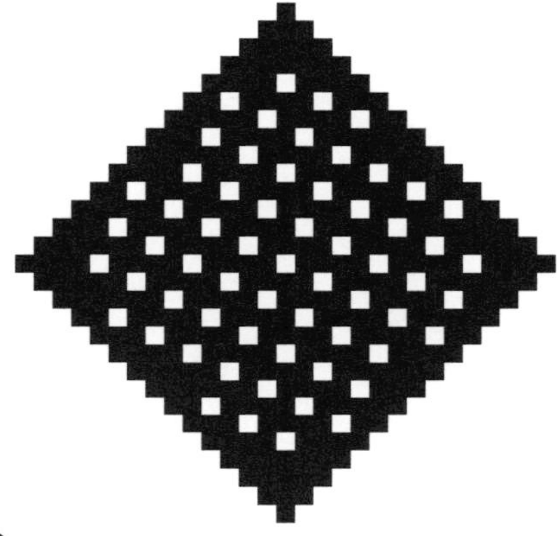

Figure 3.112
Dimensions 29 × 29

symmetry axes (see Figure 3.110), others one (see Figure 3.111). The toothed square design in Figure 3.112 has a fourfold rotational symmetry.

We may ask ourselves or our students how many different toothed square designs of certain dimensions exist that display certain characteristics. To facilitate the analysis, we may adopt the convention that two toothed square designs that are the negative of each other are instances of the same. So we may for reason of visibility only consider those toothed square designs that have all their teeth brown. How many toothed square designs of a given dimension have four axes of symmetry? Figure 3.113 shows all 16 of dimensions 7 × 7, and Figure 3.114 all 64 toothed square designs of dimensions 9 × 9 which have four axes of symmetry. How does the number $a(n)$ of distinct toothed square designs of dimensions (2n−1) × (2n−1) depend on n? How many toothed square designs of certain dimensions have two axes or one axis of symmetry? How many have fourfold rotational symmetry? For each of these sets of toothed square designs we may consider particular subsets, and generate several questions for further investigation. For example, of those with four axes of symmetry, in how many cases do the brown strands only pass over and under odd numbers of white strands? In how many cases do the brown strands never pass over more than five white strands? And so forth.

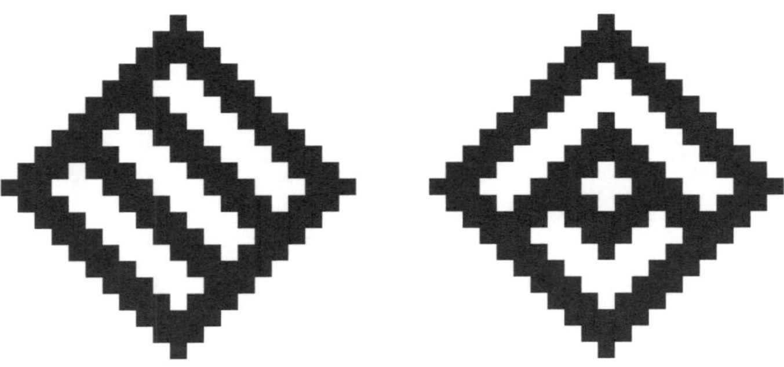

Figure 3.110
Dimensions 21 × 21

Figure 3.111
Dimensions 21 × 21

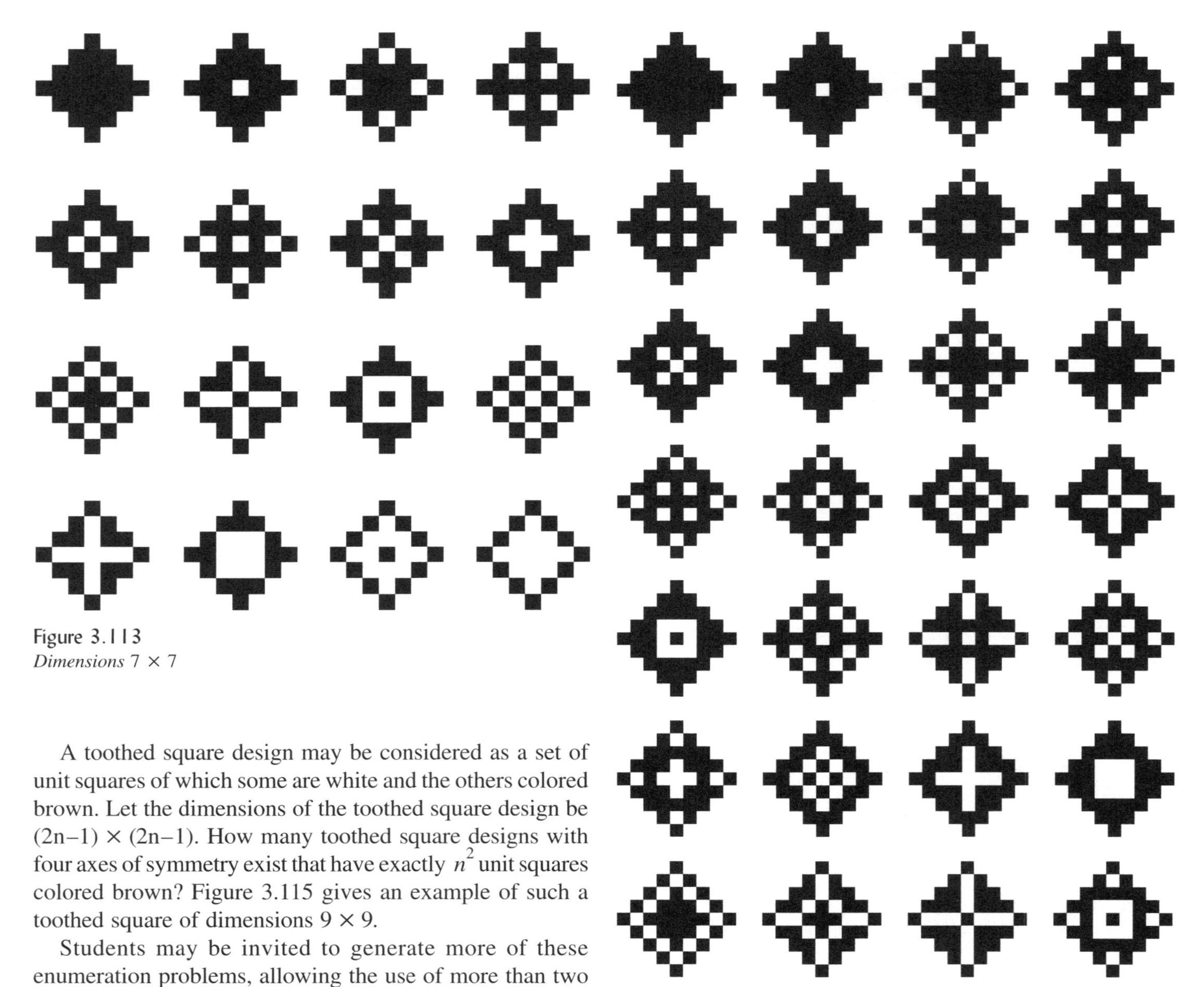

Figure 3.113
Dimensions 7×7

Figure 3.114. *Dimensions* 9×9

A toothed square design may be considered as a set of unit squares of which some are white and the others colored brown. Let the dimensions of the toothed square design be $(2n-1) \times (2n-1)$. How many toothed square designs with four axes of symmetry exist that have exactly n^2 unit squares colored brown? Figure 3.115 gives an example of such a toothed square of dimensions 9×9.

Students may be invited to generate more of these enumeration problems, allowing the use of more than two colors (brown and white).

Figure 3.114 (cont.). *Dimensions* 9 × 9

Figure 3.114 (cont.). *Dimensions* 9 × 9

Figure 3.115
5^2 *(black)* + 4^2 *(white) unit squares*

Even by odd

Toothed square designs with dimensions even by odd — as in the example already given in Figure 3.108b (dimensions 30 × 29)— appear on a series of beautifully decorated mats collected in Boma in the Congo river delta. The photograph in Figure 3.116 displays an example of such a mat. As the two dimensions of the toothed square designs on these mats are different, the designs cannot have neither four axes of symmetry nor a fourfold rotational symmetry. Figure 3.117 presents examples of such toothed square designs with two axes or one axis of symmetry. In Figure 3.118 the axial symmetry is broken by the smaller toothed square in the lower half of the design.

Figure 3.116
Mat from the lower congo

When analyzing these toothed square designs, I was puzzled by the question why did the artists of this region invent these toothed square designs with dimensions even by odd, and not odd by odd (or even by even), which would have allowed them to create designs having more symmetries. The artists preferred having oblique toothed lines with an even width, most with a width of two unit squares (see Figure 3.119).

For instance in Figure 3.117a and b these toothed lines always have width 2 or 4, and a colored strand always passes over or under an even number of white strands. The toothed

Figure 3.117
a. *Dimensions* 30×29; b. *Dimensions* 58×57

Figure 3.118
Dimensions 82 × 81

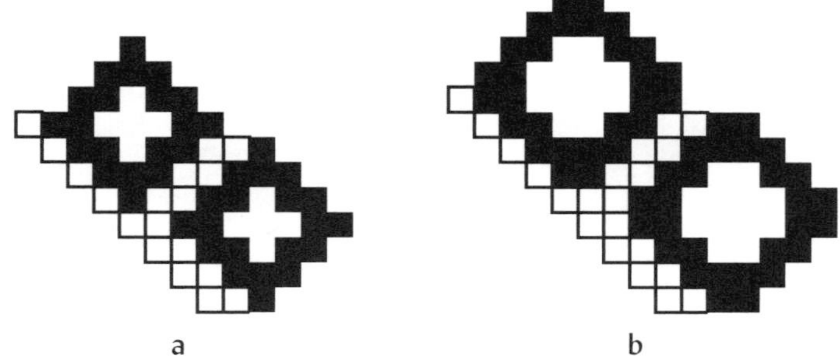

Figure 3.120

lines with width 2 are the smallest possible, and it is understandable that they were often selected when the artist liked to create an intricate design on a mat of usual size. When several toothed square designs are combined on a mat (see once more the example in Figure 3.116) in such a way that they are separated by oblique toothed lines of width 2, there are several implications. Situations like the ones represented in Figure 3.120 would obviously be ugly: neither the odd by odd toothed square designs (3.120a) nor the even by even toothed square designs (3.120b) would be nicely aligned.

Even by odd toothed square designs are the solution found by the artists as illustrated in Figure 3.121. Furthermore, when two oblique toothed lines of width 2 cross (see Figure 3.122a), immediately left of the crossing there is an opening of height 1 for a vertical white strand to pass over the central brown strand, and immediately above the crossing there is room for two white strands to pass over the brown strand. In an odd by odd or even by even crossing context (see Figure 3.122b and 3.122c), each of the oblique toothed lines

Figure 3.119

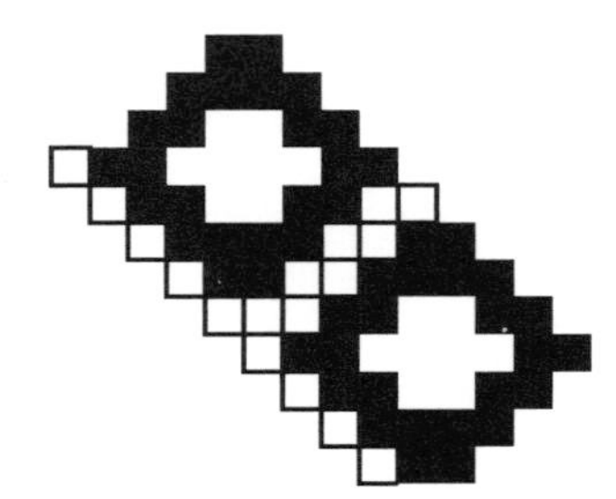

Figure 3.121

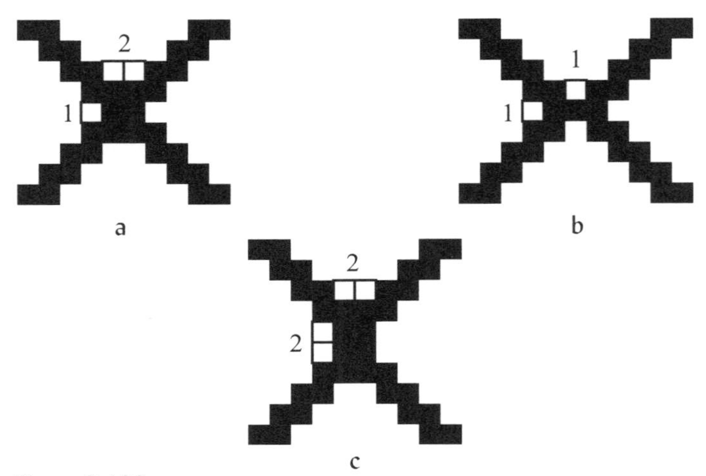

Figure 3.122

is displaced by one unit square before and after the crossing. It becomes in this way understandable why these mat weavers created even by odd toothed square designs. If they had also accepted toothed lines of width 3, as did their colleagues from which mats were collected from Cataracts (see above) both odd by odd and even by even crossings with a fourfold rotational symmetry would have been possible (see Figure 3.123).

Although toothed square designs of even by odd dimensions cannot have a fourfold rotational symmetry, the

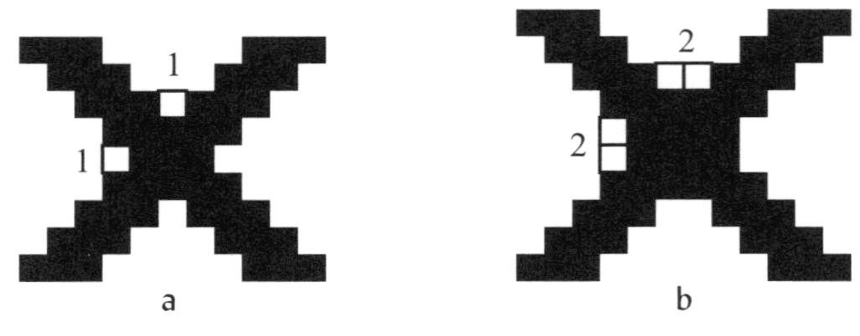

Figure 3.123

Figure 3.124
Dimensions 114 × 113

mat weavers discovered that by increasing the dimensions they could create the impression that the whole design had such a symmetry. For instance, the toothed square design of dimensions 114 × 113 shown in Figure 3.124 seems to have four axes of symmetry. The basic structure (Figure 3.125a) of the toothed square design of dimensions 90 × 89 in Figure 3.125b displays, in the same way, almost a fourfold rotational symmetry.

As in the case of the toothed square designs of odd by odd dimensions, even by odd toothed square designs may be used as a starting point for generating further geometrical problems. We may ask ourselves or our students how many different toothed square designs of certain even by odd

Figure 3.125

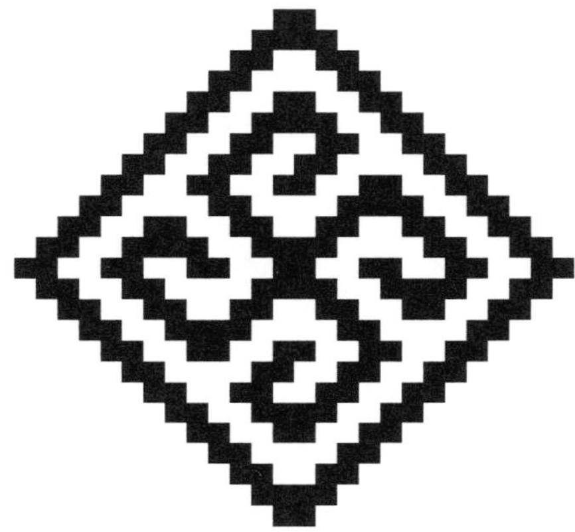

Figure 3.126
Dimensions 26 × 25

dimensions exist that display given characteristics. For example, how many toothed square designs of dimensions (2n) × (2n – 1) exist where the colored horizontal strands always pass over and under an even number of white strands?

Almost fourfold rotational symmetry

Among the attractive mats collected from Boma and from Banana on the mouth of the Congo river, there are several which display toothed square designs of even by odd dimensions, that although having twofold rotational symmetry, nevertheless give the impression of having fourfold rotational symmetry. The same designs do not have axial symmetry as it was displayed in the earlier examples we considered (Figures 3.124 and 3.125). Figure 3.126 shows one of these toothed square designs. It has dimensions 26 × 25. Two S's are crossing perpendicularly, or, describing the design in another way, it is composed of four 'arms' each consisting of a toothed line of width 2, having three clockwise 'elbows'. Figure 3.127 shows a similar design, this time of dimensions 42 × 41, and each arm has seven

Figure 3.127
Dimensions 42 × 41

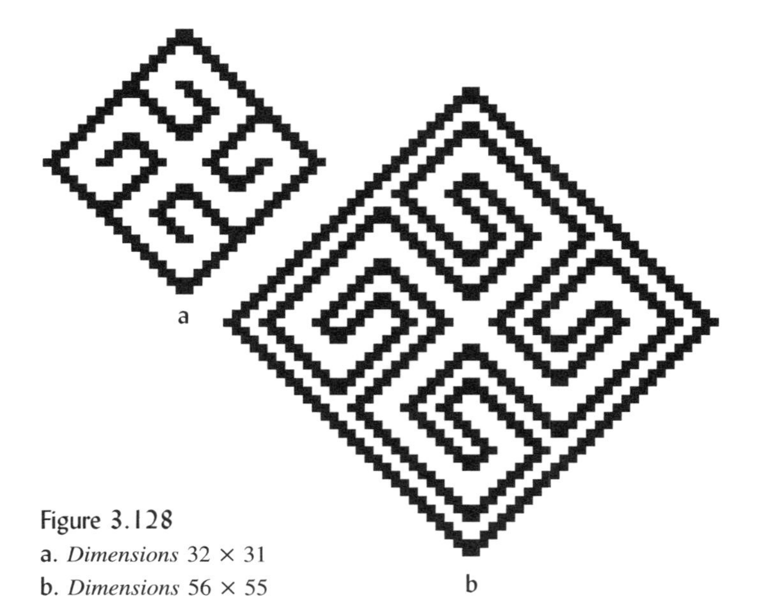

Figure 3.128
a. *Dimensions* 32 × 31
b. *Dimensions* 56 × 55

clockwise elbows. The white arms of the designs in Figure 3.128 have three anti-clockwise elbows. This time, however, the arms consist of toothed lines of width 4. The arms of the interesting design in Figure 3.129 consist of toothed lines of width 14, but instead of having thick brown arms, they are decomposed in successive closed brown and closed white toothed lines of width 2, which have 36 elbows and a central white cross with arms with three elbows.

Figure 3.130 shows a toothed rhombus design, obtained by using colored strands (in the vertical direction) of double the size of the white strands. If the same design would have been made of colored and white strands of the same width, it would have looked like the toothed square in Figure 3.131.

If the mat weavers would have admitted toothed lines with a width of 3, they could have created toothed square designs of odd by odd dimensions with (exact) fourfold rotational symmetry (see the examples in Figure 3.132). But then each of the designs would occupy more space on the mats.

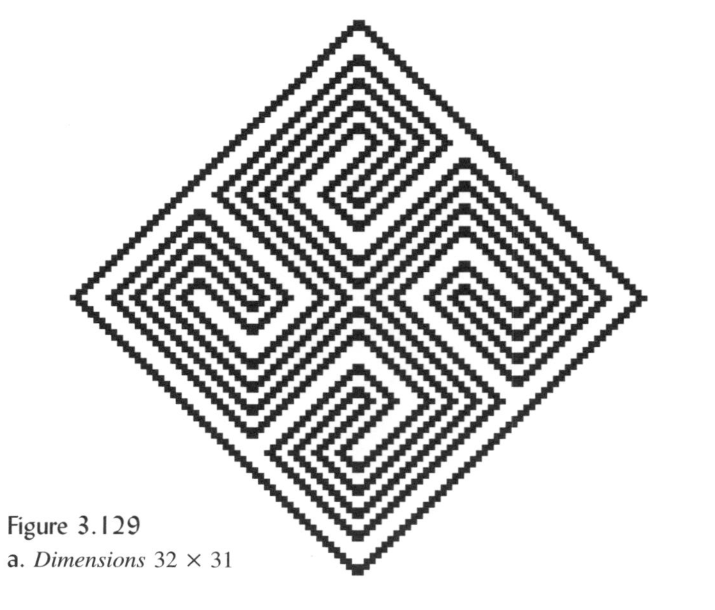

Figure 3.129
a. *Dimensions* 32 × 31

Figure 3.130

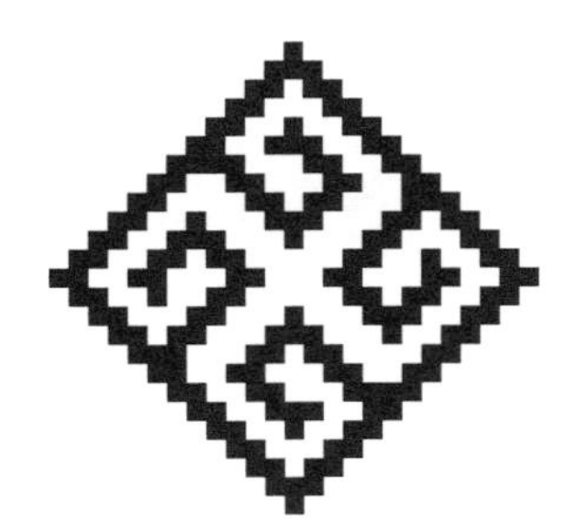
Figure 3.132

Figure 3.131

The reader is invited to generate and solve geometrical problems about the enumeration of toothed square designs of even by odd dimensions that have almost fourfold symmetry.

10 Decorative plaited strips

In this section, I will present decorative strip patterns as woven on African baskets, mats, and other objects. I will show that these African objects may serve as a starting point for interesting enumeration problems and in analyzing symmetries of strip patterns. I will consider various types of woven strip patterns: vertically-horizontally woven, diagonally woven, diagonally plaited mat strip patterns, as well as woven cylinders.

Vertically-horizontally woven strip patterns

The Bassari who live in the Senegalese-Guinean border region make waist belts, out of rectangularly cut pieces of bark, curved into a cylinder. Figure 3.133 presents a portion of a decorative strip pattern, horizontally-vertically woven around such a belt. In this representation, the vertical black [brown in the figures] and the horizontal white strands are

Figure 3.133
Bassari strip pattern

Figure 3.134

given the same width. As the direction of the individual strands is not very visible, we may simplify our representation by removing them. In this way, Figure 3.133 is transformed into Figure 3.134.

On this strip, a (rectangular) decorative motif is repeated (see Figure 3.135). Its has dimensions 8 by 7, that is, 8 vertical and 7 horizontal strands are needed to produce one copy of the motif.

Figure 3.136 presents further examples of (portions of) decorative strip patterns on Bassari belts. The dimensions of their decorative motifs are respectively 5 by 7, 4 by 7, 14 by 6, and 16 by 7.

Students may be asked to investigate how many different vertically-horizontally woven strip patterns of given dimensions m by n exist. The number depends on the criteria of acceptability of patterns. For example, there are 13 such strip patterns of dimensions 2×3 (see Figure 3.137), whereby a strip pattern like the one presented in Figure 3.138 with a not interwoven horizontal white strand is not admitted.

Using this criterion, is it possible to express generally the number of different vertically-horizontally woven strip patterns of dimensions m by n as a function of m and n?

Figure 3.135

Symmetries

Not all the vertically-horizontally woven strip patterns that I have presented have the same symmetries. The patterns in Figure 3.136a and b have a horizontal axis of symmetry, whereas the strip pattern in Figure 3.133 has both horizontal and vertical axes of symmetry. The strip patterns in Figure 3.136c and d are invariant under a glide reflection.

The study of vertically-horizontally woven strip patterns may give students the chance to discover that there exist seven symmetry classes of (one-color) strip patterns. Table 1 contains a list of these seven classes, along with the characteristics of each class.The international notation for each class appears in the second column. Since by definition, all strip patterns have translational symmetry, only other symmetries are indicated in the third column.

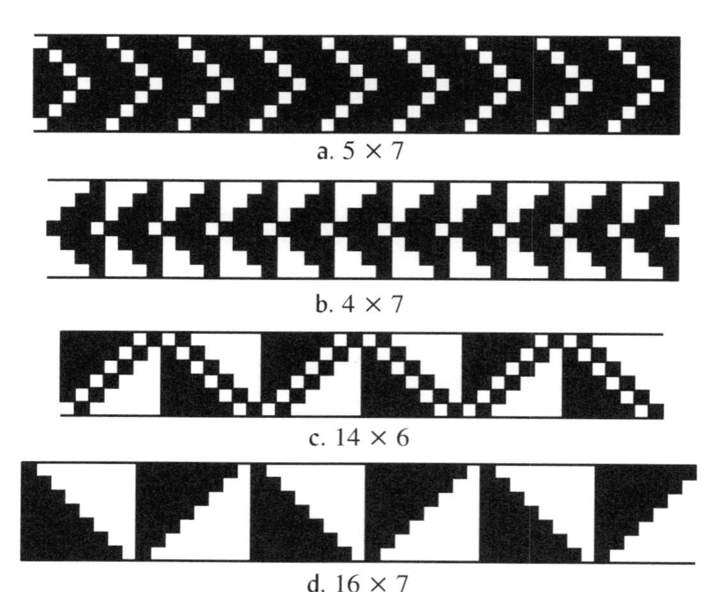

Figure 3.136.
Further examples of Bassari strip patterns

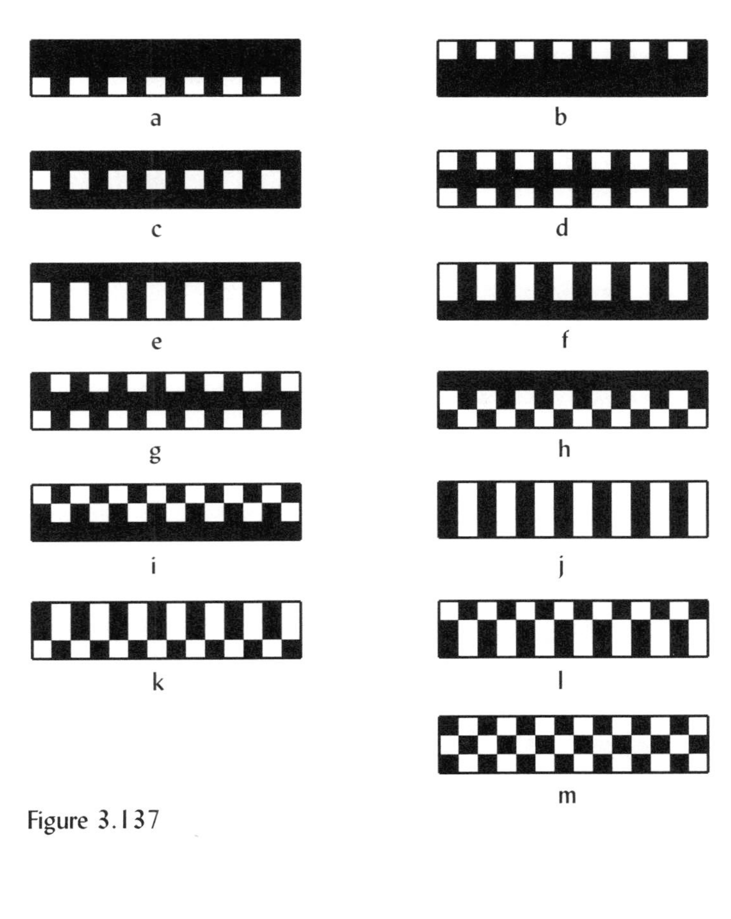

Figure 3.137

Figure 3.138

	notation*	symmetries	invariant under
1	*pmm2*	vertical, horizontal and rotational symmetry of 180°	vertical reflection, horizontal reflection and rotation through an angle of 180°
2	*pma2*	vertical, translational-reflected and rotational symmetry of 180°	vertical reflection, glide reflection and rotation through an angle of 180°
3	*pm11*	vertical symmetry	vertical reflection
4	*p1m1*	horizontal symmetry	horizontal reflection
5	*p112*	rotational symmetry of 180°	rotation through an angle of 180°
6	*p1a1*	translational-reflected symmetry	glide reflection (or reflected translation)
7	*p111*	only translational symmetry	only translation

Table 1

Students may be asked to determine whether all classes are represented under the vertically-horizontally woven strip patterns of dimensions 2×3? If not, which are the smallest dimensions for which there exist such strip patterns belonging to the seven different classes? Students may be asked to construct vertically-horizontally woven strip patterns belonging to the seven classes.

In fact, these seven classes are the seven classes of one-color strip patterns. Some of the strip patterns represented may

---

* As one sees in Table 1, in the international notation, four symbols are used to represent each of the seven classes of strip patterns. The letter *p* denotes **p**attern. If the pattern has a vertical reflection, then the second symbol is *m* (of **m**irror); if not, then the second symbol is *1*. If the pattern is invariant under a horizontal reflection, then the third symbol will be *m*; if it is invariant under a glide reflection, then the third symbol is *a*; if it has neither a horizontal reflection nor a glide reflection, then the third symbol is *1*. The last symbol is 2, if the pattern has 180° rotational symmetry (this symmetry is also called a symmetry of order 2); in all other circumstances, the last symbol is *1*.

be considered two-color strip patterns. This means that they have a symmetry that reverses or interchanges brown and white. For example, the strip patterns in Figure 3.136c and d admit vertical reflections which reverse the colors, as do the strip patterns in Figure 3.137j and m. However, these last strip patterns also admit a horizontal reflection that maintains the colors. In that sense, they belong to another symmetry class of two-color strip patterns. Students may be asked to discover all symmetry classes of two-color strip patterns[†]. For all classes, is it possible to construct vertically-horizontally woven strip patterns belonging to them? If so, which are the smallest dimensions for which there exist vertically-horizontally woven strip patterns belonging to all classes?

Figure 3.139
Example of a gipatsi *handbag*

Diagonally woven strip patterns (cf. Gerdes and Bulafo)

Traditionally only women, today also men, in the south-eastern province of Inhambane (Mozambique) weave beautifully decorated hand bags. In the Gitonga language these hand bags are called *sipatsi* (sing.: *gipatsi*). Figure 3.139 shows an example of a *gipatsi* hand bag. On the photograph the purple or dark blue strands appear as brown.

It is diagonally woven in the sense that the palm strands make an angle of 45° with the horizontal and vertical edges of the bag. Each horizontal decorative strip is visible by brown zigzag borders on a white background. Figure 3.140 displays a detail of the lower strip.

Once again since the direction of individual strands is almost not visible, we may remove them from the illustration. In this way, the decorated strip, of which Figure 3.140 shows a portion, is transformed into the strip of Figure 3.141.

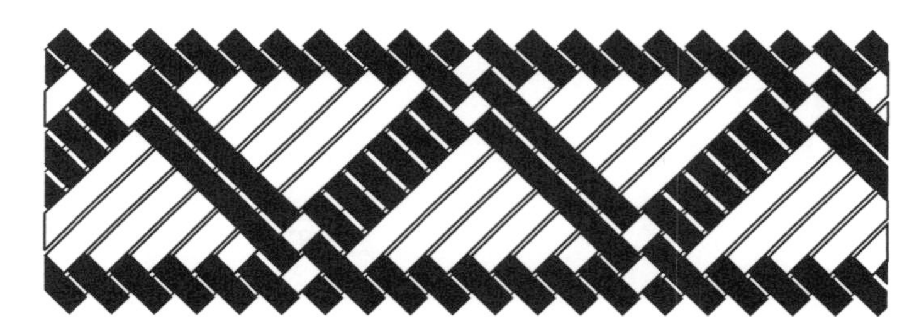

Figure 3.140

Figure 3.141

[†]The book of Washburn & Crowe (1988) may be recommended to the students for the study of symmetries of strip and plane patterns.

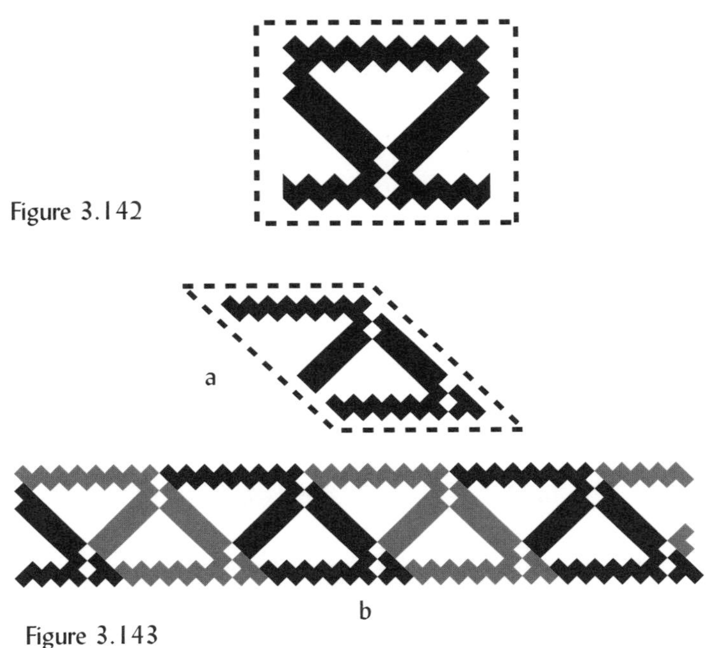

Figure 3.142

Figure 3.143

For the bag weavers, the decorative motif does not appear in a vertical position (Figure 3.142), but in an oblique or diagonal one, in agreement with the weaving direction, as in the example in Figure 3.143a. Figure 3.143b illustrates the repetition of the motif.

We may say that a brown and white toothed parallelogram (see Figure 3.144a), as in Figure 3.143, generates the respective strip pattern. In this example it has dimensions 8 by 13 (see Figure 3.144b). The first dimension may be called the *period* of the motif; the second its *diagonal height*. In other words, the period indicates how many colored strands are necessary to generate one copy of the decorative motif. As Figure 3.145 shows, differently toothed parallelograms may generate the same diagonally woven strip pattern. Naturally, they have the same dimensions.

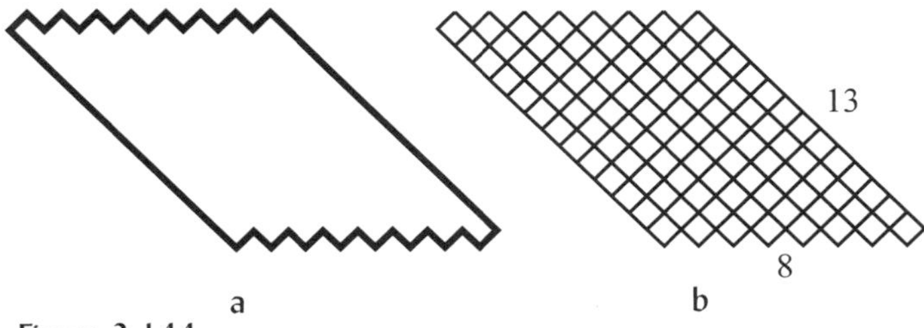

Figure 3.144

Like in the study of horizontally-vertically woven strip patterns, enumeration problems may be posed to students: How many diagonally woven strip patterns (of given dimensions $p \times d$) of the type that appear on *sipatsi* do exist?

All ornamental strips on *sipatsi* are visible: their zigzag borders are marked by the transition from brown to white. In other words, the top and bottom rows of unit squares are colored (see Figure 3.146 for the case where the diagonal height is 4). The colors of the squares between the top and

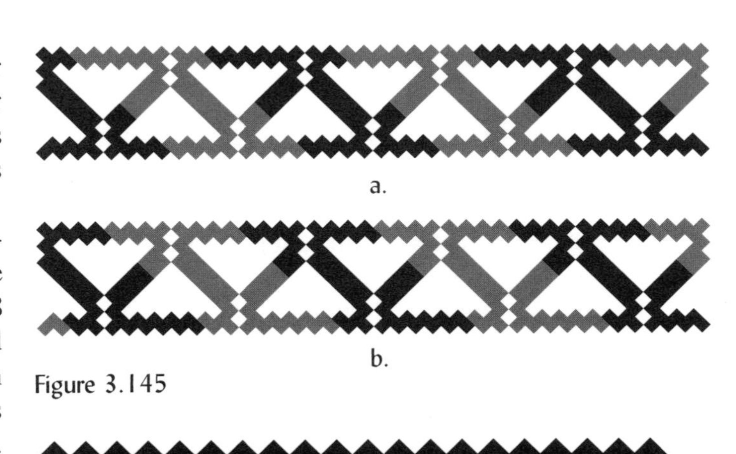

Figure 3.145

Figure 3.146

the bottom row may be freely chosen: each may be brown or may be white.

In the case where the dimensions are 3 × 3, there are only two distinct diagonally woven strip patterns. Figure 3.147 shows the two patterns (b) and motifs that generate them (a). In the case of dimensions 3 × 4, there are 20 distinct diagonally woven strip patterns of the *gipatsi* type (see Figure 3.148).

Symmetries

The creativity of *sipatsi* weavers deserves to be underlined: they have invented strip patterns of all of the seven theoretically possible classes. Figure 3.149 gives an example of each class.

Students may be asked to invent *sipatsi* strip patterns belonging to each of the seven symmetry classes. Which are the smallest dimensions for which it is possible to construct strip patterns of the *sipatsi* type belonging to all seven classes?

Diagonally plaited mat strips

Another type of diagonally woven strips appears in the plaiting of mats, as in Uganda (cf. Trowell & Wachsmann, pl. 32), in the coastal zones of Tanzania (cf. Trowell, pl. XIII), in Kenya (cf. Amin and Moll, p. 28), in Mali (cf. Jefferson, p. 165) and north Mozambique. To distinguish them from the earlier discussed diagonally woven strips let us call them *diagonally plaited mat strips*. These are long and relatively small plaited bands, that are sewn together

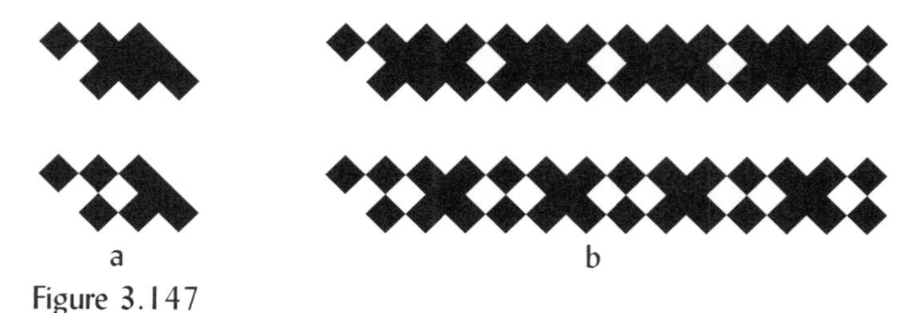

Figure 3.147

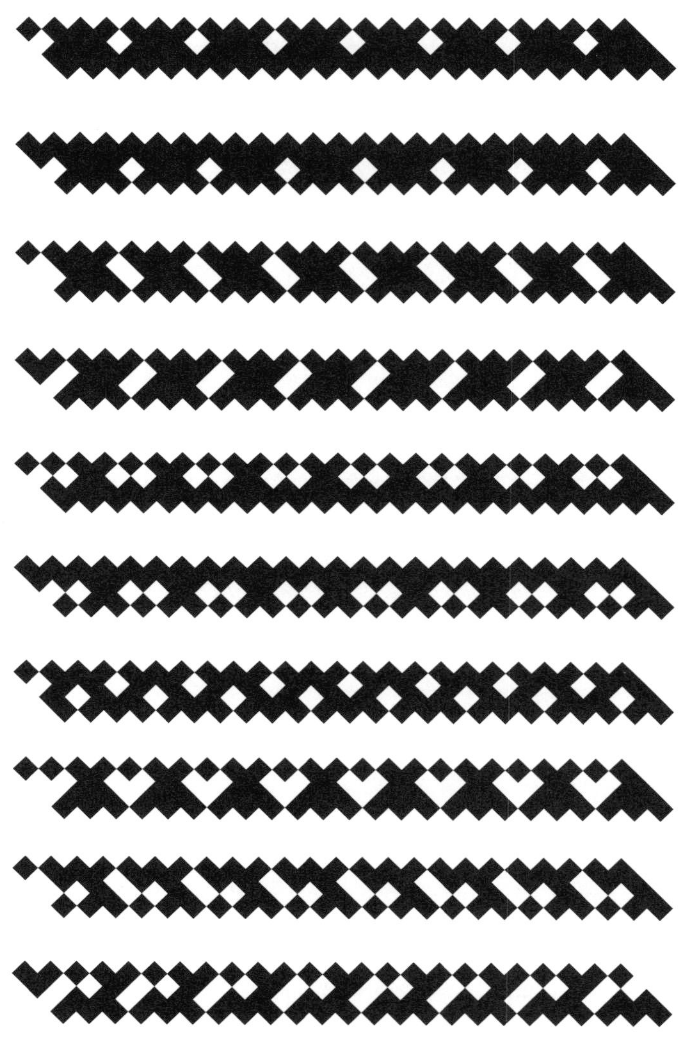
Figure 3.148a

Figure 3.148b

Figure 3.149
Examples of gipatsi *patterns with different symmetries*

to become a rectangular mat. Each strand makes an angle of 45° with the sides of the band and is folded each time it arrives at the side to continue in a direction perpendicular to the direction before the fold (see the examples in Figure 3.150, where one particular strand is darkened to show its zigzag path).

The three strips in Figure 3.150 have been woven two-over-two-under. To weave them one needs 13, 11, and 14 strands respectively. If the number of strands is even, then the strips lack rotational symmetry: the borders on the two sides have different widths.

The type of decoration that can be produced on diagonally plaited mat strips is rather different from the type of

Figure 3.150

decoration on the diagonally woven strips in the previous section. This time brown strands may cross with brown strands, and white with white. Figure 3.151 displays two examples. Both are woven with 9 strands (period and diagonal height are 9). In the first case, five white strands are followed by four brown strands; in the second case, we have 3 brown strands, followed by 2 white, 2 brown, and once more 2 white strands.

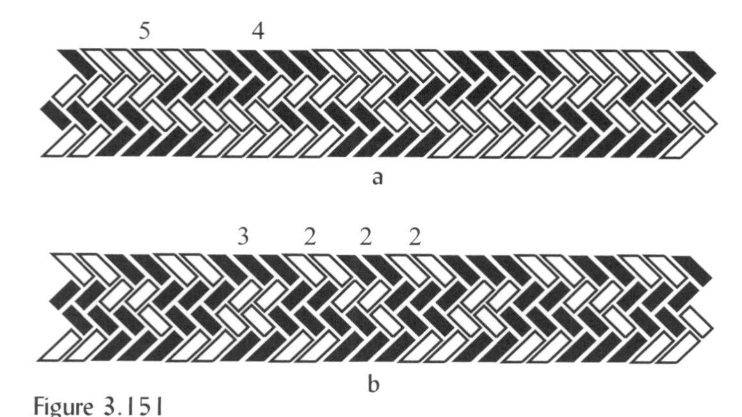

Figure 3.151

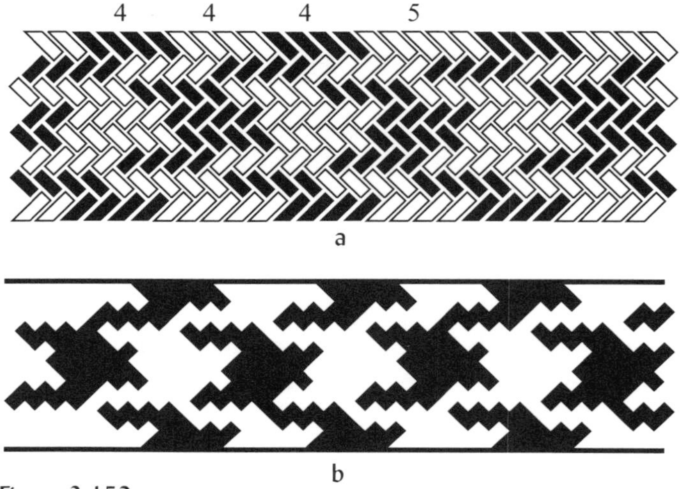

4 4 4 5

a

b

Figure 3.152.
Plaited strip from Uganda

Figure 3.152a presents part of an interesting plaited strip from Uganda. It has dimensions 17 by 17, and is invariant under a glide reflection (cf. the simplified representation in Figure 3.152b). Figure 3.153 presents two other of these plaited strips of dimensions 17×17.

Students may be invited to analyze how many possibilities exist. In other words, how many different diagonally plaited mat strip patterns of given dimensions are possible when the weaving is always over-two-under-two?

What happens if we eliminate the restriction that the plaiting should be over-two-under-two? Figure 3.154 presents four beautiful diagonally plaited mat strip patterns from the island of Zanzibar (Tanzania). All have a diagonal height of 24; their periods are respectively 4, 6, 6, and 8. Brown and white strands alternate. Which symmetries do they have?

How many diagonally plaited mat strips, in which brown and white strands alternate, of dimensions 4 by 24 exist? How many of dimensions 6 by 24? How many of dimen-

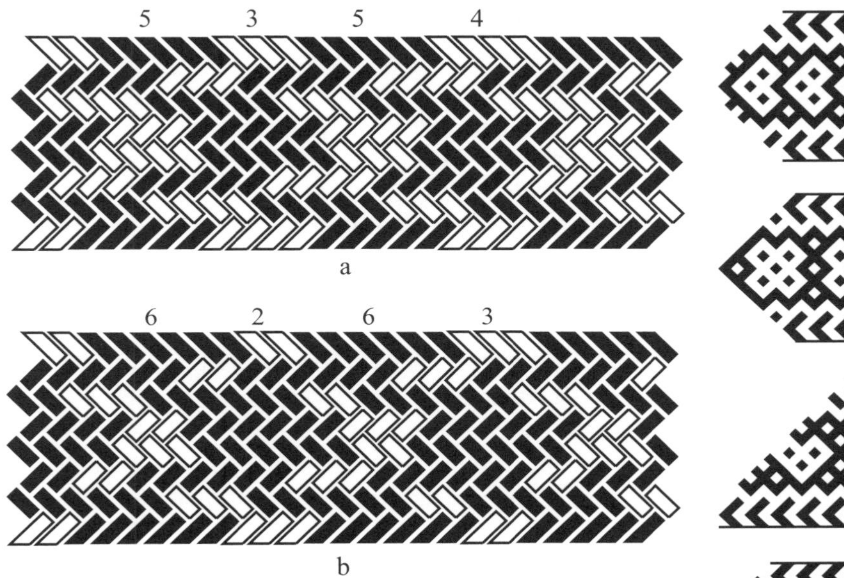

Figure 3.153.
Plaited strips from Uganda (dimensions 17 × 17)

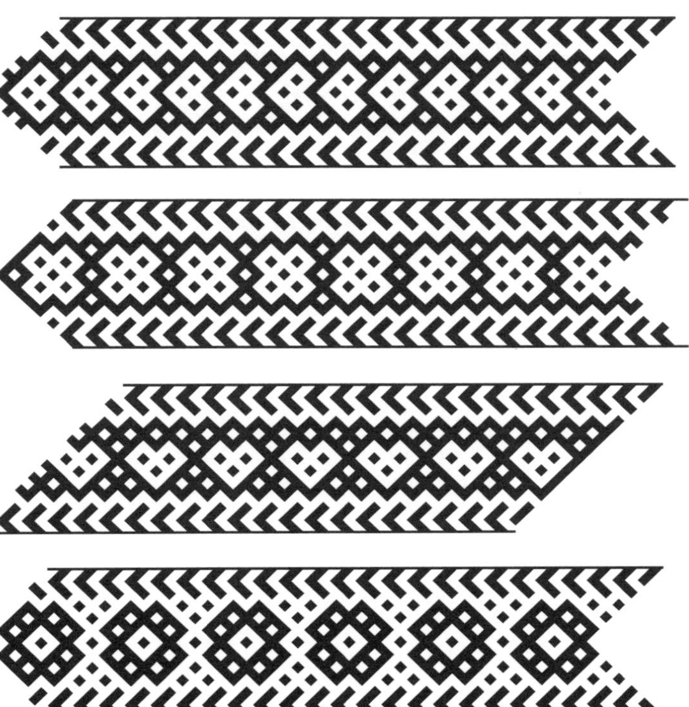

Figure 3.154
Plaited strips from Zanzibar

sions $2m$ by 24? How many of dimensions p by d? Which symmetries are possible? And so on. These are some questions students may be asked to explore. Figure 3.155 shows a Zanzibar strip of dimensions 14 × 32.

If the restriction that the brown and white strands have to alternate is taken away the answers will be different. Figure 3.156 presents another diagonally plaited mat strip from Zanzibar. Its dimensions are 14 × 28. Seven brown strands are followed by seven white strands.

Figure 3.155
Zanzibar strip of dimensions 14 × 32

Decorated woven cylinders

Usually, Mozambican weavers produce basket handles by making them out of a cylinder woven around a nucleus composed of a bundle of palm strands. At the end, the

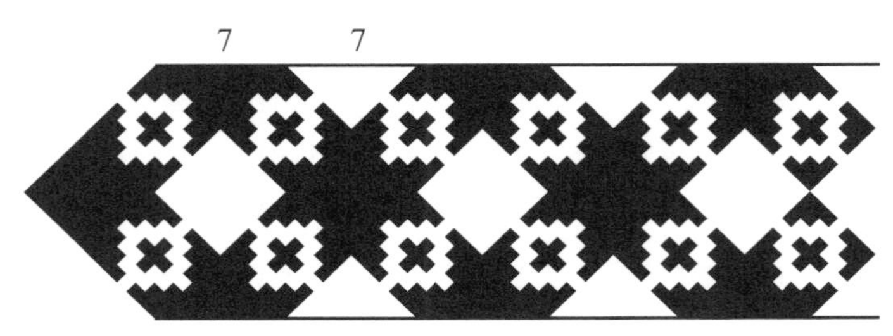

Figure 3.156
Zanzibar strip of dimensions 14×28

cylinder is bent into a handle and fastened to the basket. The weaving is normally over-one-under-one, starting with four strands (two pairs of two). The strands circle around the cylinder in spirals (helices).

Students may be asked to weave such a cylinder. Is four the minimum number of strands needed? Is it possible to weave a cylinder with five strands? With six? And so forth.

Figure 3.157
Decorated woven cylinders

Mozambican basket weavers often introduce colored strips to decorate the handles (see the examples in Figure 3.157). For instance, two of the four strands are colored, the other maintain their natural color. How many distinct cylinders are possible? And what will be the answers if only one strand is colored?

Will it be possible to use other weaves, like two-over-two-under? If so, how many strands are needed? Which decorative patterns can be introduced and which symmetries do they display? Are cylinder patterns strip patterns or plane patterns, or different? Is it possible to weave cylinders with two-color symmetry? Or, in another context, is it possible to produce shoelaces or coverings of wire cable with two- or more color symmetries? These are a few questions that may attract students and teachers to further investigation.

11 From diagonally woven baskets and bells to a twisted decahedron

Makonde artists from northern Mozambique and southern Tanzania are internationally known for their sculpture. A geometrically interesting way of weaving baskets also comes from the Makonde. Artisans weave a type of basket with standardized sizes that is used to measure the production of maize, sorghum, and beans before taking the harvest to the storage houses. The basket is called *likalala*, and has a volume of about 50 liters. The basket has the shape of a 'cylinder' with a square bottom and circular opening (see Figure 3.158).

To make the *likalala* basket, the artisan weaves a square mat with clearly visible mid lines (see *AC* and *BD* in Figure 3.159), as follows. With two sets of six strands (both sets maintain their natural yellow color; one set has been colored brown in the figures to make the weaving process more

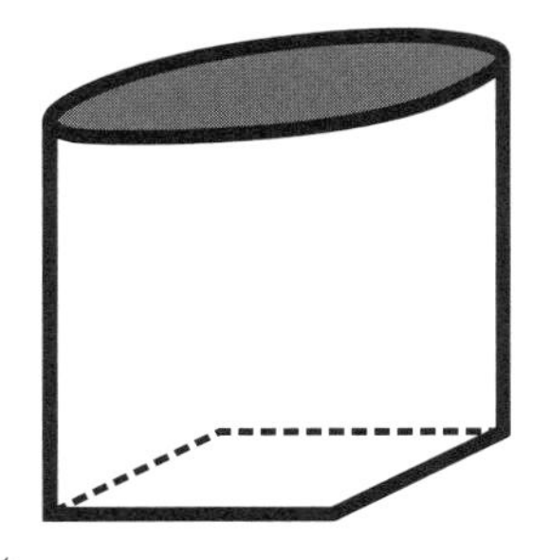

Figure 3.158
Shape of a likalala *basket*

understandable), the artisan weaves the center of the square (see Figure 3.160), called *yuyumunu*, meaning the mother who bears the whole square mat.

The artisans next weave the mid lines by introducing equal groups of four strands both at the left and at the right from the center, and 'above' and 'under' the center . Finally, they complete the square mat by weaving each of its quadrants, over-two-under-two. At the end, not only the mid lines are visible, but also the 'toothed squares' and the midpoints *A*, *B*, *C*, and *D* of the four sides of the squares (see once more Figure 3.159). At these mid points the corners of the basket will rise. The square formed by these mid points will constitute the square bottom of the basket.

Figure 3.159

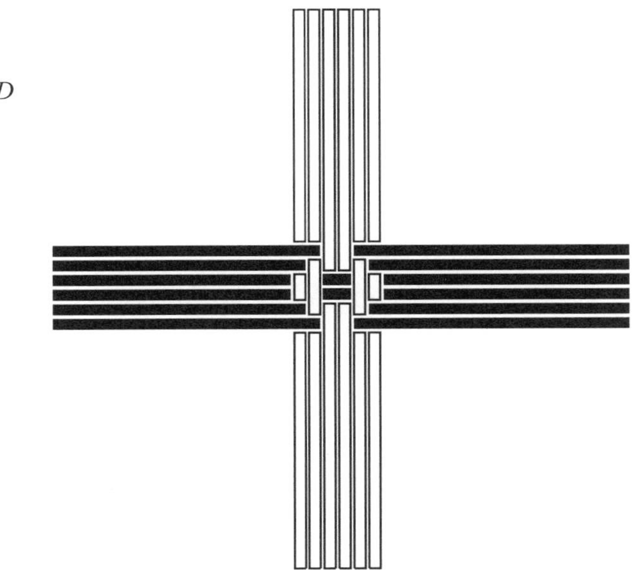

Figure 3.160

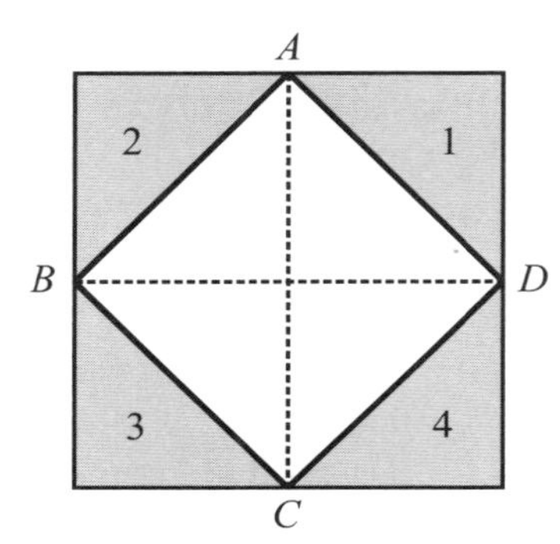

Figure 3.161

a. *woven mini-basket* b. *bottom seem from below*

Figure 3.162

The basket weaver then folds two adjacent triangles (areas '1' and '2' in Figure 3.161) upwards and weaves the outstanding (white) strands of both triangles together. Thus A becomes a corner and a part of the wall of the basket is obtained. Similarly the other two triangles are folded, and the four triangles are woven together. The artisan continues the twill weaving, whereby the strands alternately go over and under two strands, until he or she obtains a desired height, and finally fastens a circular wooden border to the basket. As a consequence of the weaving technique, the bottom square and the top circle have the same circumference.

The basic weaving technique of the *likalala* is also known in several other parts of Africa and elsewhere (cf. Gerdes, 1990b, 1999a). The 'limit' case—lowest possible number of strands—is found among the Bassari, a people (population about 10,000) of hunters who rather recently became peasants and who live in the Senegalese-Guinean border region (cf. Lestrange & Gessain). Among their baskets, they have mini-baskets that are plaited—using the over-one-under-one weave—out of only eight or even out of only four strands made of palm leaf (see Figure 3.162).

These 'baskets' were penis holders—called *epog edepog*, worn by men from the time of their circumcision onwards. The Museum of Man in Paris has several of these penis holders, collected in 1961 from the village of Etyolo. These holders do not have separate borders fastened to them, as do the *likalala* baskets from the Makonde. Instead a rim is formed by folding the outstanding strand parts and tucking them into the weave at the first opportunity.

What will happen if we do not make a rim but try to finish the basket in the same way as it started? In other words, if we try to close it with a square top. There are essentially two different possibilities: either the top square is in the same position as the bottom square (see the example in Figure 3.163a), or the 'basket' is twisted and the top square is rotated around the basket's axis about an angle of 90° (see the example in Figure 3.163b). In the first case, the height is a whole number of times the side of the square (two times in Figure 3.163a), whereas in the second case it is about a whole number and a half times the side (one and a half times the side in Figure 3.163b).

Plaited bell baskets

Bassari weavers like the second solution. In the collection of the Museum of Man is a jingle collar, called *bamboyo*,

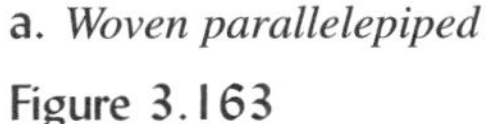

a. *Woven parallelepiped*

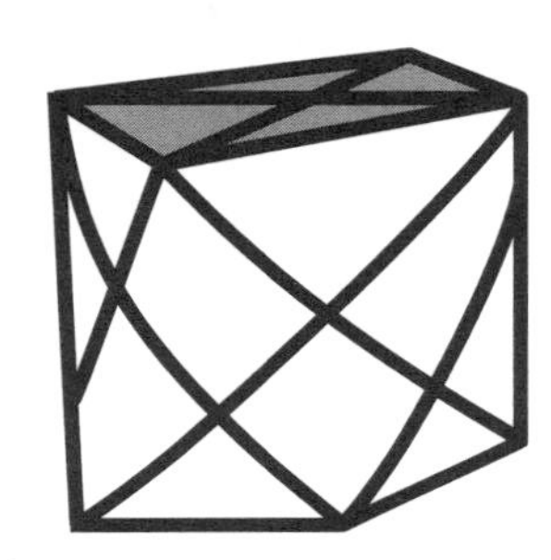

b. *Twisted Bassari mini-basket*

Figure 3.163

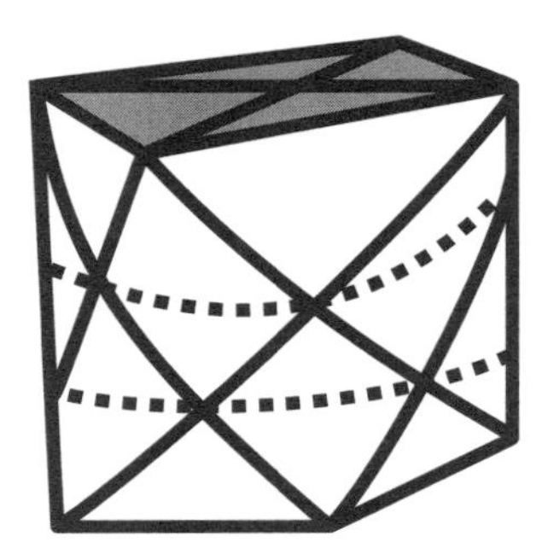

Figure 3.164

collected from Goumo, a village about 6 km from Etyolo. The collar is composed of twenty of such small closed 'baskets' as in Figure 3.163b. The sides of the squares are about 2 to 3 cm. The thread passes through the centers of the bottom and top squares and the little bell baskets contain small stones that ring when dancers wear the collar during the *omangare* feast.

The reader is invited to try to weave one of the *bamboyo* bells with four strips of flexible cardboard. Observing the end result, it may be noted that the whole little bell may be woven with only one strip! How can this be done?

On the wall of the little Bassari bell basket we have two horizontal rows of small holes (see Figure 3.164). Each hole borders a part of the four crossing strips. What if we try to weave in such a way that only one horizontal row of holes appears on the wall? The twist disappears and we obtain a woven cube (Figure 3.165a). ['Braided diagonal cube' is the expression used by Hilton and Petersen, 1988, 124–125.] It is convenient to fold the strips before starting to make the weaving easier (Figure 3.166 shows how a strip has to be folded). Four strips are needed. Figure 3.165b shows the closed path of one of the strips.

If we weave the basket in such a way that three rows of small holes appear on its wall instead of the two rows on the Bassari bell basket, the twist disappears once more, and this time we obtain a woven basket whose height is twice the length of the side of the bottom square (see once more Figure 3.163a). Two strips are needed. An interesting, more general question to investigate is how many strips (which are closed, that is, returning after some time to themselves)

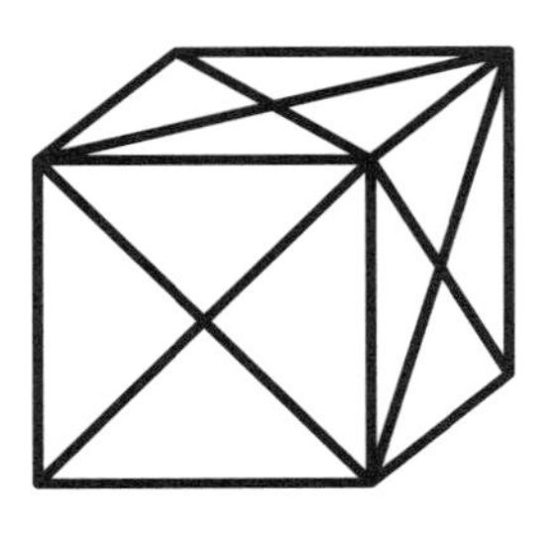

a. *Woven cube*

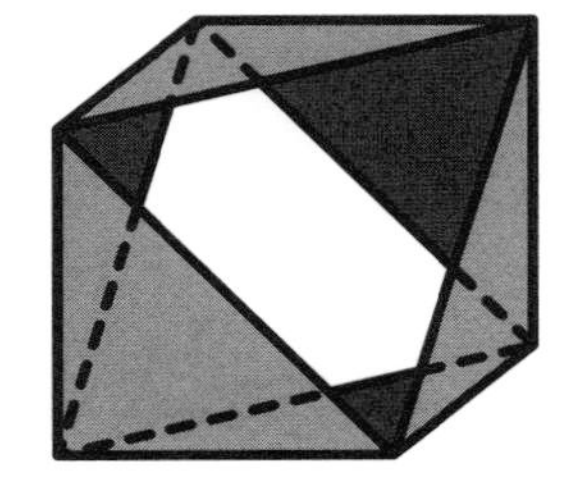

b. *One closed strip*

Figure 3.165

Figure 3.166

are needed to weave a rectangular parallelepiped of sides pa, qa, and ra, where p, q, and r denote natural numbers and using strips of the width $a\sqrt{2}/2$. In particular, for which values of p, q, and r is only one strip needed to braid diagonally the whole parallelepiped?

Let us return to the Bassari bell basket. This bell basket is really special in the sense that it can be made out of only one strip. Are there more 'baskets' with this characteristic? Will the little Bassari bell basket be the smallest closed weavable 'basket' with squares on the bottom and top that are twisted relative to each other?

Mini-*bamboyo*-bell

We saw that on the wall of the little Bassari bell basket, there are two rows of small holes that border four crossing strip parts. If there is only one such row, a braided cube is obtained. What will happen if there is *no* row of small holes?

If there is no such row, this means that after leaving the bottom square (with small holes in the center and at the vertices), the first new holes appear at the vertices of the top square. Consequently, the wall of such a woven object will be composed of isosceles right triangles of which the right angle vertex is a vertex of the upper square and the other two vertices are vertices of the lower square, or, conversely, the right angle vertex is a vertex of the lower square and the other two vertices are successive vertices of the upper square (see Figure 3.167). This 'basket' has a twisted shape since the top square is rotated around the basket axis about an angle of 45°. Let us call our 'basket' a mini-*bamboyo*-bell.

How many strips are needed to make it? Will it be rigid? Let us see. Figure 3.168 shows how to prepare a straight strip for making a mini-*bamboyo*-bell, and Figure 3.169 the final strip. Now try to weave over-one-under-one a mini-

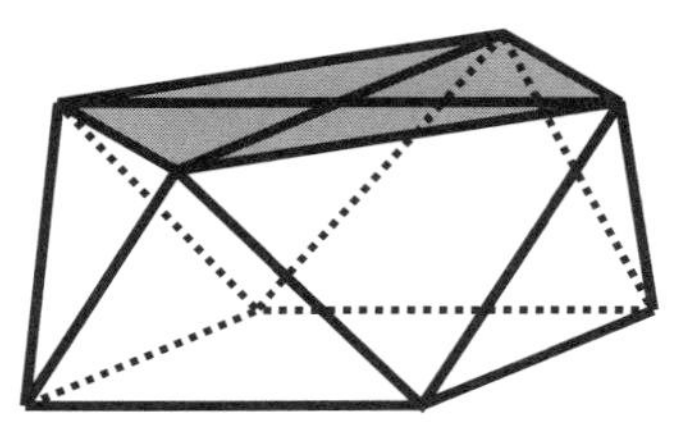

Figure 3.167
*Decahedral mini-*bamboyo *bell*

1) Begin with a long strip of cardboard paper:

2) Fold straight back (the shadowed part represents the back of the cardboard.

3) Unfold

4) Fold down:

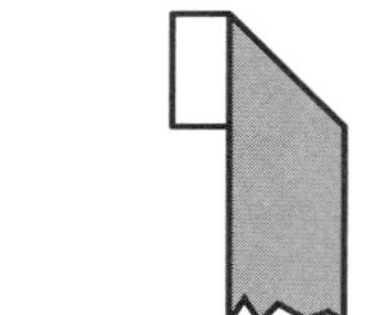

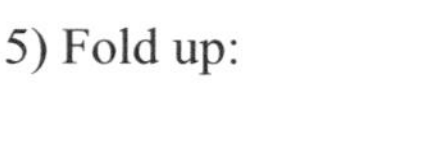

5) Fold up:

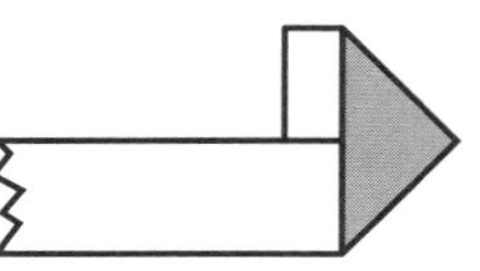

Figure 3.168
*Preparing a strip for weaving the mini-*bamboyo *bell*

bamboyo-bell. Like the Bassari bell basket it is made out of only one strip, and, on the other hand, maybe surprisingly, it is an extremely rigid 'basket'.

What can be said about its shape? It is a decahedron with congruent squares on the top and bottom, and eight

6) Unfold:

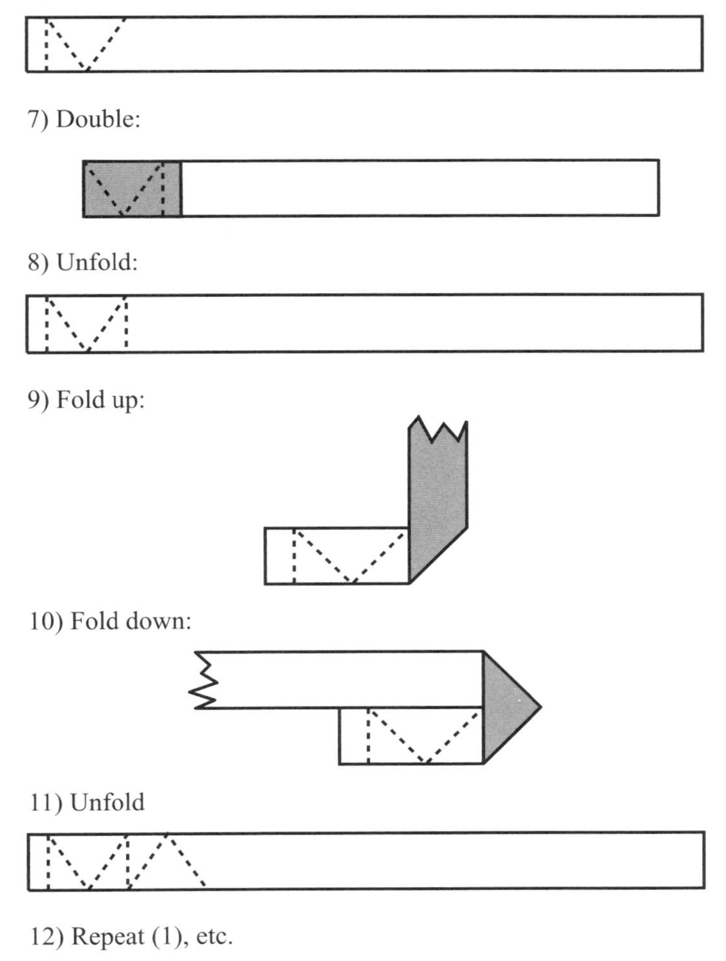

12) Repeat (1), etc.

Figure 3.169

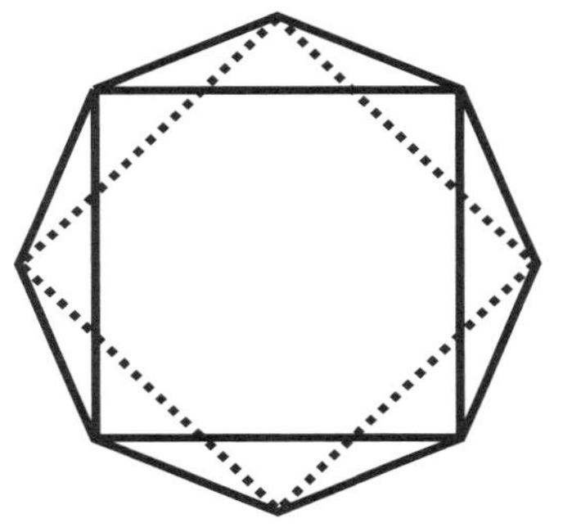

Figure 3.170

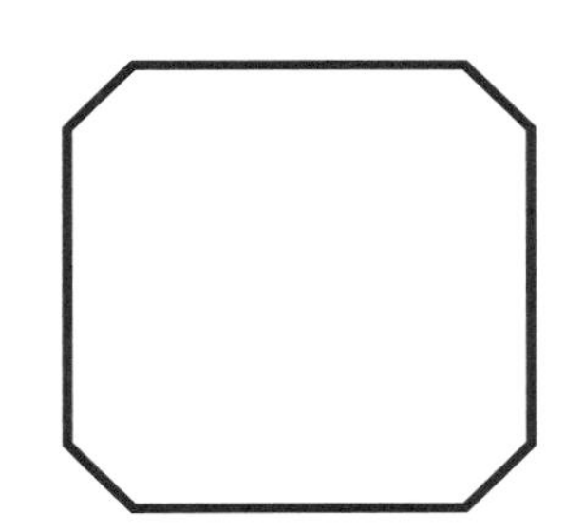

Figure 3.171

congruent triangles, each with the size of a quarter of the squares. Thus the surface area is $4a^2$, if a is the side of the squares (or $a\sqrt{2}/2$ the width of the initial strip). Figure 3.170 presents a view from above of the decahedron.

Each horizontal section of the decahedron is a semi-regular octagon like in Figure 3.171, having the midsection regular. All octagons have the same circumference, $4a$, as the perimeter of the top and bottom squares.

What is the volume of this decahedron? The volume can be determined in various ways. The reader is invited to find his or her own ways, before reading on.

Volume of decahedral mini-bamboyo-bell

Our decahedron may be considered as a truncated pyramid of which four congruent wedges have been cut off (see Figure 3.172).

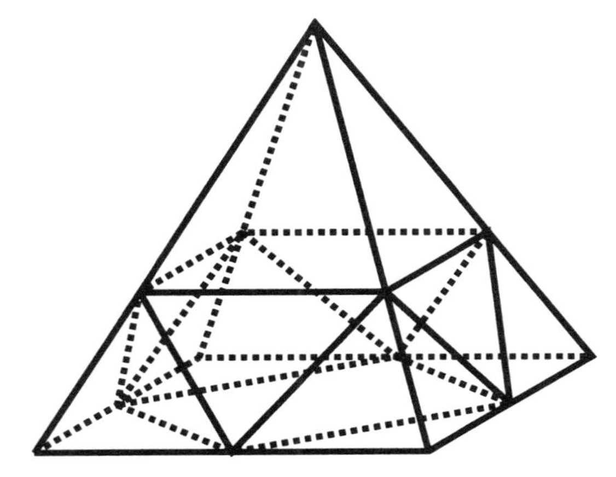

Figure 3.172

Since the bottom square of the truncated pyramid has the same length as the diagonal of the decahedron's squares, its volume is given by

(1) $$\frac{h}{3}\left[\left(a\sqrt{2}\right)^2+a^2\sqrt{2}+a^2\right]=\frac{ha^2}{3}\left[3+\sqrt{2}\right].$$

The volume of each of the wedges is given by

(2) $$\frac{h}{3}\times\frac{1}{2}\left[\frac{a\sqrt{2}}{2}\right]^2=\frac{h}{3}\times\frac{a^2}{4}.$$

Therefore the volume of the decahedron is equal to

(3) $$\frac{ha^2}{3}\left[3+\sqrt{2}\right]-4\times\frac{h}{3}\times\frac{a^2}{4}=\frac{ha^2}{3}\left(2+\sqrt{2}\right)\cdot$$

Now we must express the height h in terms of a. Let us consider a vertical section of the decahedron, that passes through its axis and two of its vertices (see Figure 3.173).

Using the theorem of Pythagoras, we find

(4) $$h^2=\left[\frac{a}{2}\right]^2-\left[\frac{a\sqrt{2}}{2}-\frac{a}{2}\right]^2=-\frac{a^2}{2}+\frac{a^2\sqrt{2}}{2},$$

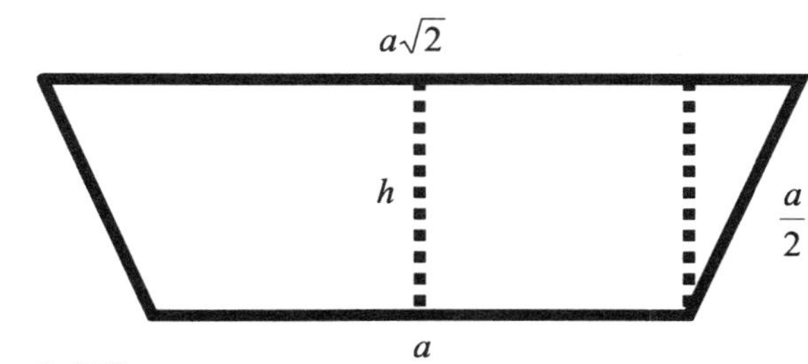

Figure 3.173

leading to

(5) $$h=a\sqrt{\frac{\sqrt{2}-1}{2}}.$$

This corresponds to h $\approx 0.455a$.

Substitution in (3) gives us the volume of the decahedron:

(6) $$\begin{aligned}V&=\frac{a^3}{3}\sqrt{\frac{\sqrt{2}-1}{2}}\left(2+\sqrt{2}\right)\\&=\frac{a^3}{3}\frac{\sqrt{\sqrt{2}-1}}{\sqrt{2}}\sqrt{2}\left(\sqrt{2}+1\right)\\&=\frac{a^3}{3}\sqrt{\left(\sqrt{2}-1\right)\left(\sqrt{2}+1\right)\left(\sqrt{2}+1\right)}\\&=\frac{a^3}{3}\sqrt{\sqrt{2}+1}\approx 0.5179a^3.\end{aligned}$$

An alternative way to calculate the volume is to use integral calculus. Let $A(y)$ be the area of a semi-regular horizontal section, where y represents the distance between the section and the base of the decahedron. Then the volume of the decahedron is given by

(7) $$V=\int_0^2 A(y)\,dy.$$

It may be easier to express the area of a horizontal section

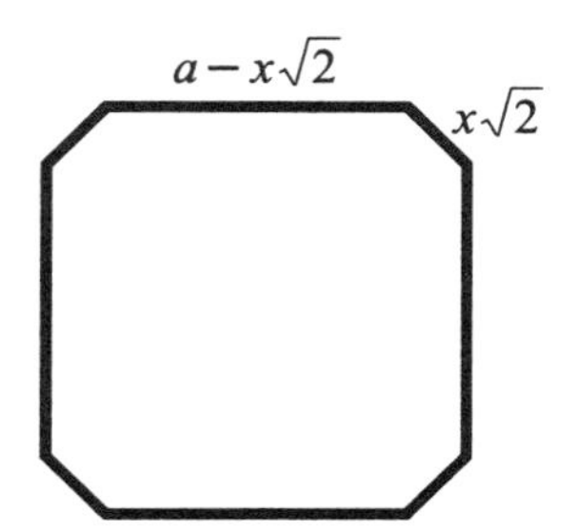

Figure 3.174

in function of x, whereby x represents the distance from the section to the base measured along an edge of the decahedron. As $dy=\sqrt{\sqrt{2}-1}\,dx$, it follows that

$$V=\sqrt{\sqrt{2}-1}\int_0^{a\sqrt{2}/2} A(x)\,dx.$$

Once determined $A(x)=a^2+(4-2\sqrt{2})ax+(4-4\sqrt{2})x^2$ (see Figure 3.174), the final result (6) is obtained.

How can the concept of the decahedral mini-*bamboyo*-bell be generalized? By changing the ratio of height to side of the squares (see the example in Figure 3.175). By changing the angle of the twist. By changing to squares with different sizes. By changing the squares into other regular polygons, etc. In each case, questions related to areas and volumes may be formulated and analyzed. Which of these figures admit circumscribed spheres?

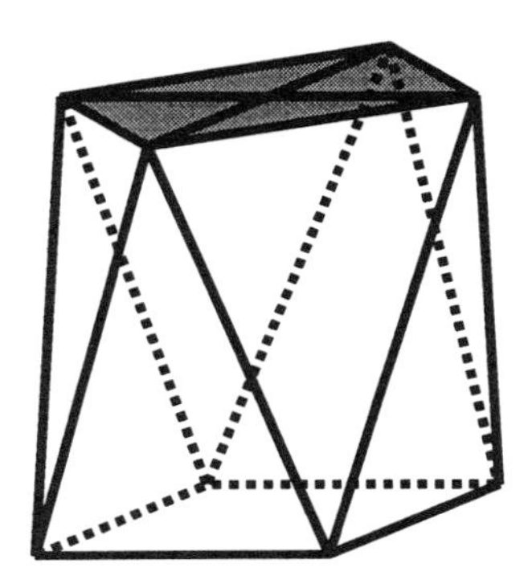

Figure 3.175

Looking back

In this section it was shown that an analysis of the weaving of Makonde baskets and Bassari bells may lead students both to the weaving of a cube, of parallepipeds, and of a twisted decahedron, and to consider several questions related to areas and volumes of the geometrical forms involved.

Bibliography

Amin, Mohamed & Moll, Peter (1983), *Portraits of Africa*, Harvill Press, London.

Bruins, E. and Rutten, M. (1961), *Textes Mathématiques de Suse*, Paris.

Changuion, Paul; Matthews, Tom and Changuion, Annice (1989), *The African Mural*, Struik, Cape Town & New Holland, London.

Coart, E. (1927), Les nattes, in: *Annales du Musée du Congo Belge: Ethnographie et Anthropologie*, Série III: Notes analytiques sur les collections du Congo Belge, Tome II: Les industries indigènes, Fascicule 2: 195–202, Pl. XXII–LXIV.

Dunsmore, S. (1983), *Sepak Raga (Takraw) — The Southeast Asian Ball Game*, Sarawak Museum, Kuching

Etienne-Nugue, Jocelyne (1982), *Crafts and the Arts of Living in the Cameroon*, Louisiana State University Press, Baton Rouge.

Faublée, J. (1946), *Ethnographie de Madagascar*, Musée de l'Homme, Paris.

Gerdes, Paulus (1985a), Conditions and strategies for emancipatory mathematics education in underdeveloped countries, *For the Learning of Mathematics*, Montreal, Vol. 5, No.1, 15–20.

—— (1985b), Three alternate methods for obtaining the ancient Egyptian formula for the area of a circle, *Historia Mathematica*, New York, Vol. 12, 261–268.

—— (1988), On culture, geometrical thinking and mathematics education, *Educational Studies in Mathematics*, Dordrecht /Boston, 1988, Vol.19, No.3, 137–162. Reproduced in: Bishop, Alan (Ed.), *Mathematics Education and Culture*, Kluwer Academic Publishers, Dordrecht/Boston, 1988, 137–162; and in: Powell, Arthur B. and Frankenstein, Marilyn (Eds.), *Ethnomathematics: Challenging Eurocentrism in Mathematics Education*, State University of New York Press, 1997, 223–247.

—— (1990a) (Ed.), *Matemática? Claro! - 8a Classe*, Instituto Nacional do Desenvolvimento da Educação, Maputo.

—— (1990b), *Ethnogeometrie: Kulturanthropologische Beiträge zur Genese und Didaktik der Geometrie*, Franzbecker Verlag, Bad Salzdethfurth.

—— (1991), Fivefold Symmetry and (basket)weaving in various cultures, in: Hargittai, Istvan (Ed.), *Fivefold Symmetry in a Cultural Context*, World Scientific Publishing Co., Singapore, pp. 243–259.

—— (1995), *Women and Geometry in Southern Africa*, Universidade Pedagógica, Maputo.

—— (1996), *Femmes et Géométrie en Afrique Australe*, L'Harmattan, Paris.

—— (1998a), *Women, Art and Geometry in Southern Africa*, Africa World Press, Trenton NJ / Asmara (Eritrea)

—— (1998b), Molecular Modeling of Fullerenes with Hexastrips, *The Chemical Intelligencer*, New York, Vol. 4(1), 40–45.

—— (1998c), On some geometrical and architectural ideas from African art and craft, in: Kim Willimas (Ed.), *Nexus II: Architecture and Mathematics,* Edizioni dell'Erba, Firenze, pp. 75–86.

—— (1999a), *Culture and the Awakening of Geometrical Thinking. Anthropological, Historical, and Philosophical Considerations. An ethnomathematical study* MEP-Publications (University of Minnesota), Minneapolis (Short English language edition of 1990b).

—— (1999b), Geometrical Models of Fullerenes with Hexastrips, *The Mathematical Intelligencer*, Vol. 21(1) 6–12, 27.

—— (1999c), On mathematical ideas in cultural traditions of central and Southern Africa, in: Helaine Smith (Ed.), *Mathematics across cultures: a history of non-western mathematics,* Kluwer, Dordrecht (in press).

—— (1999d), Gerade und Ungerade—Zu einigen mathematischen Aspekten der Mattenflechterei der yombe-Frauen am unteren kongo, in: Erhard Scholz (Ed.), *Festband zum 60sten Geburtstag von Harald Scheid,* Wupertal (in press).

Gerdes, Paulus and Bulafo, Gildo (1994): *Sipatsi: Technology, Art and Geometry in Inhambane*, Universidade Pedagógica, Maputo.

Golomb, Solomon (1994), *Polyominoes: Puzzles, Patterns, Problems, and Packings*, Princeton University Press, Princeton.

Hilton, Peter and Pedersen, Jean (1988), *Build your own polyhedra*, Addison-Wesley, New York.

Jefferson, Louise (1974), *The Decorative Arts of Africa*, Collins, London.

Kroto, H.W., Heath, J.R., O'Brian, S.C., Curl, R.F. and Smalley, R.E. (1985), C_{60}: Buckminsterfullerene, *Nature*, 318, 162–163 (Reproduced in Aldersey-Williams, H. *The most beautiful molecule: The discovery of the Buckyball*, John Wiley & Sons, New York, 1995).

Lestrange, Marie-Thérèse de and Gessain, Monique (1976), *Collections Bassari du Musée de l'Homme*, Musée de l'Homme, Paris.

Meurant, G. and Thompson, R.F. (1995), *Mbuti Design — Paintings by Pygmy Women of the Ituri Forest*, Thames and Hudson, London.

NTTC (1976), *Litema, Designs collected by students of the National Teacher Training College of Lesotho*, NTTC, Maseru (Lesotho).

Schildkrout, Enid and Keim, Curtis (Eds.) (1990), *African Reflections: Art from Northeastern Zaire*, American Museum of Natural History, New York.

Somjee, S. (1993), *Material Culture of Kenya*, East African Educational Publishers, Nairobi.

Stuhlman, Franz (1910), *Handwerk und Industrie in Ostafrika: Kulturgeschichtliche Betrachtungen*, Hamburg.

Torday, Emil and Joyce, T. A. (1922), *Notes ethnographiques sur des populations habitant les bassins du Kasai et du Kwango oriental*, Musée Royal du Congo Belge, Bruxelles.

Trowell, Margaret (1960), *African Design*, Praeger, New York.

Trowell, Margaret and Wachsmann, K. (1953), *Tribal Crafts of Uganda*, Oxford University Press, London.

Washburn, Dorothy and Crowe, Donald (1988), *Symmetries of Culture: Theory and Practice of Plane Pattern Analysis*, University of Washington Press, Seattle.

Weule, Karl (1908), *Wissenschaftliche Ergebnisse meiner Ethnographischen Forschungsreise in den Südosten Deutsch-Afrikas*, Berlin.

Wyk, Gary van (1998), *African Painted Houses. Basotho Dwellings of Southern Africa,* Harry Abrams Publ., New York.

4

The 'sona' sand drawing tradition and possibilities for its educational use

The theme of the fourth chapter is the geometry of the *sona* sand drawing tradition among the Chokwe in southern-central Africa. As slavery and colonial domination disrupted and destroyed so many African traditions, the *sona* tradition with its strong geometrical component virtually disappeared. Consequently, after an introduction on the Chokwe people, I present in Section 1 elements of a reconstruction of the sand drawing tradition. In Section 2, I present examples of an educational exploration of *sona*. Section 3 constitutes an excursion, underscoring the mathematical potential of the *sona* tradition. Here I contribute to developing the geometry of Lunda-designs and Lunda-patterns, which I discovered in the context of analyzing a particular class of *sona*. Chapter 4 can be considered a summary of my books *Sona Geometry* (1993/4, 1994, 1995, 1997b), *Lusona: Geometrical recreations of Africa* (1991, 1997a) and *Lunda Geometry* (1995).

Introduction: The Chokwe and their sand drawing tradition

The *sona* sand drawing tradition was cultivated principally by the Chokwe (also written as Tshokwe, Tchokwe, Cokwe and Quiocos), but also by related peoples such as the Ngangela and Luchazi. These peoples live in eastern Angola, northwest Zambia, and neighboring zones of Congo/Zaire (see Figure 4.1). A cultural characteristic they have in common is their important male initiation rites called *mukanda*, which last between six and eight months.

The Chokwe, with a population of about one million (1991), inhabit predominantly the northeastern part of Angola, a region called Lunda, and neighboring zones from Congo/Zaire and Zambia. It was about 1600 when Lunda immigrants arrived in central Angola, coming via Congo/Zaire from the Lake Tanganyika area. They adopted the name Chokwe after a tributary of the Lungwe-Bungo that flows into the Zambeze River.

Traditionally, Chokwe women worked in agriculture, and men devoted themselves to hunting. Artisans occupied themselves with iron making, painting, sculpture, and furniture making, and weaving of mats, baskets, and so on. Their

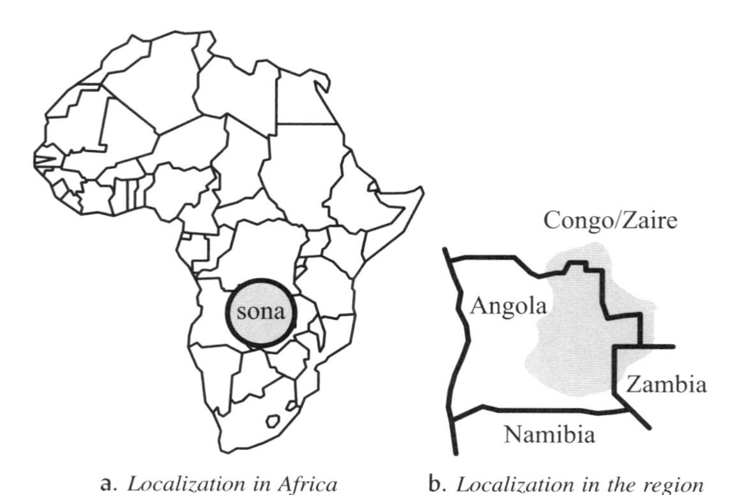

a. *Localization in Africa* b. *Localization in the region*

Figure 4.1
Regional distribution of the sona drawing tradition

art and craft achieved a high degree of perfection. However, colonial penetration and later occupation, at the end of the nineteenth century, provoked a cultural decline.

The sona sand drawing tradition

When Chokwe men were meeting at their village central or at their hunting camps, sitting around a fire or in the shadow of leafy trees, they enjoyed telling stories and illustrating them with drawings in the sand. These drawings are called *sona* (singular: *lusona*). Today the word *sona* refers also to writing. Many of the sand drawings or *sona* belong to an old tradition. They refer to proverbs, fables, games, riddles, animals, and so forth and play an important role in transmitting knowledge and wisdom from one generation to the next.

Young boys enjoyed making sand drawings with their fingers. They learned the meaning and execution of the easier drawings during their period of intensive schooling, the *mukanda* initiation rites. The more difficult *sona* were

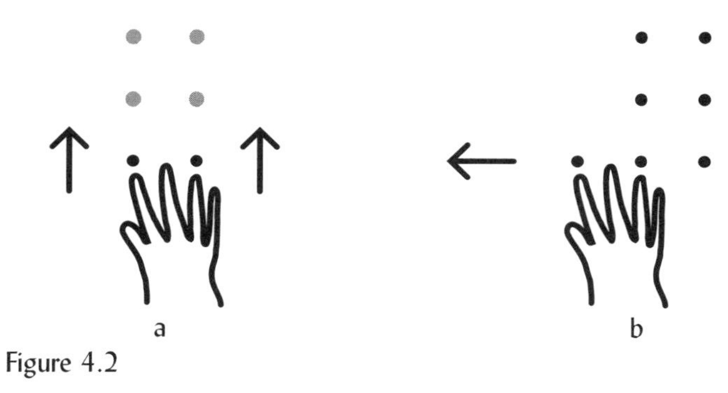

Figure 4.2

only known by those story tellers, who were real *akwa kuta sona* (those who know how to draw). These drawing experts passed on the knowledge about the meaning and execution techniques of *sona*, to their male descendants. They were highly esteemed, forming part of an elite in Chokwe society.

To facilitate memorizing their standardized drawings the *akwa kuta sona* invented the following mnemonic device: after cleaning and smoothing the ground, a drawing expert first sets out, with the tips of the index and middle or ring fingers of his right-hand an orthogonal net (grid or array), of equidistant points or dots. Figure 4.2 shows how the points are impressed in the sand.

Often an additional series of points is then marked in the centers of the unit squares of the first grid (see Figure 4.3a).

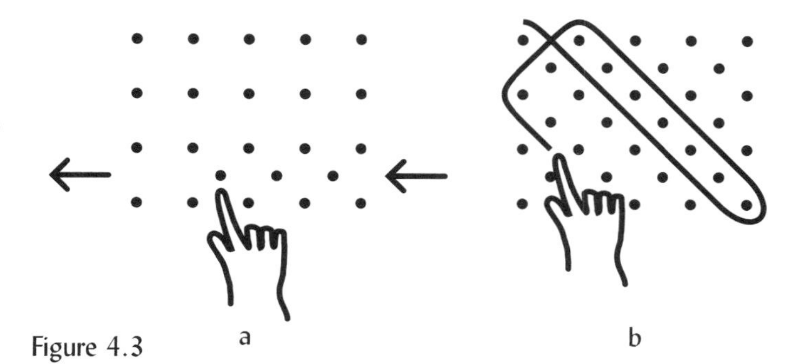

Figure 4.3

After counting and checking the various points in each row, the drawer traces a line figure with the tip of his right index finger. Normally the drawing is composed of one or more lines that embrace the points of the net. Figure 4.3b illustrates the execution of a curve.

Both the number of rows and the number of columns of the array of points as well as the rules for executing drawings depend on the motif to be represented. By applying their method—an example of the use of a coordinate system—the *akwa kuta sona* reduced, in general, the memorization of a complete design to that of a few numbers (the **dimensions** of the lattice of points) and of a **geometric algorithm** for the construction of the curve(s).

The lines had to be executed with a smooth and continuous motion. Stopping halfway would be interpreted by the bystanders as a lack of knowledge. Experts executed the drawings swiftly. As a rule, if the figure was fairly intricate, the drawing was made in silence. But, if it was relatively simple, the tale was told as the drawing proceeded. Once drawn in the sand, the designs were generally erased.

Easier *sona* were also, for instance, painted on house walls, whereby the lines embraced red points. Figure 4.4 shows an Angolan scarf decorated with small *sona*, a sketch made by the Italian priest Cavazzi during his visit to Angola in the seventeenth century.

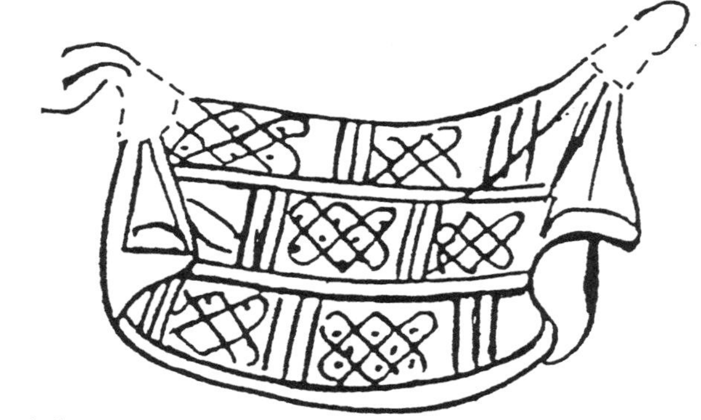

Figure 4.4
Sketch of a 17th century Chokwe scarf

With slave trade and later colonial occupation, the *sona* drawing tradition began to disappear. Some missionaries and ethnographers collected *sona* and saved them from oblivion. The largest collection of *sona* and the most important was published by the Portuguese ethnographer Fontinha. His book contains 287 different drawings he collected in the 1940s and 1950s. Already at that time it was extremely difficult to collect so many *sona*. It was rare to find people who knew more than half a dozen *sona*.

The simpler patterns were known by a large number of men. However, the secret of the more complicated ones was known by only some older *akwa kuta sona*. Fontinha observed that "with each day that passes and with each elder that dies, so too disappear precious witnesses of their collective past" (Fontinha, 1983, p. 39).

During nearly half a century among the Ngangela, the American missionary Pearson found only four men who had real knowledge of *sona* sand drawings. According to his informants, these 'sandgraphs' had been handed down through a few men over many generations (Pearson, 1977). Hamelberger (1952) and Dos Santos (1961) published collections of *sona* from the Chokwe, and Kubik (1988) from the Luchazi. Kubik considers *sona* as the traditional library of these peoples, drawn "to convey to the male community ideas about existing institutions, to stimulate fantasy, abstract logical thinking and even meditation" (Kubik, 1987, p. 58). At the same time, drawing *sona* constituted a game, a form of entertainment and recreation.

As an example of stories and their accompanying *sona*, the following boxes contain the stories of "The lion and the boy" and "The rooster and the jackal".

The Lion and the Boy

A boy and a lion grew up together and were always friends. One day, they went hunting together. It was the turn of the boy to kill a stag, but he lost his senses. Immediately, the lion made a fire and prepared a medicine to reanimate the boy. They returned home with the slaughtered animal, and there was a feast, that further consolidated still more the friendship that linked the two.

Some time afterwards, they went hunting again and this time the lion went off to kill a buffalo. The lion fell down on the ground, feigning to have fainted. The boy supposed that the lion was dead and thought immediately that he could become the most famous hunter of the region if he would prepare an amulet with the eyes, tip of the nose, and parts of the ears of the lion. As soon as the boy tried to cut the animal the lion stood up and having observed the treachery of his friend, he killed the boy.

Ever since then, the two big hunters—men and lions —were never seen together again (Fontinha, p. 230).

In the *lusona*, the boy and the lion are represented by the two bigger dots. The monolinear drawing has been obtained from a 3×5 plaited-mat design by joining additional loops.

The Rooster and the Jackal

The rooster Kanga and the jackal Mukuza wanted to marry the same woman. Both contacted her father with a proposal of marriage. He asked for payment in advance and they immediately agreed. Suddenly, there ran a rumor that the promised woman had died. Kanga started to cry inconsolably, whereas Mukuza only regretted having lost his advanced payment. Then the father, who intentionally had spread the rumor to see who would be worthy of his daughter, gave her to the rooster who had demonstrated his sincerity (Fontinha, p. 234).

In the *lusona* the two upper bigger dots represent the rooster and the jackal. The lower bigger dot represents the woman.

1 Some geometrical aspects of the sona sand drawing tradition

Symmetry and monolinearity as cultural values

More than 80 percent of the reported *sona* are symmetrical, having at least one axis of symmetry (see the examples in Figure 4.5). Frequent are *sona* with double symmetry, that is, with two perpendicular axes of symmetry (see Figure 4.6). *Sona* with only a rotational symmetry of 180° or 90° are less common (see Figure 4.7). The high frequency of sand drawings with one or more axes of symmetry constitutes an expression of the importance of (axial) symmetry within the culture of the Chokwe and related peoples.

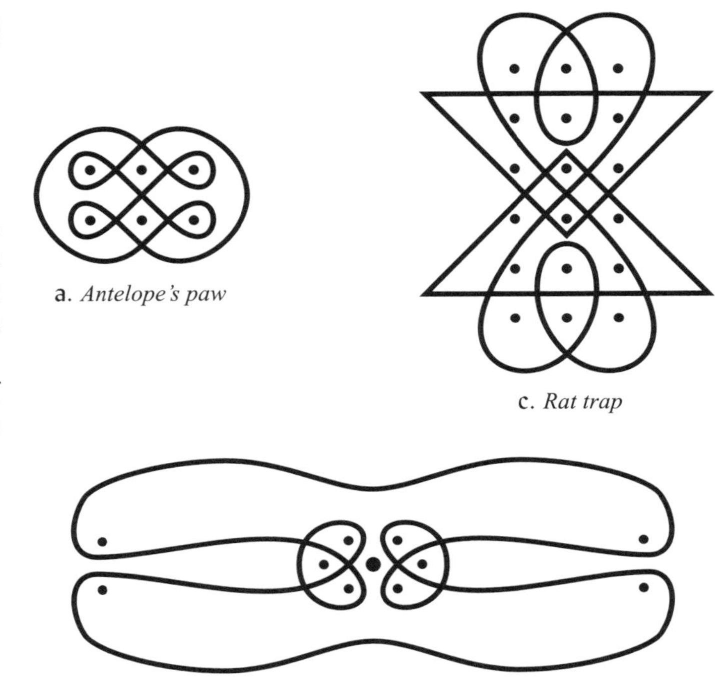

a. *Antelope's paw*

c. *Rat trap*

b. *Spider in the middle of its cobweb*

Figure 4.6

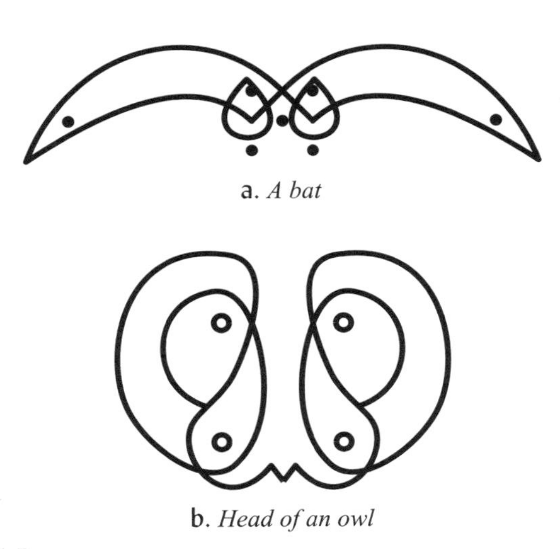

a. *A bat*

b. *Head of an owl*

Figure 4.5

About three-quarters of the reported *sona* are monolinear. In other words, they are composed of only one line; the line can intersect itself, but a segment of the line can never be traced over again* (see the examples in Figures 4.5, 4.6, 4.7a, b; Figure 4.6c is composed of three lines [3-linear]).

* The concept of monolinearity is not the same as traceability or an Euler graph, in graph theory. In graph theory, tracing a graph permits two segments to be tangent at a point. In contrast, when drawing a *lusona*, two segments (of the same line or of distinct lines) are not allowed to be tangent, but they may intersect.

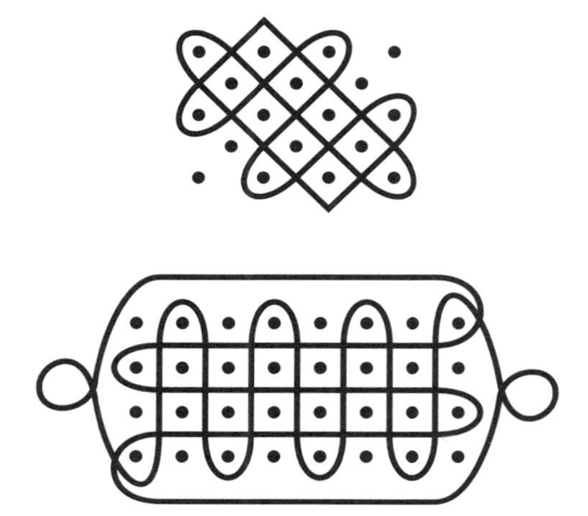

a. *Two representations of the* tchela *board game*

b. *Men-lions that, stealthily, plan their intrigues*

Figure 4.7

Some designs are made of two monolinear drawings: both lines are traced at the same time, each by one hand (see the examples in Figure 4.8). Monolinearity had a high cultural value.

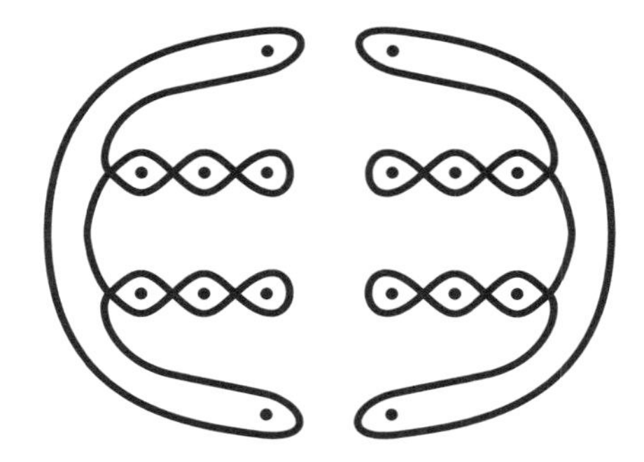

a. *People collecting mushrooms*

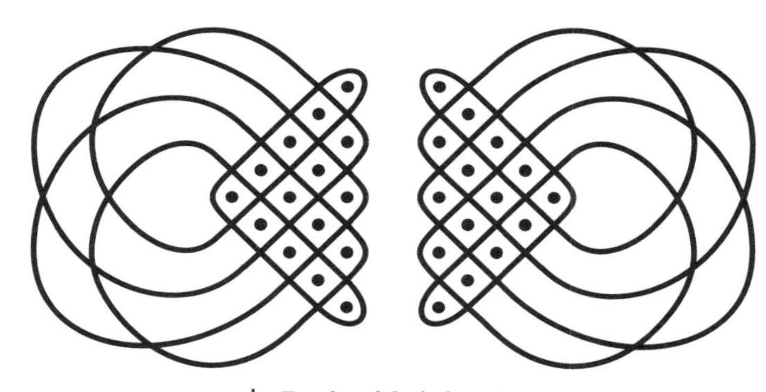

b. *Trunks of the* kajana *tree*

Figure 4.8

Symmetry and monolinearity constituted complementary ideals. The *akwa kuta sona* drawing experts preferred designs that were simultaneously symmetrical and monolinear. However, the two values are not always compatible. For instance, the *lusona* in Figure 4.9a has double symmetry, but is composed of two lines. When one transforms it a little bit in such a way that it becomes monolinear (Figure 4.9b), it looses one symmetry axis.

Sometimes *sona* are not symmetrical as result of the meaning they have to transmit, as the next example will show. The monolinear drawing with fourfold symmetry in

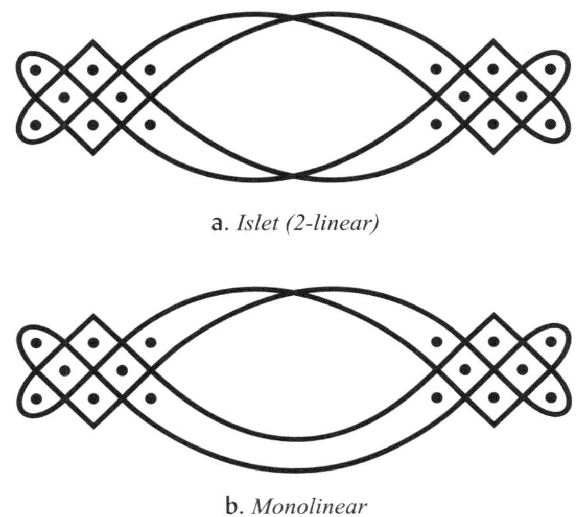

a. *Islet (2-linear)*

b. *Monolinear*

Figure 4.9

Figure 4.10b

Figure 4.10a represents for the Chokwe a plant, from which one extracts a venom that is rubbed on weapons; for the Ngangela it represents a sculptured chair made from a single block of wood. Derived from this *lusona* is another without rotational symmetry, but still monolinear (see Figure 4.10b). It illustrates the following fable (Dos Santos, 1961, p. 87):

> Sambálu, the rabbit (positioned at point *B*), discovered a salt mine (point *A*). Immediately, the lion (point *C*), the jaguar (point *D*) and the hyena (point *E*) came to demand possession and affirm the rights of the strong. Then the rabbit quickly made a fence to isolate the mine from all usurpers [In fact, only from *B* can one go to point *A*, without going beyond the line that represents the fence]. The fable affirms the inviolable rights of the weak.

Geometric algorithms

Sona may be classified according to their dimensions and the algorithms used for their construction. In the following, I present some examples of algorithms invented by the

Figure 4.10a

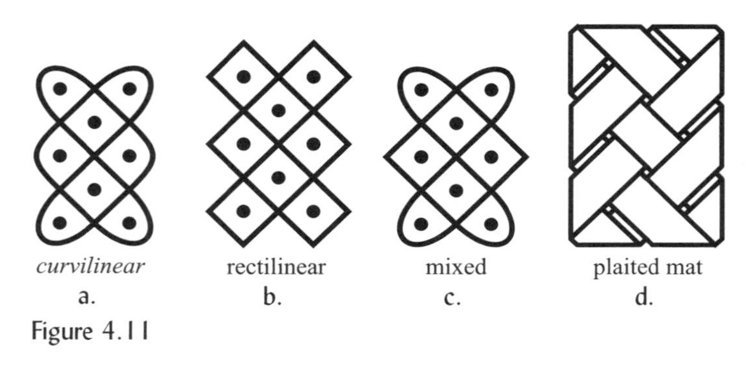

Figure 4.11

Plaited-mat designs

The most frequently used algorithm is the one derived from mat weaving: The lines of the *sona* correspond to the strands of plaited mats, making a 45° angle with its side. Thus the three *sona* in Figure 4.11a, b, and c correspond to the small mat in Figure 4.11d. The bends at the borders may be smooth (curvilinear), rectilinear, or mixed whereby some bends are smooth and others rectilinear. The drawings in Figure 4.11 have dimensions 3 by 2 (three main rows two points each).

Figure 4.12 presents examples of *sona* belonging to the class of plaited-mat designs. Sometimes the form of the lines is adapted or auxiliary lines are added to make the drawings more expressive, as the examples in Figure 4.13 show.

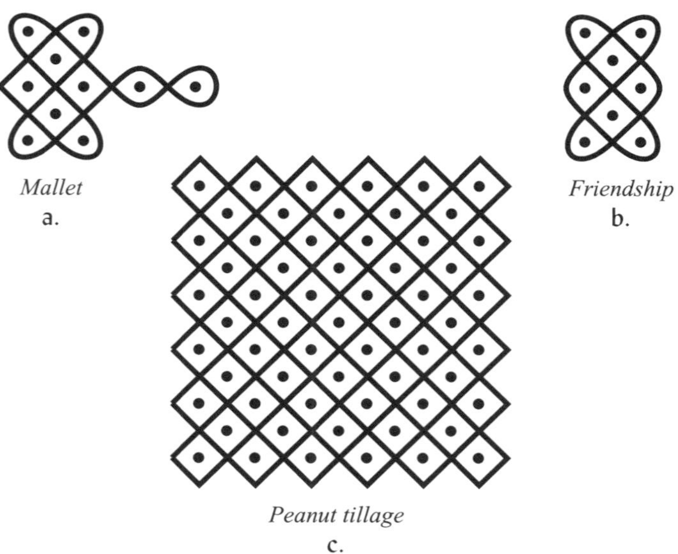

Figure 4.12

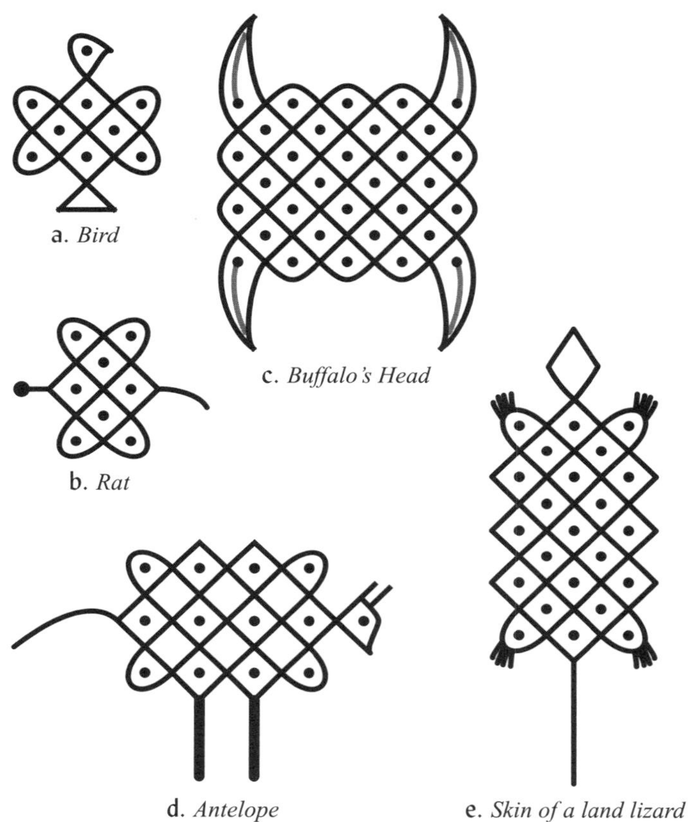

Figure 4-13

Figure 4.14

Chased-chicken-path algorithm

Both monolinear Chokwe *sona* with twofold symmetry, reproduced in Figures 4.14a and b, represent the trajectory described by a wild chicken when pursued. They are drawn with the same geometrical algorithm, not taking into account that the first is curvilinear and the second mixed. The dimensions of their grids of reference points are, however, different (5×6 and 9×10). Among the Ngangela, another monolinear *lusona* has been collected that satisfies the algorithm (see Figure 4.14c), but now the dimensions are 3×8. The drawing shows the path of the *ngonge* insect as it slowly eats its way around the inner bark of a tree, cutting off the flow of sap. In time it destroys the tree.

Figure 4.15 shows this really beautiful algorithm 'at work' in the case of the dimensions 9×10. It is difficult to imagine how one of the *akwa kuta sona* drawing experts might have invented this algorithm. In some other cases it is possible to reconstruct the possible path of invention, as will be shown in the next section.

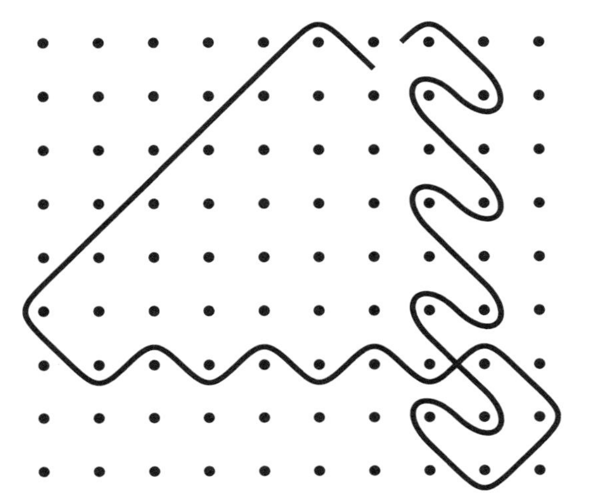

Figure 4.15

Figure 4.16

Loop algorithms

The algorithm used to draw the *lusona* in Figure 4.16 may have been invented from a zigzag line (see Figure 4.17a) to which some loops were applied (like in Figure 4.17b), obtaining Figure 4.17c, and after extension of the line at its extremities the final drawing is obtained (Figure 4.16).

In a similar way, the monolinear *lusona* in Figure 4.18a might have been obtained from a double zigzag (Figure 4.18b).

The algorithm used to draw the monolinear representation of a fire (Figure 4.19a) and several other *sona*, seems to have been invented by joining loops to a monolinear plaited-mat design (see Figure 4.20).

b.

a. *End of struggle between two rival chiefs*

Figure 4.18

a. *Fire*

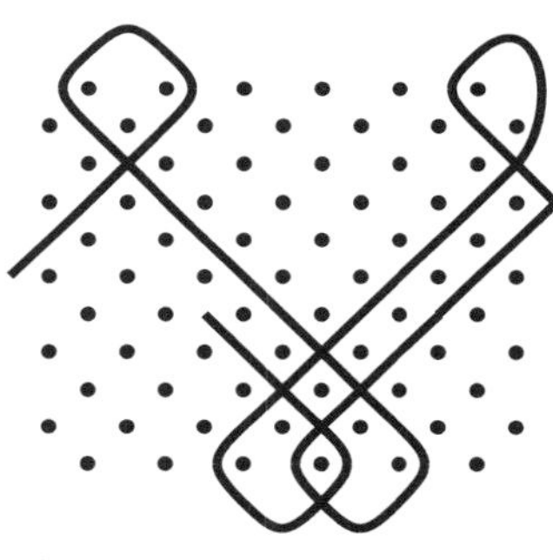

b. *step in the execution of the drawing*

Figure 4.19

a

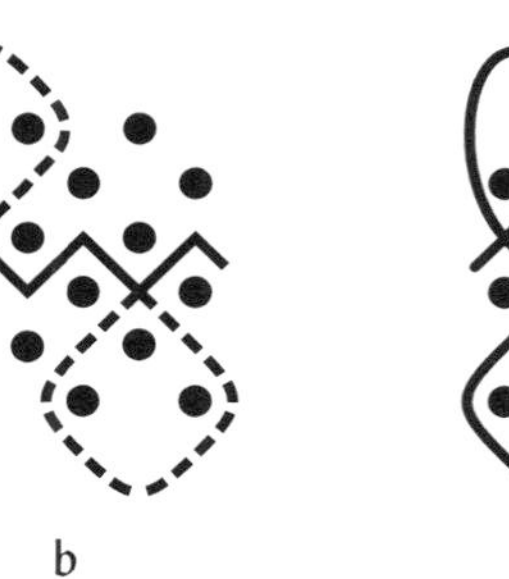

b

c

Figure 4.17

a. *Plaited-mat design of dimensions* 3 × 7

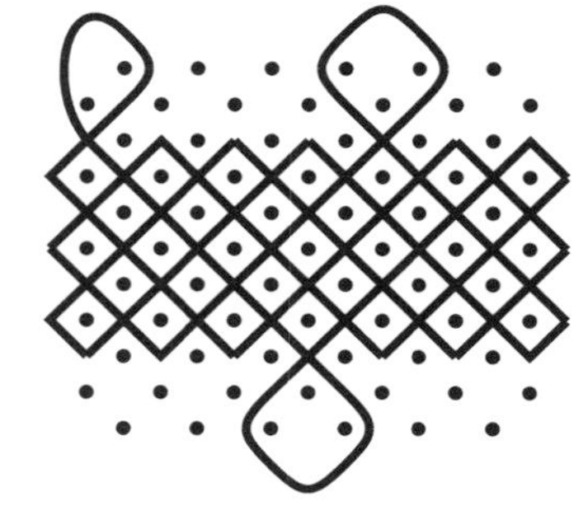

b. *Joining loops*

Figure 4.20

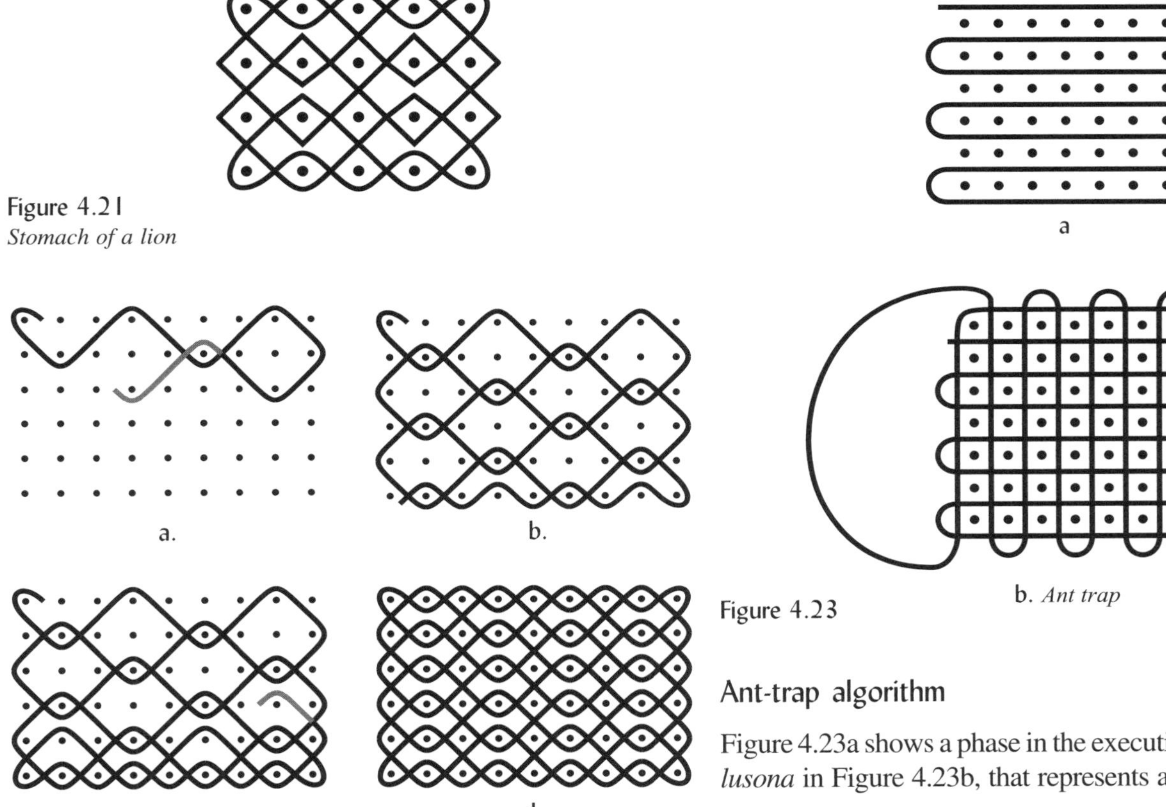

Figure 4.21
Stomach of a lion

a. b. c. d.

Figure 4.22

a

b. *Ant trap*

Figure 4.23

Lion's stomach algorithm

The *lusona* in Figure 4.21 is monolinear, enjoying a double symmetry. Its dimensions are 4 by 5. The construction of the smooth version with dimensions 6 × 9, applying the same algorithm, is illustrated in Figure 4.22. It is interesting to observe how the line 'inverts' after having traveled half of the course.

Ant-trap algorithm

Figure 4.23a shows a phase in the execution of the monolinear *lusona* in Figure 4.23b, that represents a trap to catch ants.

Spider's algorithm

Figure 4.24 shows a monolinear *lusona*, representing a spider. It enjoys a fourfold rotational symmetry. Smaller versions are displayed in Figure 4.25.

Systematic construction of monolinear figures

The *akwa kuta sona* invented several methods to construct monolinear figures, as will be illustrated in the following by the examples of triangular designs and the use of chain rules.

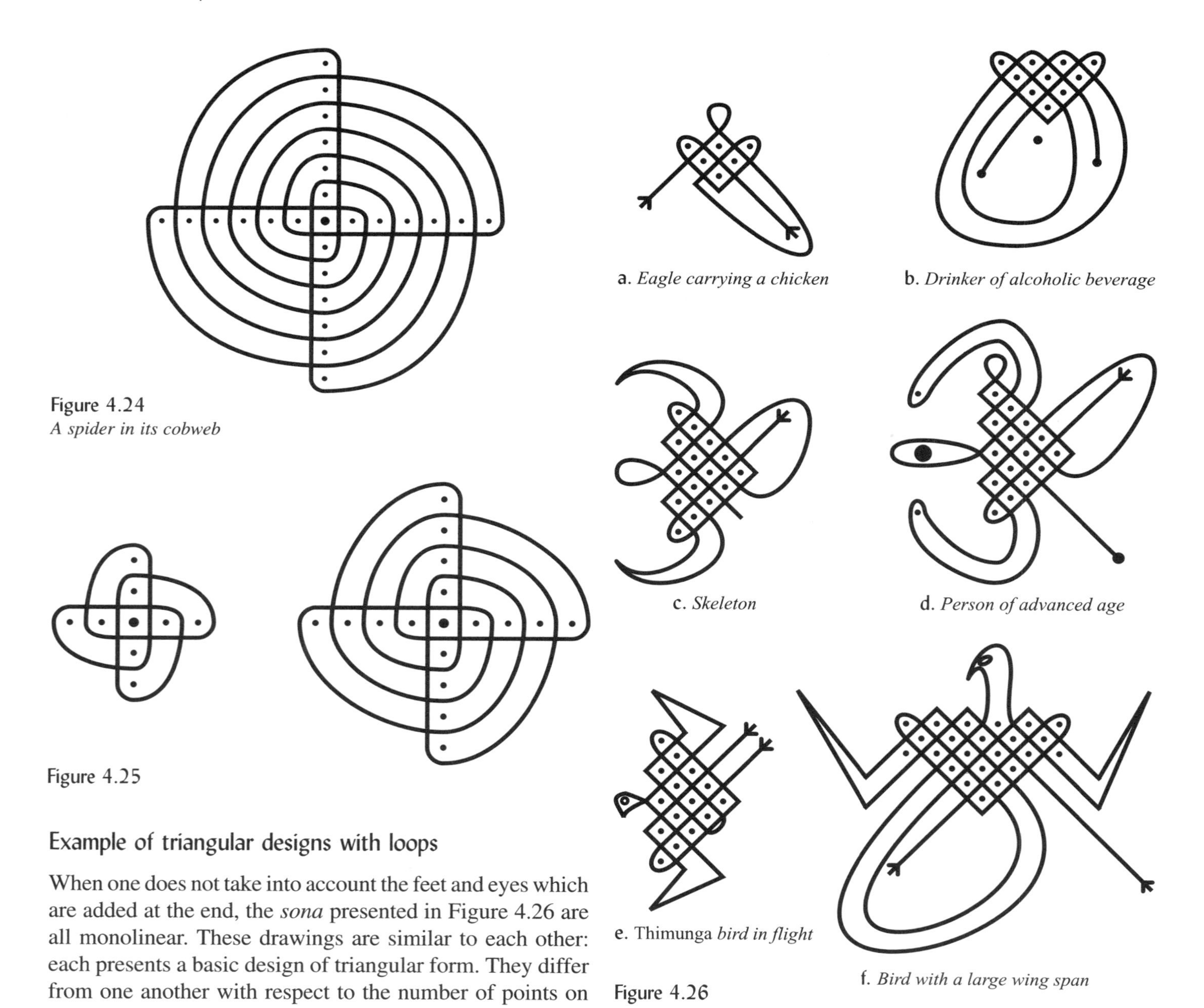

Figure 4.24
A spider in its cobweb

Figure 4.25

a. *Eagle carrying a chicken*

b. *Drinker of alcoholic beverage*

c. *Skeleton*

d. *Person of advanced age*

e. Thimunga *bird in flight*

f. *Bird with a large wing span*

Figure 4.26

Example of triangular designs with loops

When one does not take into account the feet and eyes which are added at the end, the *sona* presented in Figure 4.26 are all monolinear. These drawings are similar to each other: each presents a basic design of triangular form. They differ from one another with respect to the number of points on each side: 3 (a), 4 (b), 5 (c, d, e), and 6 (f), respectively. It is highly probably that the *akwa kuta sona* who invented these *sona* began with triangular designs and transformed them into monolinear designs with the help of one or more loops. The monolinear designs so obtained were adapted topologically so that the drawing experts could express the ideas they wanted to transmit through the *sona*.

Figure 4.27 shows the construction of the basic design for the *lusona* in Figure 4.26b. Figure 4.28 illustrates some possibilities for the construction of the basic designs with five points on each side (cf. also Figure 4.8b). The sand

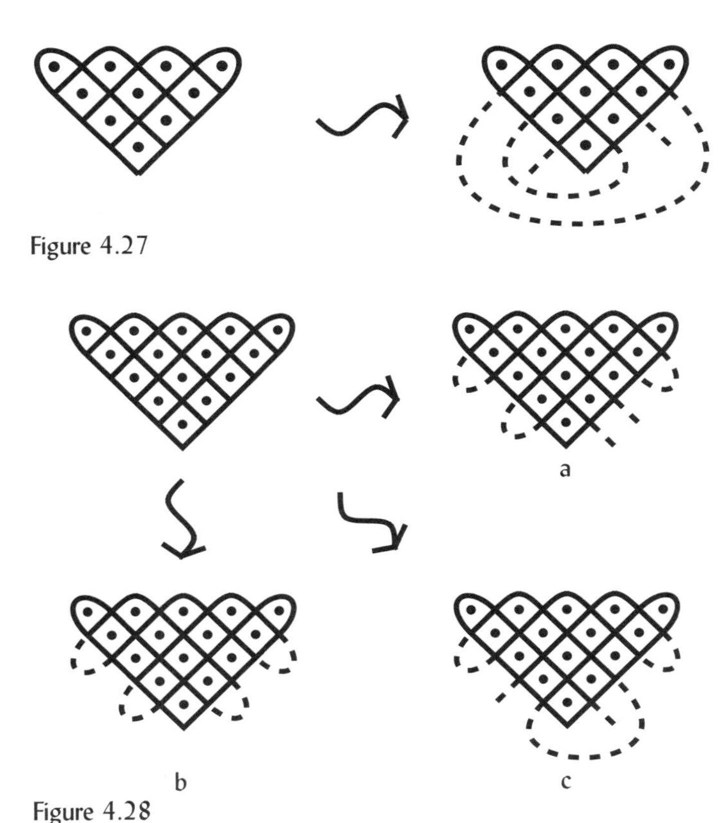

Figure 4.27

Figure 4.28

Figure 4.29

drawings in Figures 4.26d and e are variants of the same basic solution. Another solution (b) is (topologically) found twice in the *lusona* called *Kalunga*, representing God (see Figure 4.29).

General chain rules

The *sona* drawing experts knew several rules to 'chain' monolinear designs to obtain new and bigger monolinear designs. They often applied these rules.

A frequently used rule concerns the creation of a new, closed monolinear design by joining two (or more) open(ed) monolinear designs in such a way that an endpoint of the first *lusona* connects to an endpoint of the second *lusona* (first step). After connecting the two remaining endpoints (second step), a closed monolinear design appears (see Figure 4.30).

Figure 4.31 gives an example where two copies of a plaited-strip design are chained to a closed monolinear *lusona* (not taking into account the auxiliary vertical line in the center).

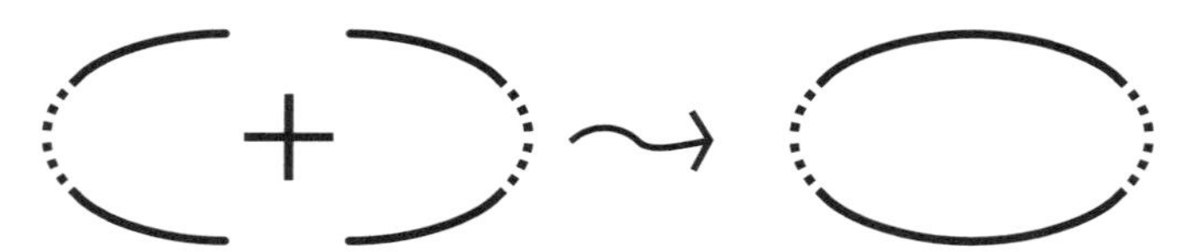

a. *two open monolinear designs* b. *one closed monolinear design*

Figure 4.30
Schematic representation of the first chain rule

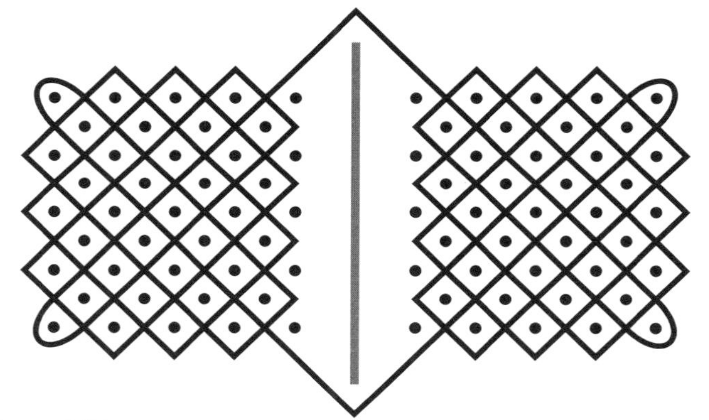

Figure 4.31
Battlefield

Figure 4.32
Firewood

To construct the monolinear *lusona* in Figure 4.32, six monolinear open 'cells' are joined together to form a bigger closed monolinear design.

a. *sona element*

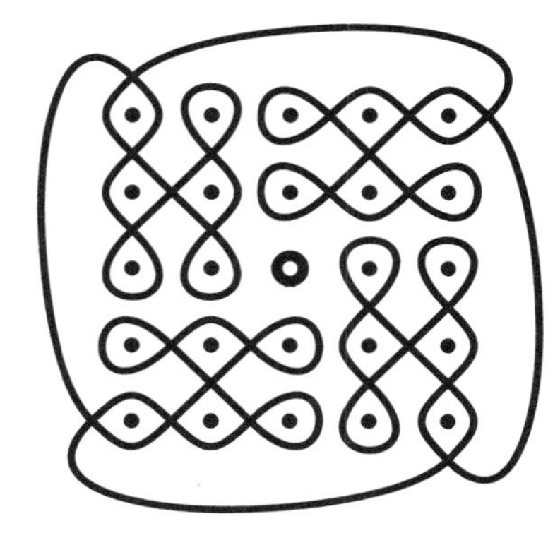

b. *Forest where the* gundu *bird abounds*

Figure 4.33

One of the applications of the first chain rule is found in the construction of monolinear *sona* with rotational symmetry. In Figure 4.33 four copies of the same *sona* element are joined (see Figure 4.7b for another example).

The second rule the *akwa kuta sona* were using consists of the 'fusion' of two monolinear designs to obtain a new monolinear design (see Figure 4.34).

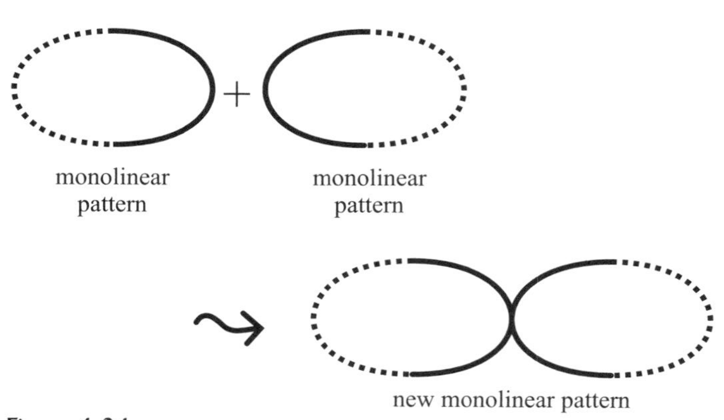

Figure 4.34
Schematic representation of the second chain rule

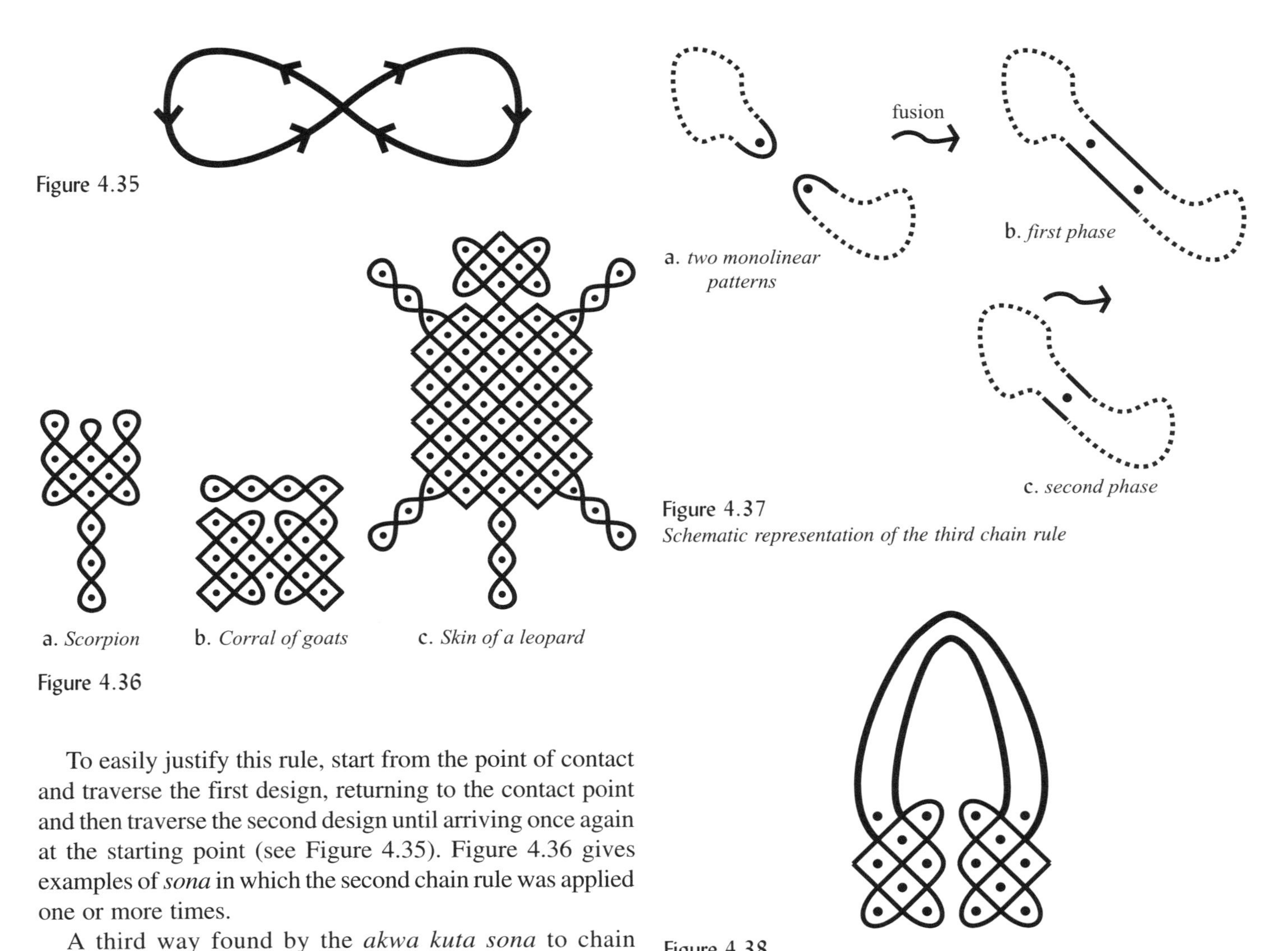

Figure 4.35

a. *Scorpion* b. *Corral of goats* c. *Skin of a leopard*

Figure 4.36

Figure 4.37
Schematic representation of the third chain rule

Figure 4.38
Double iron bell (a symbol of royalty)

To easily justify this rule, start from the point of contact and traverse the first design, returning to the contact point and then traverse the second design until arriving once again at the starting point (see Figure 4.35). Figure 4.36 gives examples of *sona* in which the second chain rule was applied one or more times.

A third way found by the *akwa kuta sona* to chain monolinear designs is shown schematically in Figure 4.37. One may complete it in one or two successive phases. During the first phase, the lines are fused somewhere on the boundary. Afterwards, two boundary grid points can be fused into a single point.

Figure 4.38 presents an example whereby two plaited-mat designs are connected. Only the first phase is completed.

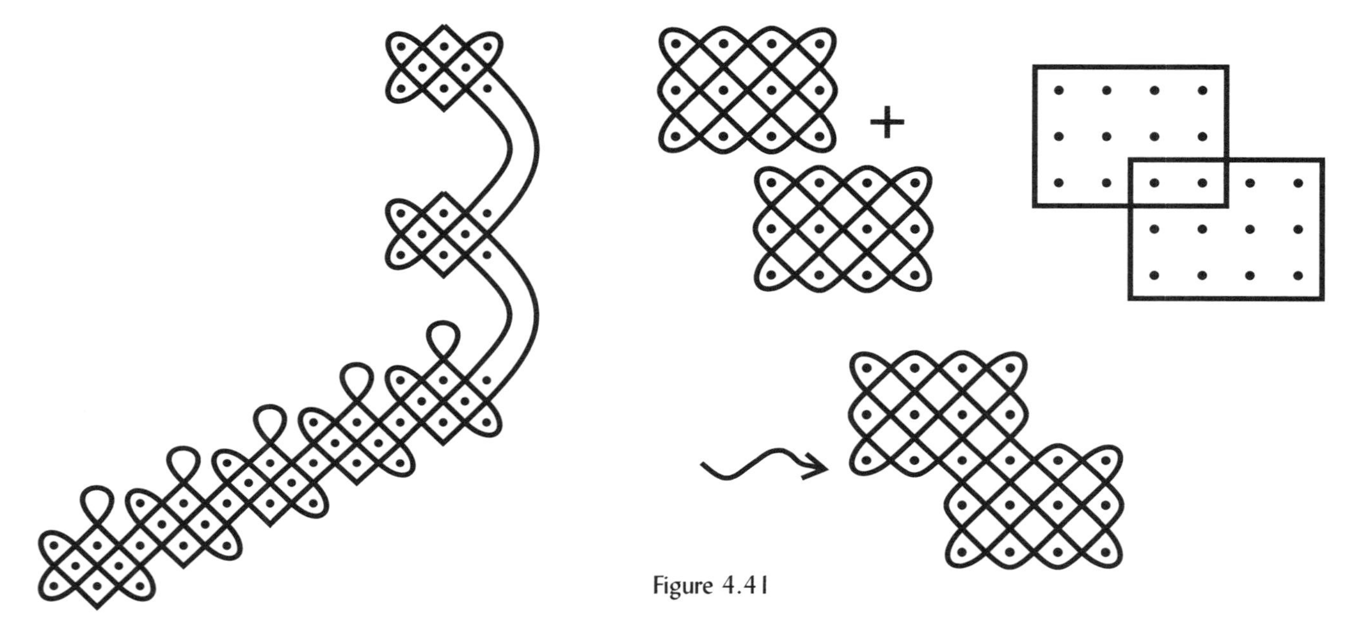

Figure 4.39

Figure 4.41

In constructing the *lusona* shown in Figure 4.39, the third chain rule is applied six times, advancing twice only until the first phase. The second phase is completed four times (see schematically the execution of one such a fusion in Figure 4.40).

Chain rules for plaited-mat designs

The *sona* drawing experts found also one (or more) specific rule(s) to chain monolinear *sona* of the plaited-mat design type to bigger monolinear designs. Figure 4.41 presents schematically such a chain rule in the case of chaining two

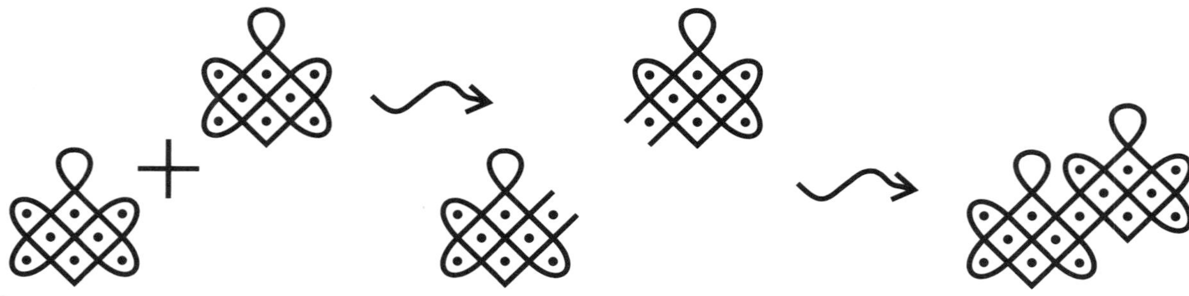

Figure 4.40

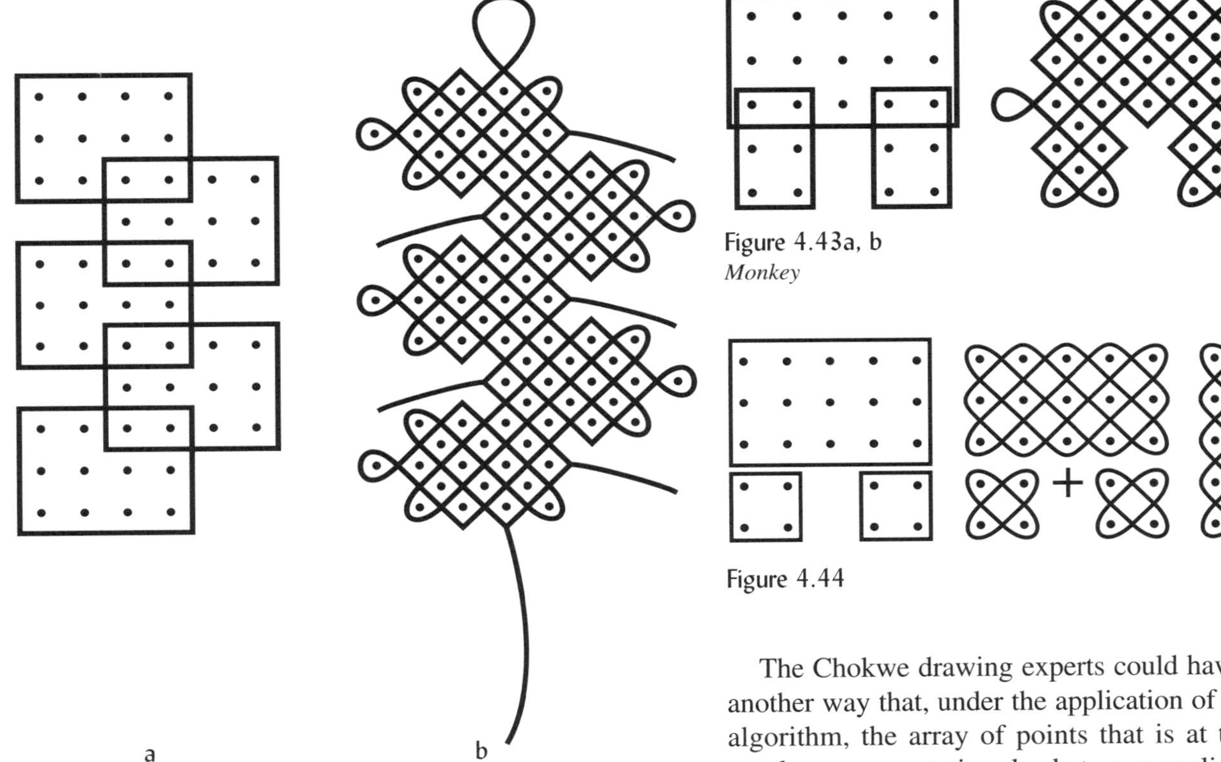

Figure 4.42

Figure 4.43a, b
Monkey

Figure 4.44

3×4 plaited-mat designs, leading to a *lusona* that represents an animal that died inside a rock.

In the representation of a leopard with five cups (see Figure 4.42b) this rule has been applied four times (Figure 4.42a). If we ignore the tails that have been added at the end, the drawing is monolinear.

The same rule may have been applied to discover the basic design, that, with head and tail added, is the *lusona* that represents a monkey (Figure 4.43).

The Chokwe drawing experts could have discovered in another way that, under the application of the plaited-strip algorithm, the array of points that is at the basis of the monkey representation, leads to a monolinear design (see the scheme in Figure 4.44): joining square grids to the initial array does not change the number of lines necessary to embrace the points of the final array.

2 Examples of mathematical-educational exploration of sona

As so many *sona* are symmetrical, they may obviously be used in the mathematics classroom to discover several symmetry properties. For instance, the property that points

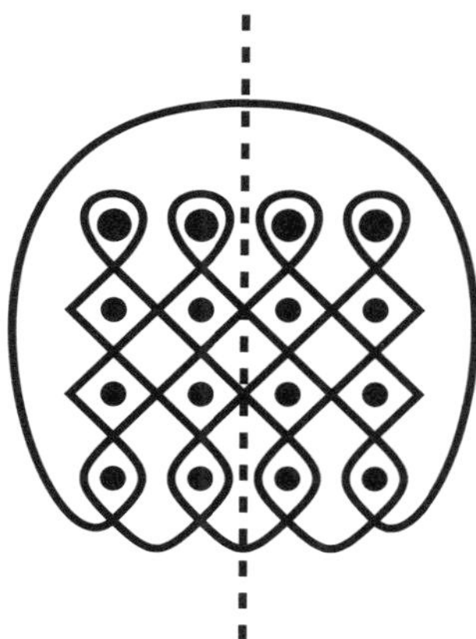

Figure 4.45
Village defended by a fence

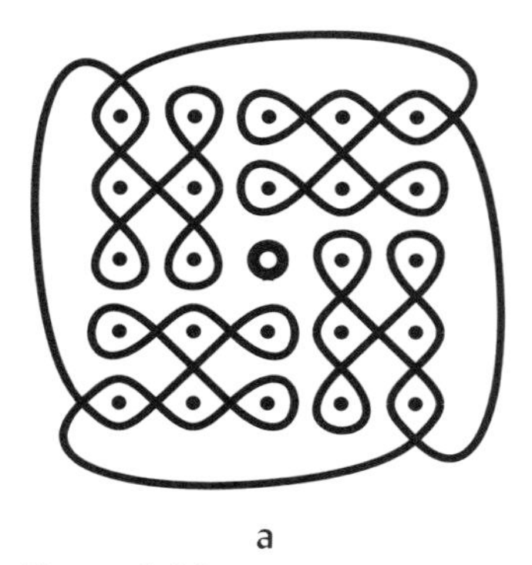

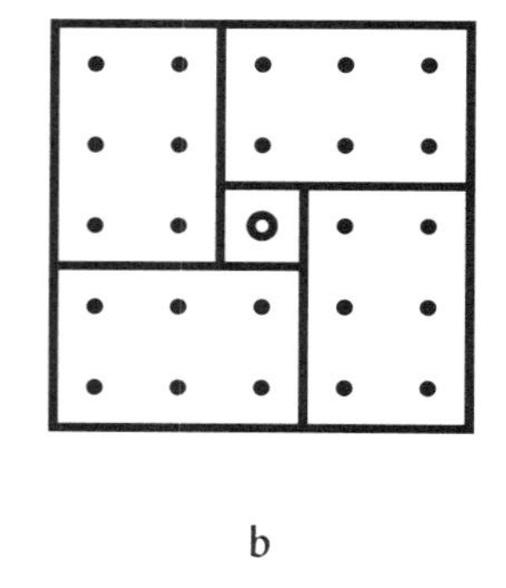

Figure 4.46

which correspond under a reflection are at the same distance from the symmetry axis (see the example in Figure 4.45). In the case of the grid points, these distances can be counted as so many times (half) the unit distance between horizontally neighboring grid points. However, *sona* provide many other connections to mathematics education. I will suggest ways to establish connections between *sona* and the mathematics from primary to university levels.

Arithmetical relationships and progressions

(cf. Gerdes, 1988, 1990)

Each *sona* drawing redistributes, we may say, the points of the reference net. For example, in order to represent the forest with many *qundu* birds (cf. Figure 4.33) one starts with a 5×5 grid singling out the central dot. Around the center are drawn four copies of the same *sona* element, each embracing 2×3 or 6 dots. In other words $5 \times 5 = 4 \times 6 + 1$ (see Figure 4.46b). A possible extrapolation is that the square of an odd number is equal to four times a number plus one.

The following examples show that *sona* may be used as a starting point for the study of sums of sucessive terms of arithmetical progressions.

In the case of the antelope's representation (cf. Figure 4.13d), the *lusona* redistributes obliquely the points of the initial three rows of four points each. This invites one to notice how many points there are between each pair of neighboring oblique lines. As Figure 4.47 displays, there are $(1 + 2 + 3) + (3 + 2 + 1)$ points. In order words, the redistribution leads to the establishment of the relationship:

$$3 \times 4 = (1+2+3)+(3+2+1) = 2 \times (1+2+3).$$

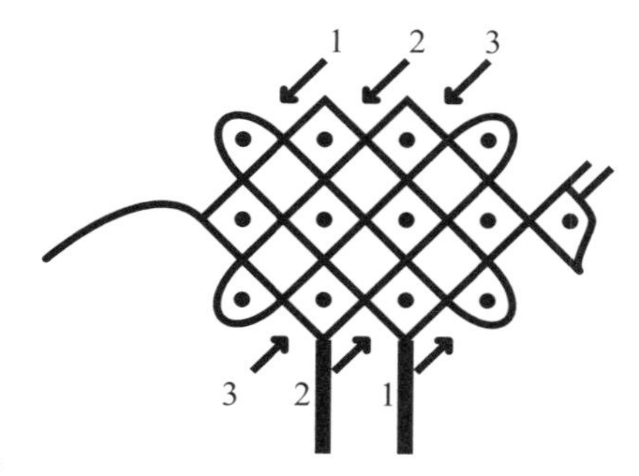

Figure 4.47

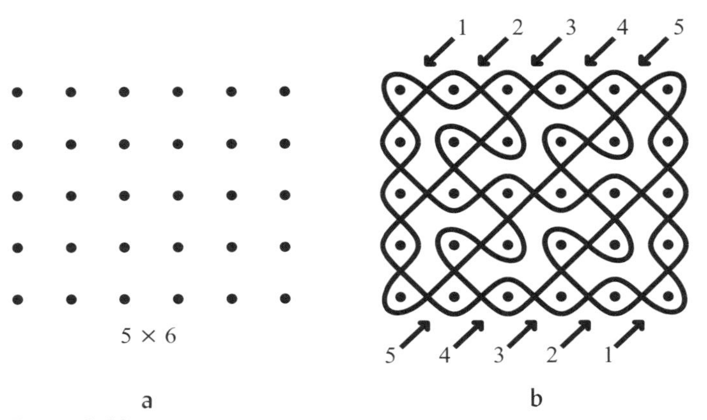

Figure 4.48

An analogous consideration of a chased-chicken-path design (cf. Figure 4.14a) leads to (see Figure 4.48):

$$\begin{aligned} 5 \times 6 &= (1+2+3+4+5) + (5+4+3+2+1) \\ &= 2 \times (1+2+3+4+5). \end{aligned}$$

Extrapolation on the basis of these (and other) examples may lead to:

$$n \times (n+1) = 2 \times (1+2+3+\cdots+n).$$

Alternatively

$$(n+1)^2 - (n+1) = 2 \times (1+2+3+\cdots+n),$$

may be extrapolated from

$$6^2 - 6 = 2 \times (1+2+3+4+5),$$

suggested (see Figure 4.49) by the *Kalunga* representation in Figure 4.29.

The *lusona* representation of three bats in Figure 4.50 is like the antelope's representation (Figure 4.47) based on a plaited-mat design of dimensions 3 × 4, but this time additional grid points have been added. Looking to the number of points in the oblique direction (Figure 4.51a), one sees that their total number corresponds to $(1+3+5) + (5+3+1)$. The total number of grid points is also equal to the number of principal grid points plus the number of

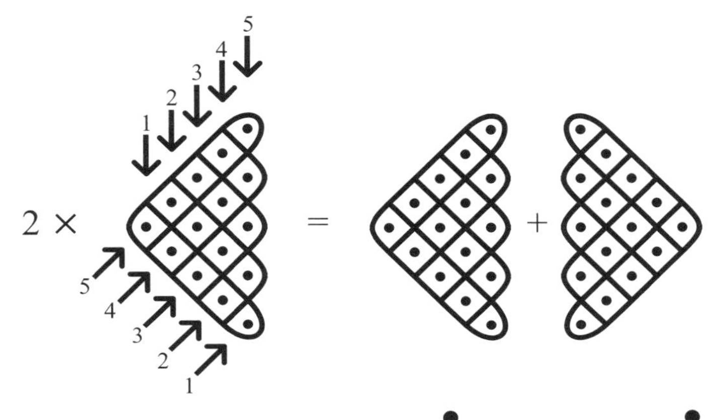

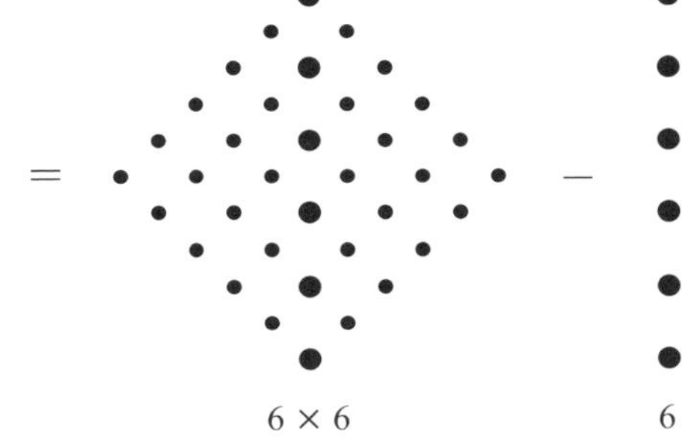

Figure 4.49

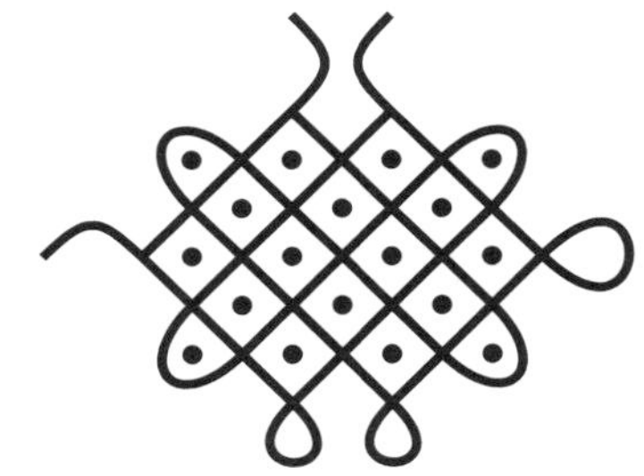

Figure 4.50
Three bats (mother and two cubs)

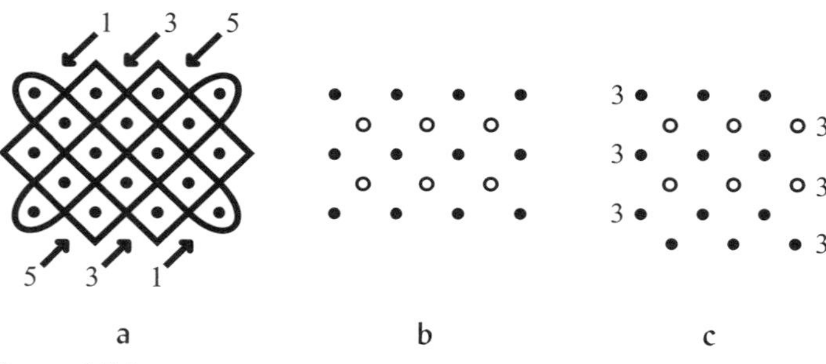

Figure 4.51

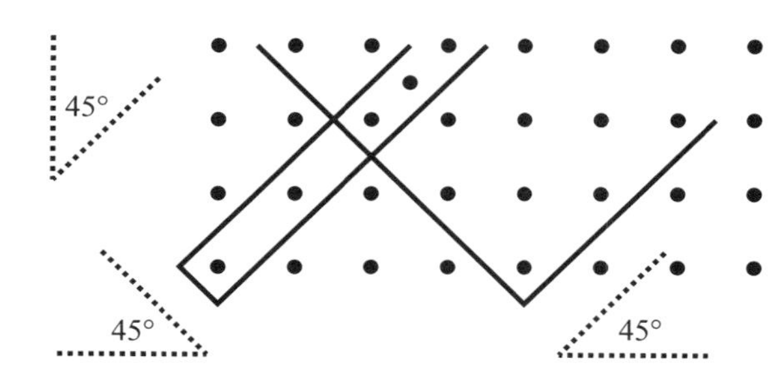

Figure 4.53

additional points (Figure 4.51b). This may be evaluated geometrically as twice 3×3 by rotating the last column of principal points to a horizontal position (Figure 4.51c).

The same considerations are possible with other plaited-mat designs of dimensions $m \times (m+1)$ with additional points. Figure 4.52 displays the case 4×5 (cf. Figure 4.13c). Extrapolation gives that the sum of the first n odd numbers is n^2.

The number of lines in a rectangular plaited-mat design (cf. Gerdes, 1988, 1993; Jaritz)

A possibly interesting topic for an exploration at the lower secondary level, would be the investigation of how many lines are there in a plaited-mat design. Or, in an alternative formulation, how many lines that satisfy the plaited-mat algorithm does one need to embrace all the points of a reference grid. First of all, the pupils may try to define the common characteristics of these lines: they form both before and after a reflection, angles of 45° with the sides of the reference frame (see Figure 4.53).

In order to represent an antelope's head, one starts with a 2×4 reference frame and needs 2 lines to embrace all grid points (see Figure 4.54).

Drawing the *sona* in Figure 4.12 and 4.13 gives some experimental data, summarized in Table 4.1, where r denotes the number of rows and c the number of columns of the rectangular point grid. Let $f(r, c)$ be the number of lines as a function of r and c.

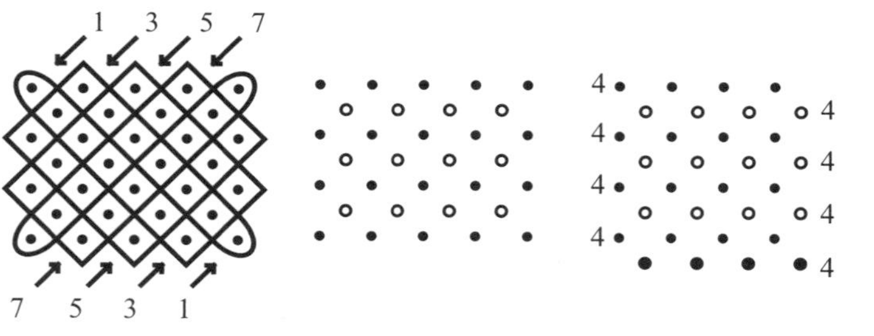

Figure 4.52

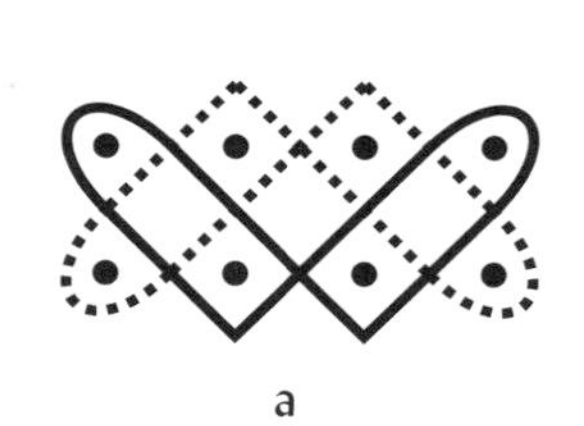

a

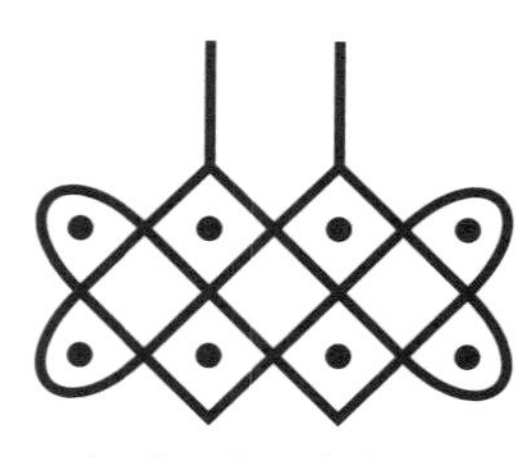

b. Antelope's head

Figure 4.54

r \ c	1	2	3	4	5	6	7	8	9	10
1										
2			1	2						
3		1		1						
4					1					
5			1							
6						6				
7										
8										
9										
10										

Table 4.1

r \ c	1	2	3	4	5	6	7	8	9	10
1	1	1	1	1	1	1				
2	1	2	1	2	1	2				
3	1	1	3	1	1	3				
4	1	2	1	4	1	2				
5	1	1	1	1	5	1				
6	1	2	3	2	1	6				
7										
8										
9										
10										

Table 4.2

Rotating a plaited-mat design through an angle of 90° does not change the number of lines. Thus $f(r, c) = f(c, r)$, that is the table has a symmetry axis. After having obtained some further experimental data, like, for instance, the ones in Table 4.2, the pupils may try to discover numerical patterns in the rows, columns and diagonals. They may try to guess still unknown values, and verifying them by drawing the lines and counting them. So they may come to formulate some conjectures. For instance, the pupils may observe that

- the only numbers in the second row are 1 and 2;
- the numbers on the principal diagonal are 1, 2, 3, 4, 5, 6,...;
- the numbers in the fourth row are the divisors of 4: 1, 2, and 4;
- and the numbers in the sixth row are the divisors of 6;
- on the two diagonals immediately next to the principal diagonal, only the number 1 appears;
- also in the direction perpendicular to the principal diagonal are several diagonals with only 1's, etc.

It may be conjectured, for example, that

- $f(4, c)$ is a divisor of 4;
- $f(6, c)$ is a divisor of 6;
- $f(r, c)$ is a divisor of r;
- $f(r, r) = r$;
- $f(r, r + 1) = 1$;
- $f(r, r - 1) = 1$;
- $f(r, c)$ is a common divisor of r and c;
- when the greatest common divisor of r and c is equal to 1, one plaited-mat line alone embraces all the grid points (in other words, the plaited-mat design is monolinear);
- when $r + c$ is equal to a 5, then $f(r, c) = 1$;
- when $r + c$ is equal to a 7, then $f(r, c) = 1$.

Some guesses may be wrong. For instance, in the case of the conjecture

- when $r + c$ is equal to an odd number, then $f(r, c) = 1$,

one may find as a counterexample $r = 3$, and $c = 6$ with $f(3, 6) = 3$. This may lead to further observation and testing and a refinement of the conjecture to

- when $r + c$ is equal to a prime number, then $f(r, c) = 1$.

Once the conjectures are found and tested, the question arises of how to be sure of them. In other words, how to prove (each of) them? For example, how to prove that

- $f(r, c)$ is the greatest common divisor (gcd) of r and c,

Other conjectures might be of help?

It is easy to prove that in the case of a square grid, each line has to come to its beginning after one turn around the square's center (see Figure 4.55, where an auxiliary square of size $r = c$ has been drawn around the square grid).

As we saw (Figure 4.44), the Chokwe *akwa kuta sona* were probably aware of the fact that 'applying a square' to a rectangular grid does not change the number of plaited-mat lines needed to embrace all reference points. This is a direct consequence of the first chain rule and the result $f(r, r) = r$.

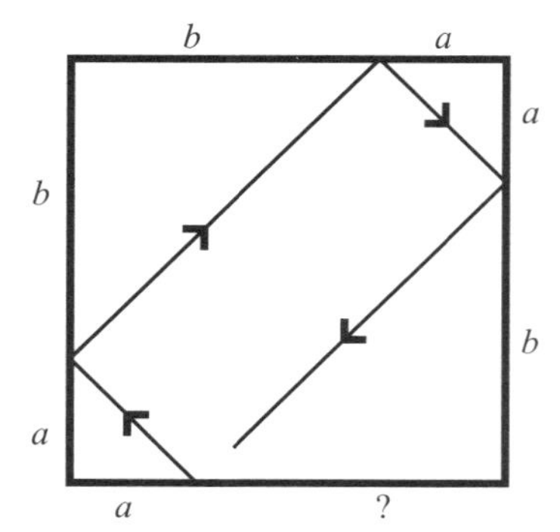

Figure 4.55

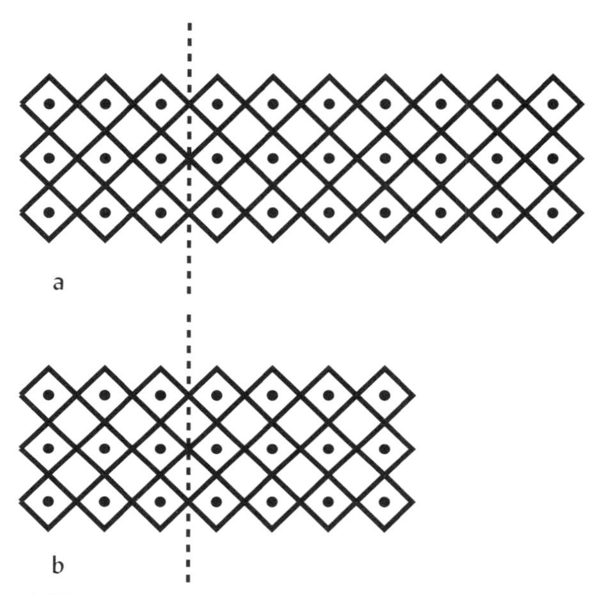

Figure 4.56

May this be used to evaluate $f(r, c)$ in other cases, that is when $r \neq c$? For example, $f(3, 10) = ?$ (see Figure 4.56a). Cutting off the 3×3 plaited-mat square on the left, does not change the number of lines, that is $f(3,10) = f(3, 10\text{-}3) = f(3,7)$. Proceeding in the same way (Figure 4.56b, etc.), one finds $f(3,7) = f(3,4) = f(3,1) = f(2,1) = f(1,1) = 1$. Similarly, $f(26,34) = f(26,8) = f(18,8) = f(10,8) = f(2,8) = \cdots = f(2,2) = 2$. In other words, we have found an algorithm ('cutting off squares') that permits the determination of $f(r, c)$, even if one does not know that $f(r, c) = \gcd(r, c)$.

Figure 4.57 displays the idea of the 'cutting off squares' schematically in the 3×10 case.

Another way to evaluate $f(r, c)$ may be discovered by the pupils by observing each of the plaited-mat lines in a concrete example. Let us consider the case of dimensions 3×6, and draw each of the lines separately. There are 3

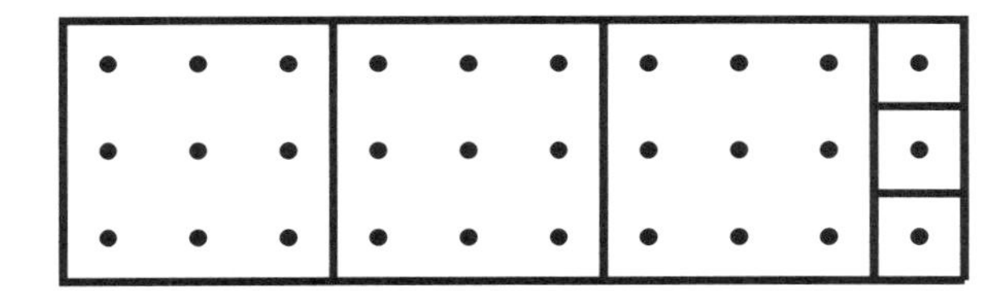

Figure 4.57

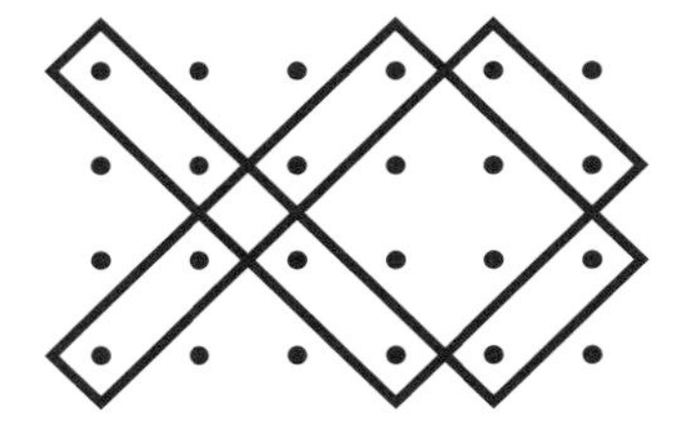
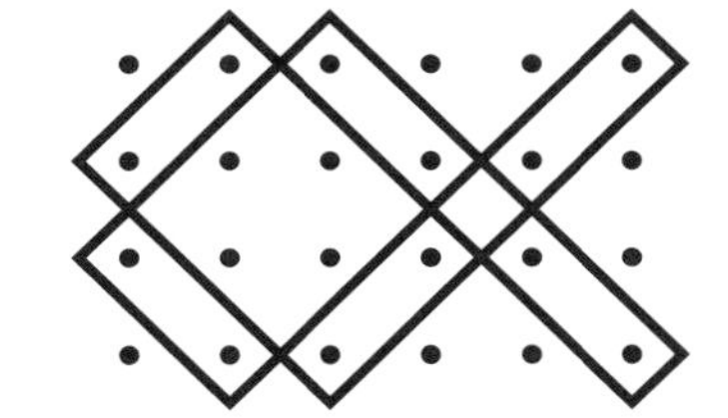

Figure 4.59

lines (see Figure 4.58). Each of these plaited-mat lines embraces two points of the first row; each embraces one point of the first column. Comparison with other examples, may lead to the formulation of some hypotheses. In the case of dimensions 4×6, each plaited-mat line embraces 3 points of the first row and 2 points of the first column (see Figure 4.59).

As distinct lines embrace different points of the first row, and each of the lines embraces the same number of points of this row, the total number of lines $f(r, c)$ should be a divisor of the number of points of the first row, that is of c. Similarly, $f(r, c)$ should be a divisor of r.

In other words, extrapolation of the basis of the analysis of several concrete examples, leads to

$f(r, c)$ is a divisor of r,
$f(r, c)$ is a divisor of c,

that is, $f(r, c)$ is a common divisor of r and c. Comparing the obtained values for $f(r, c)$ with the common divisors of r and c, may conduct the pupils to: $f(r, c)$ is the greatest common divisor of r and c [$\gcd(r, c)$].

Cautious observation of the behaviour of the plaited-mat lines which start in the top left corner of the point grids (see the example in Figure 4.60) may lead the pupils to discover the following interesting conjecture:

$f(r, c) = \gcd(r, c)$
= minimum number of points "embraced" by "branches" of a plaited-mat line that passes through a grid of dimensions $r \times c$.

Geometrical Determination of the Greatest Common Divisor of two Natural Numbers

From either these two or other ways to evaluate $f(r, c)$, it is not difficult for the pupils to arrive at the geometric equivalent of Euclid's arithmetical algorithm for the determination of the greatest common divisor of two natural numbers.

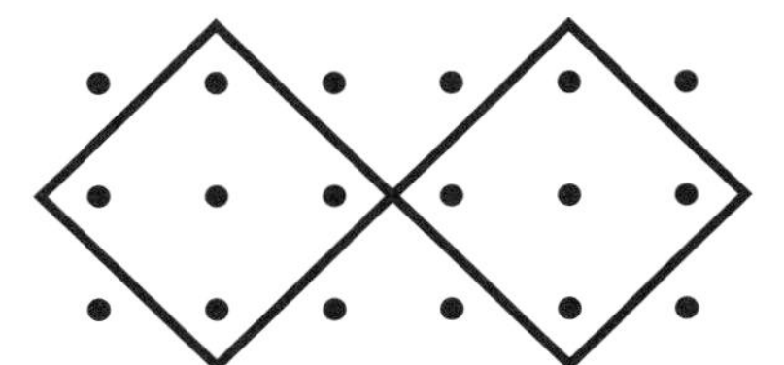
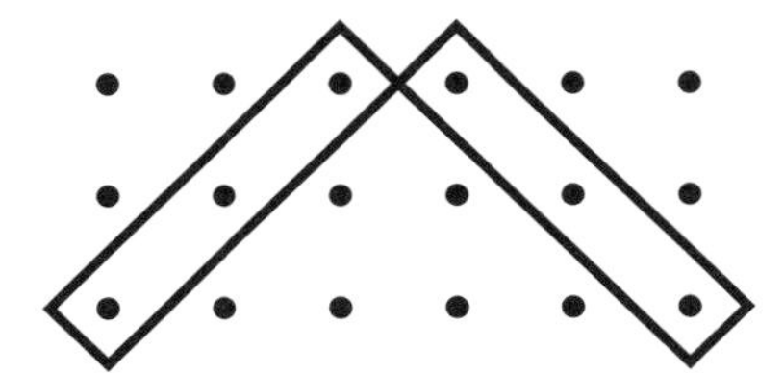

Figure 4.58

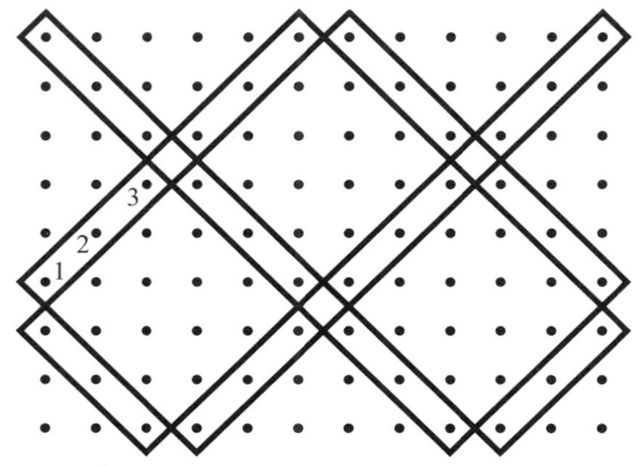

Figure 4.60

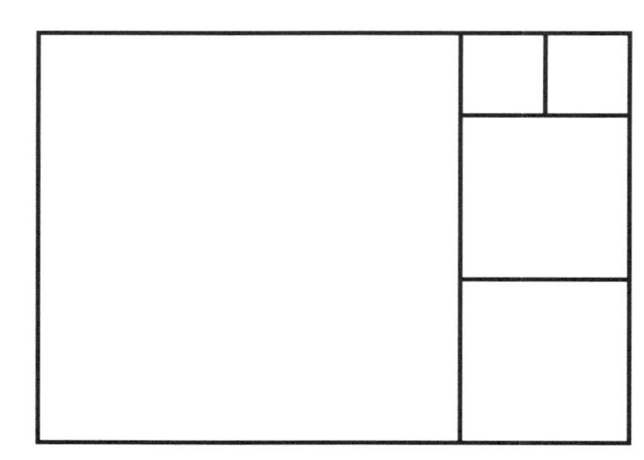
Figure 4.61b

In the first case, Figure 4.61 illustrates phases of the possible discovery process through the example $(r, c) = (15, 21)$.

For the second case, Figure 4.62 illustrates the possible successive steps in this discovery process. In Figure 4.62b, the single plaited-mat line has been substituted by a polygonal line. At this stage, the gcd(r, c) may be geometrically interpreted in the following way (cf. Figure 4.62c):

gcd(r, c) = length of the side of the greatest square with which it is possible to fill up a $r \times c$ rectangle (same units of length) .

Figure 4.62d shows that it is not necessary to draw the whole polygonal line of 4.62b in order to find gcd(r, c). It is sufficient to consider a reduced polygonal line, that diagonally "cuts off squares" from the original $r \times c$ rectangle.

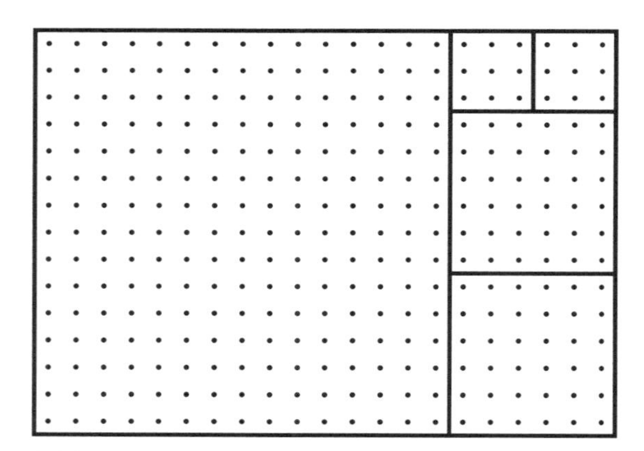
Figure 4.61a

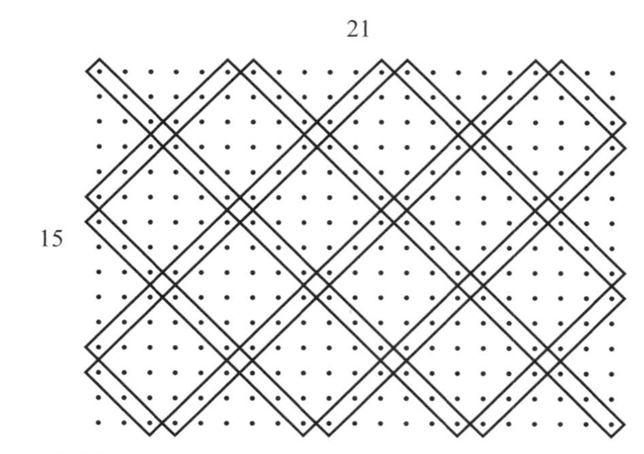

Figure 4.62a

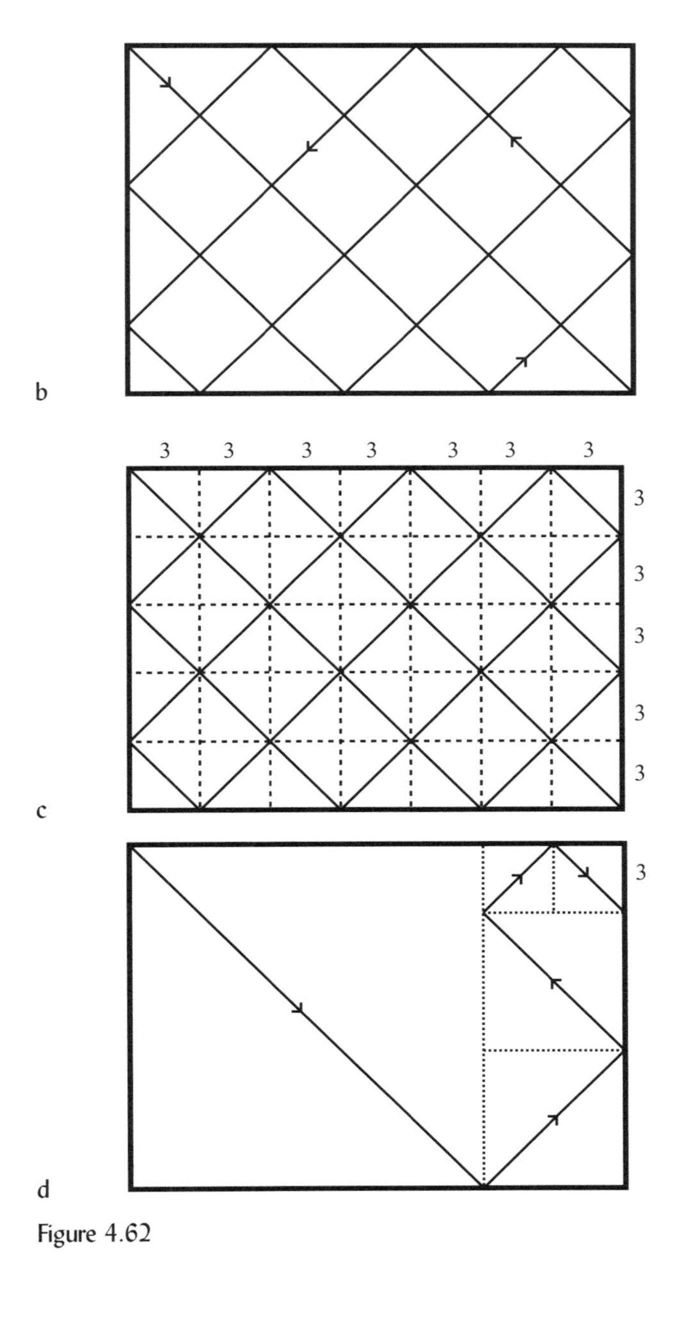

Figure 4.62

The number of "lion's stomach" lines

(cf. Gerdes, 1993, 1993a)

The consideration of algorithms other than the plaited-mat one, leads to new questions about the relationship between the number of lines, the algorithm for their construction and the dimensions of the point grids. For some algorithms it is relatively easy to establish this relationship. The "lion's stomach" algorithm (cf. Figure 4.21) is such a one.

When one uses this algorithm in the case of a grid of dimensions 7 × 3, one needs three lines in order to embrace all points of the grid (see Figure 4.63). How does the number of lines depend on the dimensions r and c of the grid?

First students may discover that when the number of columns is even, the drawing does not look quite like a lion's stomach (see the example in Figure 4.64). Therefore only odd values of c have to be considered. The number of rows r has to be at least 2.

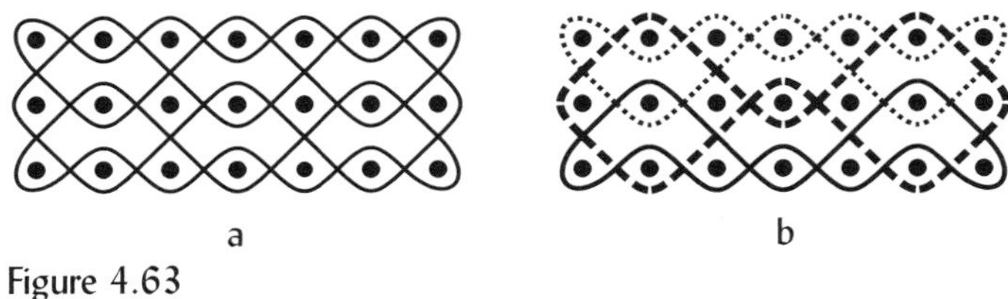

Figure 4.63

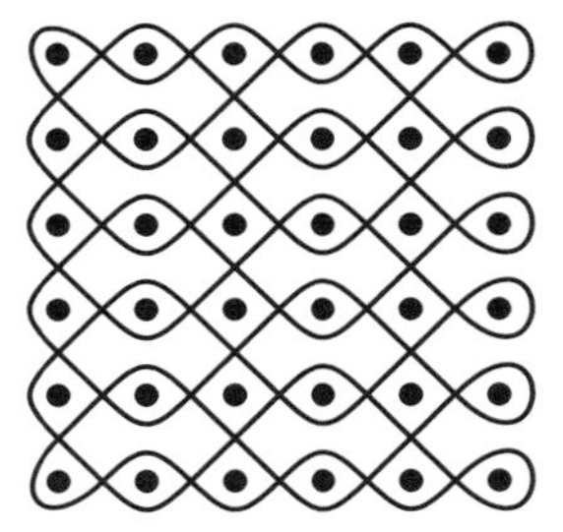

Figure 4.64

r \ c	3	5	7	9	11	13	15	17
2	2	1	2	1	2			
3	3	1	3	1				
4	4	1	4					
5	5							
6	6			1				
7								
8								
9								

Table 4.3

Students may experiment with concrete values of r and c, draw the corresponding figures, count the lines and record the data collected in a table (Table 4.3). Extrapolation on the basis of these experimental data may lead to Table 4.4 and to the formulation of a conjecture such as:

r \ c	3	5	7	9	11	13	15	17
2	2	1	2	1	2	1	2	1
3	3	1	3	1	3	1	3	1
4	4	1	4	1	4	1	4	1
5	5	1	5	1	5	1	5	1
6	6	1	6	1	6	1	6	1
7	7	1	7	1	7	1	7	1
8	8	1	8	1	8	1	8	1
9	9	1	9	1	9	1	9	1

Table 4.4

Figure 4.65

> the number of "lion's stomach" lines necessary to embrace all points of a $r \times c$ grid is equal to 1 if $c - 1$ is a multiple of 4 and equal to r if $c - 1$ is not a multiple of 4.

Now the students may test their conjecture. For instance, is the hypothesis verified in the case $r = 4$, $c = 13$ (see Figure 4.65)? Once sure of the conjecture, the question arises, how to prove it?

In the following example, it may be a little more difficult to find a conjecture.

The number of "chased-chicken" lines

(cf. Gerdes, 1993, 1993a)

Figure 4.14 displays examples of monolinear "chased-chicken" designs of dimensions 5×6, 9×10, and 3×8, respectively. When one uses a grid of dimensions 10×5 and applies the same algorithm (Figure 4.15), one needs three of such "chased-chicken" lines to embrace all points of the grid (see Figure 4.66).

Students may investigate the question, How are the number of "chased-chicken" lines related to the dimensions of the rectangular grid? [$g(r, c) = ?$].

In order to get a design that looks like Figure 4.14, the number of rows r has to be odd, and the number of columns c has to be even. The students may now experiment with

Figure 4.66

concrete values of r and c, draw the figures, count the “chased-chicken” lines and record the data collected (Table 4.5).

Observing the data in the table, the students may arrive at conjectures, such as

- (at least) the contents of first rows are periodic. In the first row 1 and 2 are repeated (period = 2); in the second row 3, 1, 1 seem to repeat (period = 3);
- on the principal diagonal, only the number 1 appears, or in other words
- when the number of rows is one less than the number of columns, then the corresponding “chased-chicken” design is monolinear;
- just below the principal diagonal appear successive natural numbers 3, 4, 5, 6, …;
- disregarding the first number in the first column, the first column and the second row are equal;
- disregarding the first number in the second column, the second column and the third row are equal;
- disregarding the first row, the table is symmetric.

Extrapolation on the basis of the collected experimental data and of the observed regularities, may lead to the Table 4.6. Now the students may test their conjecture. For instance, is the hypothesis verified in the case $r = 17$, $c = 14$? At the same time they may analyze questions such as

- What relationship, if any, exists between the numbers in the third row and its period? What happens in the case of the fourth row? What happens in general?;
- What relationship, if any, exists between the periods, r and c?

r \ c	4	6	8	10	12	14	16	18
3	1	2	1	2	1	2	1	
5	3	1	1	3	1	1	3	
7	1	4	1	2	1	4		
9	1	1	5	1				
11	3	2	1	6				
13	1	1						
15								
17								

Table 4.5

r \ c	4	6	8	10	12	14	16	18
3	1	2	1	2	1	2	1	2
5	3	1	1	3	1	1	3	1
7	1	4	1	2	1	4	1	2
9	1	1	5	1	1	1	1	5
11	3	2	1	6		2	3	2
13	1	1	1	1	7	1	1	1
15	1	4	1	2	1	8	1	2
17	3	1	1	3	1	1	9	1

Table 4.6

The students may discover that the numbers in any row or column are divisors of the respective period. Furthermore, since every number in the table belongs at the same time to a column and to a row, it is a divisor of both periods. They may now conjecture that

$$g(r, c) \text{ is a common divisor of } \frac{r+1}{2} \text{ and } \frac{c+2}{2}.$$

or still further

$$f(2m, 2n{+}1) \text{ is the greatest common divisor (gcd) of } \frac{r+1}{2} \text{ and } \frac{c+2}{2}.$$

Comparing Tables 4.5 or 4.6 with Table 4.2 (which refers to the problem of counting the number of plaited-mat lines), may accelerate arrival at the last conjecture.

Once more, the next question is, "How can we prove the conjecture?" I have analyzed this question with a group of pre-service mathematics teachers in Mozambique. As variations on this theme, other challenging questions for further investigation arose such as

- What characterizes non-rectangular grids of monolinear "chased-chicken" designs? In other words, under what conditions is a 'chased-chicken' pattern monolinear? (see Figure 4.67 for an example)
- Figure 4.68 shows a variation of a "chased-chicken" design where the successive vertical broad zigzags of the "chased-chicken" line are (bilaterally) symmetrical instead of parallel. Under what conditions are such designs monolinear? How will the number of lines depend on the dimensions of the rectangular grid?

Figure 4.67

Algorithms and monolinear designs

In a similar manner (to that of the previous section) we may analyze other algorithms such as those used in the case of the "fish trap" (Figure 4.16), the "end of a struggle" (Figure 4.18), and "fire" (Figure 4.19) representations. For what grids, will the applications of the various algorithms lead to monolinear designs? Or more generally, how does the number of lines (if it possible to have more than one) depend on the dimensions of the grids?

Figure 4.68

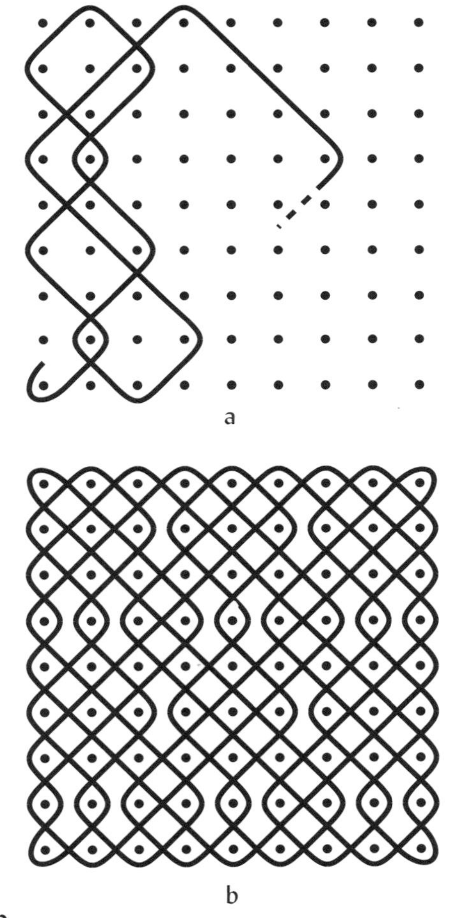

a

b

Figure 4.69

The reader is invited to invent his/her own algorithms and analyze questions such as the above. Figure 4.69a shows an example of the use of a *sona* inspired algorithm at work; the corresponding design is monolinear (Figure 4.69b).

Recreations of the type "Find the missing designs"

(cf. Gerdes, 1991, 1997)

The *akwa kuta sona* drawing experts experimented with algorithms and dimensions. They found series of *sona* constructed with the same algorithms. Their activity suggests educationally interesting activities, where each time some elements of a series are given and the student/reader has to find others. The given designs are in the style of the *sona*; all are monolinear. One has to discover the construction rule for these drawings and then draw the missing designs, applying the same construction rule and using the correct dimensions. Figures 4.70 to 4.73 present examples of this

Figure 4.70

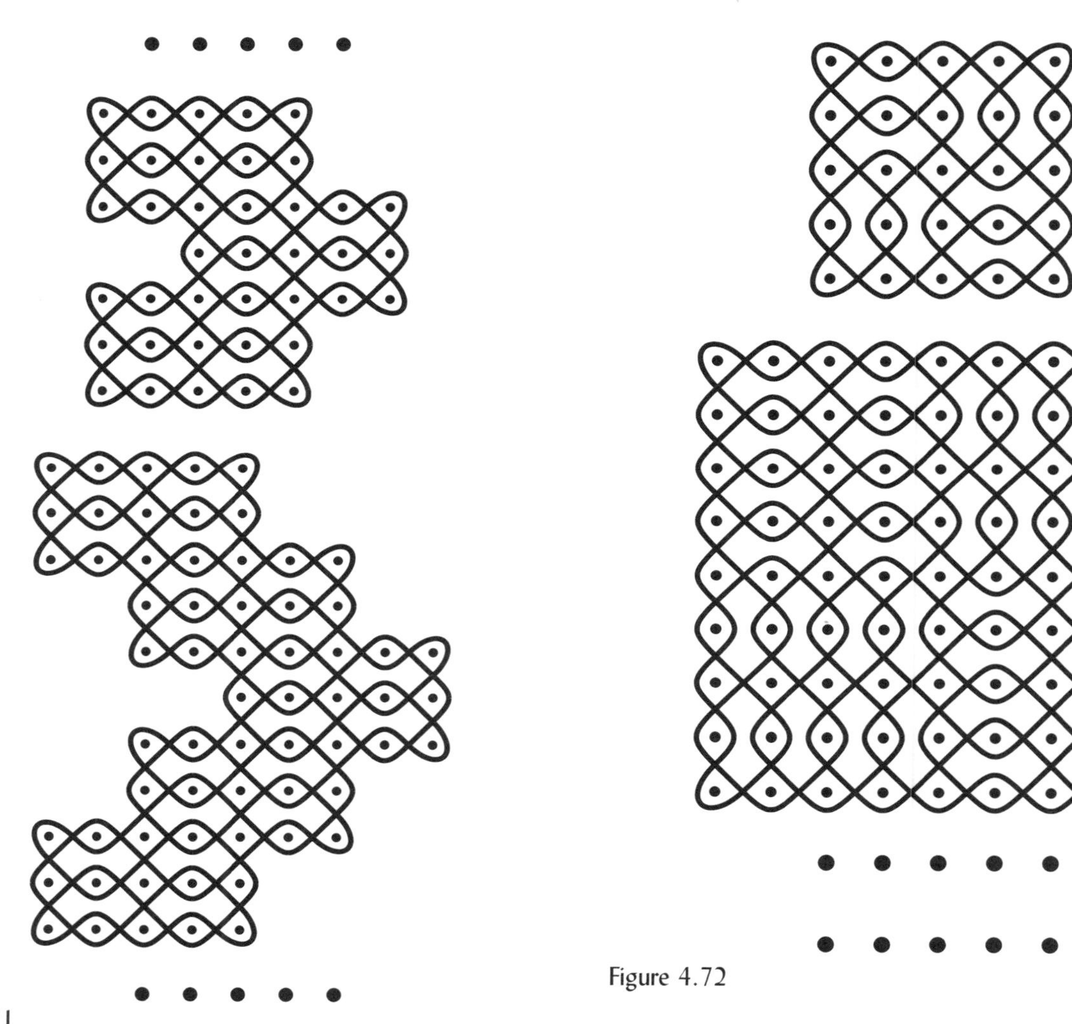

Figure 4.71

Figure 4.72

type of 'Find the missing designs' activities. Further examples may be found in the book *Lusona: Geometrical Recreations of Africa*, based on problems presented to a seminar of mathematics teacher-students at Mozambique's Universidade Pedagógica. Participation in the seminar was voluntary and normally it was difficult to end the two-hour session of the seminar since participants 'lost their notion of time' and went well over time.

Figure 4.73

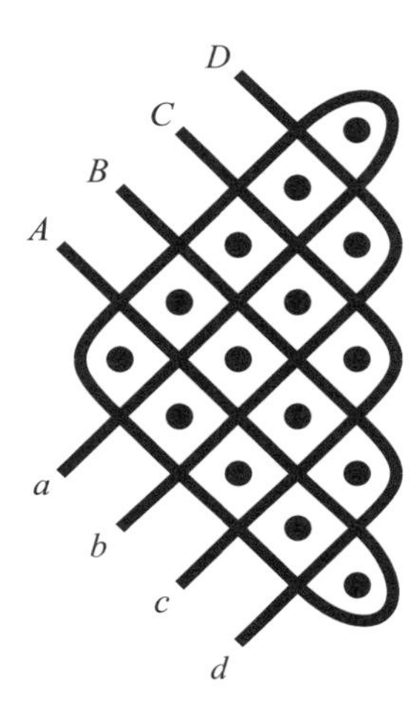

Figure 4.74

Systematic construction of monolinear designs

(cf. Gerdes, 1993, 1993a)

The systematic construction of monolinear designs by the *akwa kuta sona* drawing masters, discussed in chapter 17, may be explored educationally in diverse ways, as the following examples may illustrate.

First example: Triangular designs

Looking to the symmetrical and monolinear, left half of the *lusona* in Figure 4.8b, it may be said that it may be obtained from the triangular design in Figure 4.74 by joining the endpoints *a, b, c, d, A, B, C,* and *D* with loops *Ad, Bc, Cb,* and *Da*.

Many questions for further reflection and analysis arise, such as:

- Are there other ways for joining these endpoints such that symmetrical, monolinear designs appear? How many?
- How many possibilities for monolinear drawings exist if there are *n* endpoints on each side of the 'triangular design'?
- How many of them are symmetrical?

Second example: Formations of flying birds

The representation of three flying ducks in Figure 4.75 is both monolinear and symmetrical. Is it possible to draw them so that they fly in a V-formation? One may start with drawing the basic structure (see Figures 4.76a and b) and

Figure 4.75

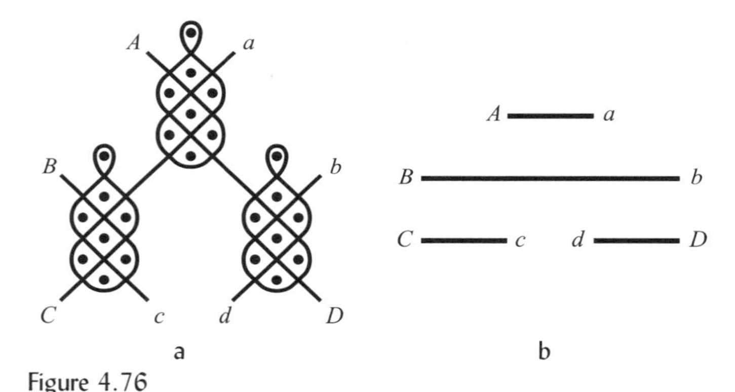

Figure 4.76

then analyze how to join, symmetrically, the loops to produce a final drawing that is monolinear.

Figure 4.77 shows a solution. Is it the only solution? Which are the solutions if one tries to represent similarly 6, 10, or more ducks in a flying V-formation? And so forth.

Chain rules

It is not difficult to understand why the general chain rules (the *akwa kuta sona* had discovered) and the second chain rule for plaited-mat designs ('applying the squares') are true. But what about the first chain rule for plaited mat

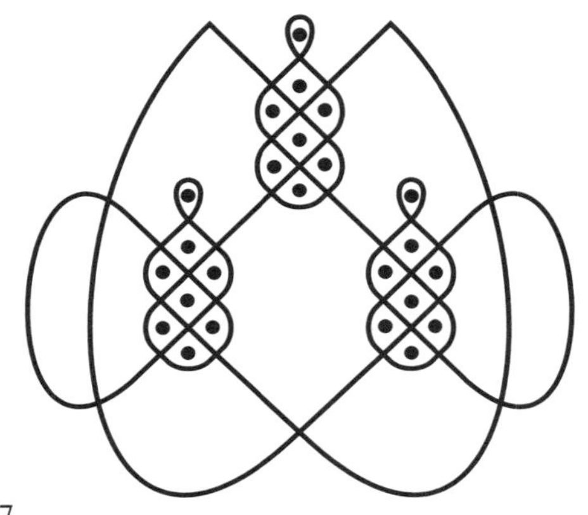

Figure 4.77

designs (as illustrated in Figure 4.41)? How do we prove the theorem embedded in the rule?

Many possibilities for further exploration exist. For example, one may try to discover other general chain rules, or other specific rules for the chaining of plaited-mat designs; prove them and see if they may be used to explain the mono- or polylinearity of designs. For instance, to embrace all points of the (polyominal) grid in Figure 4.78, How many plaited-mat lines are necessary?

Another group of questions consists of the examination of the existence of specific chain rules for other algorithms, like those of the lion's stomach and of the chased-chicken.

3 Excursion: Generation of Lunda-designs

(cf. Gerdes, 1996, 1997)

As an example of the fertility of the *sona* for further educational investigation and mathematical research, I would like to present the Lunda-designs. I discovered these in the context of analyzing a class of *sona*.

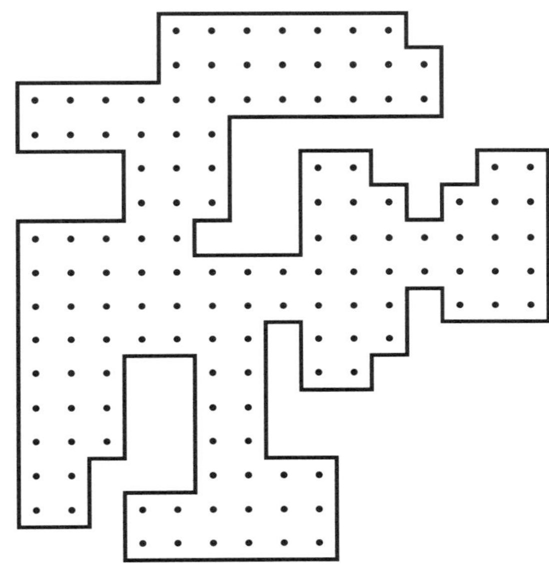

Figure 4.78

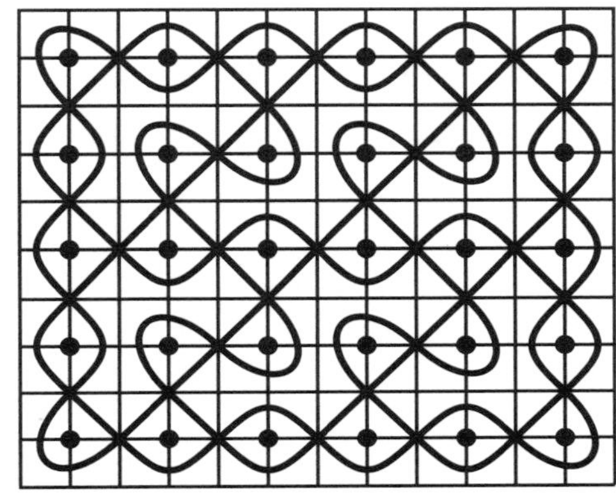

Figure 4.79

Discovering an interesting property

To facilitate the execution of the *sona* sand drawings I was analyzing, I used to draw them on squared paper with a distance of two units between two successive grid points. In this way, a monolinear design such as the 'chased-chicken path' (see Figure 4.79) passes exactly once through each of the small squares inside the rectangle surrounding the grid.

This gives the possibility of enumerating the small squares, **1** being the number attributed to the small square where one starts the line, and **2** the number of the second unit square through which the curve passes, and so on successively until the closed curve is complete. Figure 4.80 shows the counting in the case of the 'chased-chicken' path (dimensions 5×6), when we start at the rectangle's centre.

The path followed by the 'chased chicken' is aesthetically pleasing and the design displays a rotational symmetry of 180°. This leads to the following question: How is the beauty and the symmetry of this sand drawing reflected in the enumeration of the small squares? For instance, what happens with the numbers of two small squares, which correspond under a rotation of 180°? Moreover, will the numerical rectangle be interesting, for example, is it 'magic', in the sense of having the same sum of the numbers for each row?

It turns out that the numerical rectangle is 'magic' modulo 4. Figure 4.81 shows the enumeration modulo 4 of the little squares in the case of the 'chased-chicken' design with dimensions 3×4, when we start counting at the rectangle's lower left corner. The sum of the numbers of each row is 0 mod 4; the sum of the numbers in each column is 1 mod 4.

Moreover, a beautiful surprise appears: the placing of 0, 1, 2, 3 is alternately clockwise and anti-clockwise around the grid points; between four neighboring grid points there are always four equal numbers (see Figure 4.82). Figure 4.83a shows the transition to modulo 2. Coloring the unit squares with number 1 black, and the ones with number 0 white, a black-and-white design is obtained (Figure 4.83b).

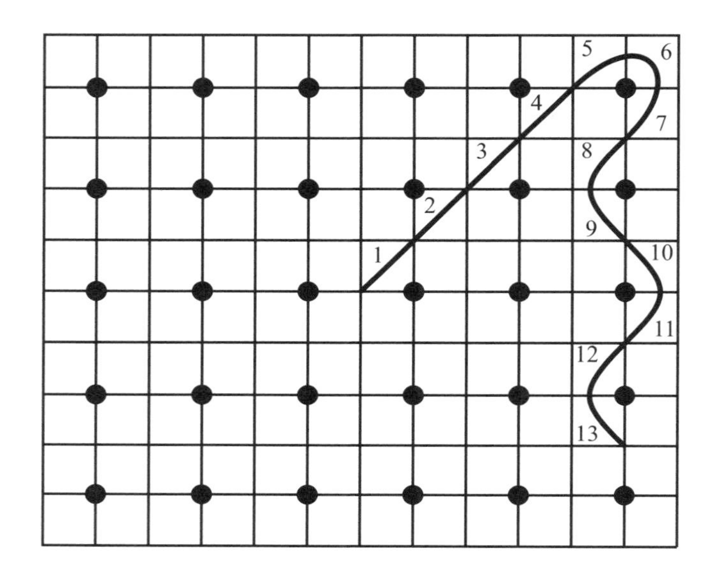

106	105	69	70	102	101	33	34	98	97	5	6
107	68	104	103	71	32	100	99	35	4	96	7
67	108	76	75	31	72	40	39	3	36	8	95
66	109	77	30	74	73	41	2	38	37	9	94
110	65	29	78	62	61	1	42	58	57	93	10
111	28	64	63	79	120	60	59	43	92	56	11
27	112	84	83	119	80	48	47	91	44	12	55
26	113	85	118	82	81	49	90	46	45	13	54
114	25	117	86	22	21	89	50	18	17	53	14
115	116	24	23	87	88	20	19	51	52	16	15

Figure 4.80

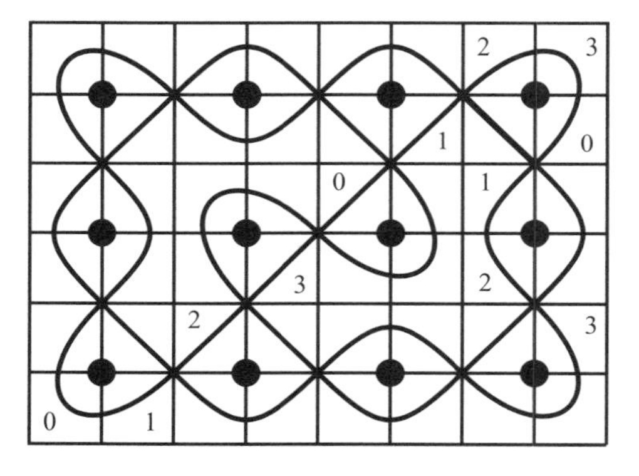

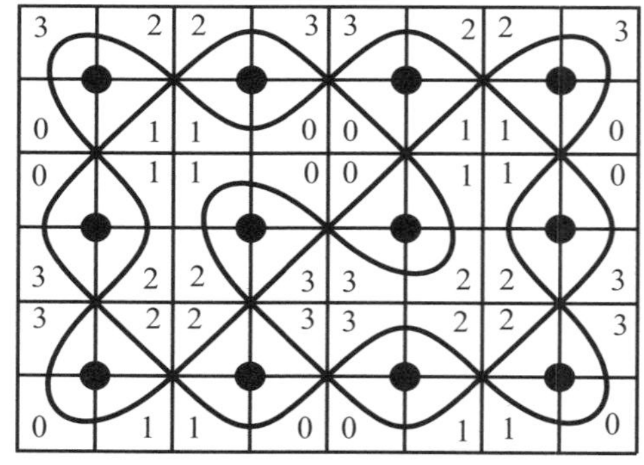

Figure 4.81

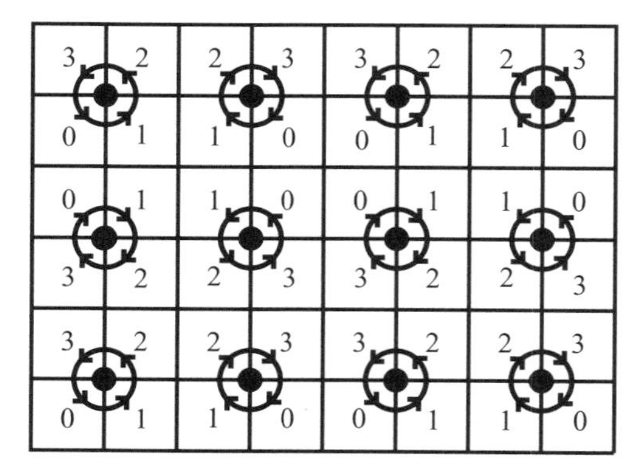

Figure 4.82

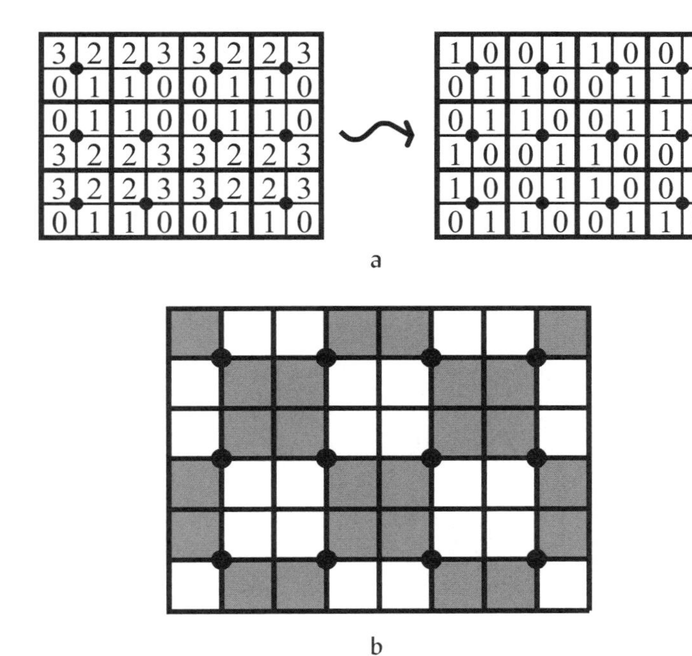

Figure 4.83

Will something similar happen with 'chased-chicken' designs of higher dimensions? And with other monolinear sand drawings such as the 'lion's stomach' (see Figure 4.21)?

Indeed, something similar happens. This follows from the fact that both the 'chased-chicken' and the 'lion's stomach' designs are regular mirror curves.

Mirror curves (cf. Jablan)

The 'chased-chicken' design may be considered as a mirror curve, that is:

- it is the smooth version of the polygonal path described by a lightray emitted from the starting place S (see Figure 4.84) at an angle of 45° to the rows of the grid;*
- as the ray travels through the grid it is reflected by the sides of the rectangle and by the "double-sided mirrors" it encounters in its path. The mirrors are placed horizontally or vertically, midway, between two neighboring grid points, as in Figure 4.85.

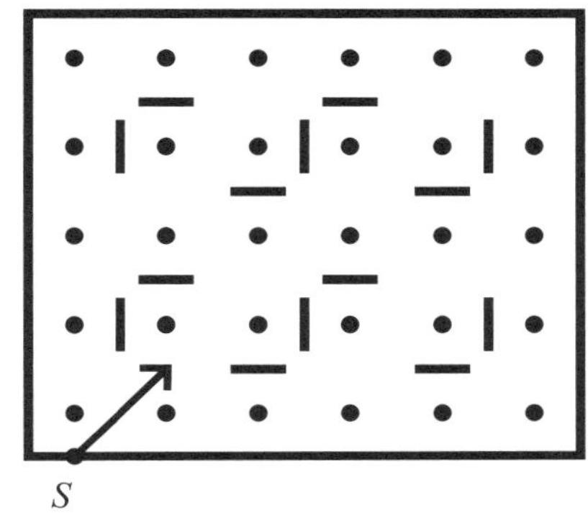

Figure 4.84
Emission of a lightray from the point S

In Figure 4.86 the mirrors for the 'lion's stomach' design and the first design in figure 4.73 are shown. As in the case of the 'chased-chicken' design, the mirrors appear only in

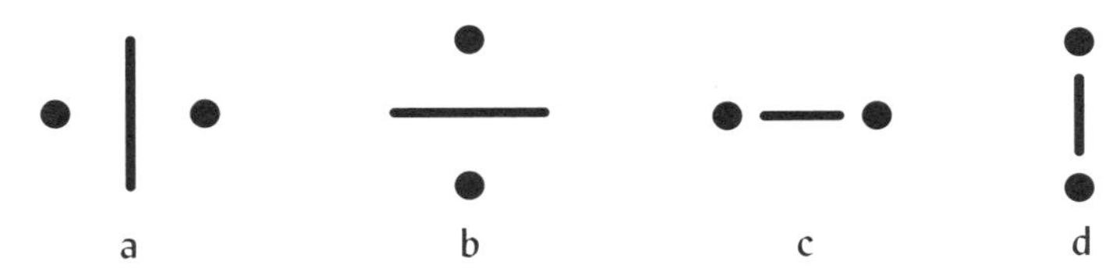

Figure 4.85
Possible positions of mirrots in the center between neighboring grid points

* When introducing a coordinate system, in such a way that the points of the lowest row of the grid have the coordinates (1,1), (3,1), (5,1), etc., the coordinates of the starting place S are (1,0).

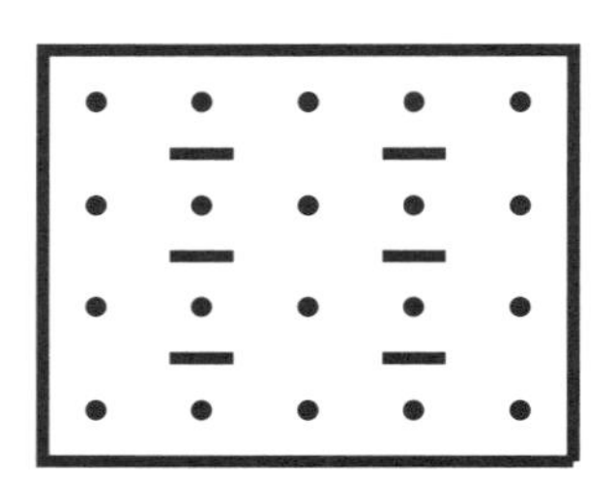

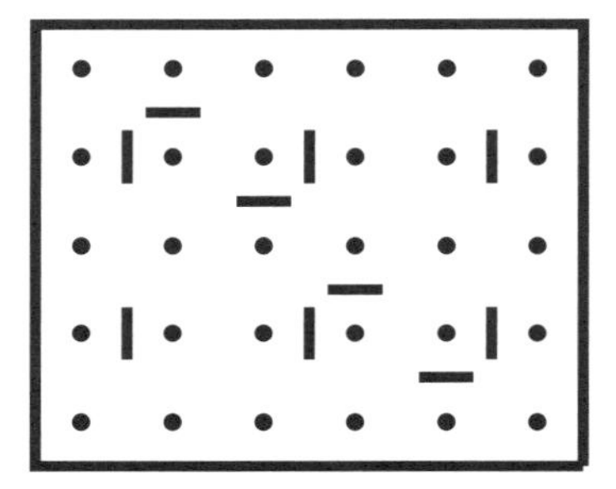

Figure 4.86

the positions (a) and (b) illustrated in Figure 4.85. When this happens, the mirror curve will be called *regular*.

Regular mirror curves have the interesting enumeration properties modulo 4 and 2 of the previous section. The reader is invited to find proofs for the theorems (cf. the proofs in Gerdes, 1993, ch. 6 and in Gerdes, 1996). When one or more mirrors appear in the position (c) or (d) the result is different, however, as the example of the black-and-white coloring of a non-regular mirror curve in Figure 4.87 illustrates.

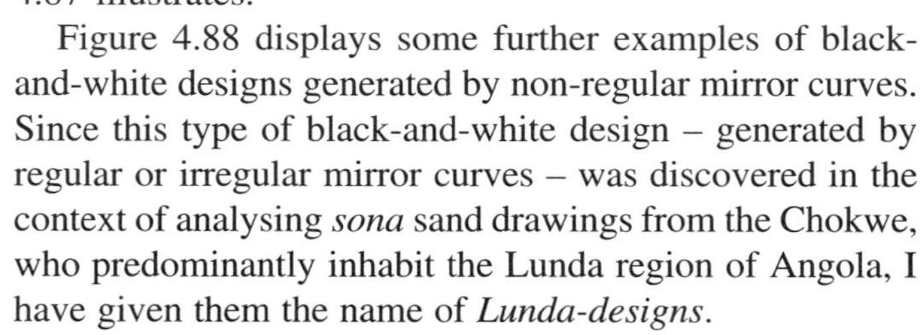

Figure 4.88 displays some further examples of black-and-white designs generated by non-regular mirror curves. Since this type of black-and-white design – generated by regular or irregular mirror curves – was discovered in the context of analysing *sona* sand drawings from the Chokwe, who predominantly inhabit the Lunda region of Angola, I have given them the name of *Lunda-designs*.

Searching for the common characteristics of Lunda-designs, I observed and proved the following symmetry properties:

i) In each row there are as many black unit squares as there are white unit squares;

ii) In each column there are as many black unit squares as there are white unit squares;

iii) Along the border each grid point always has one black unit square and one white unit square associated with it (see Figure 4.89);

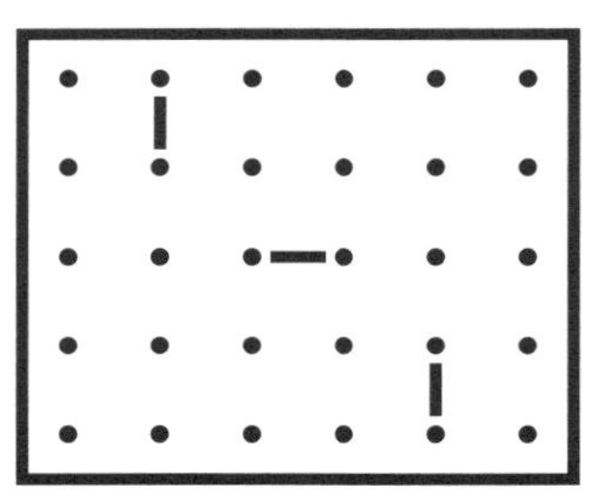

a. *Mirror design*

b. *Corresponding mirror curve*

c. *Starting the coloring*

d. *Final brown-and-white pattern (with the grid points marked)*

e. *Final brown-and-white pattern (without the grid points)*

f. *Final brown-and-white pattern (without the rectangle border)*

Figure 4.87

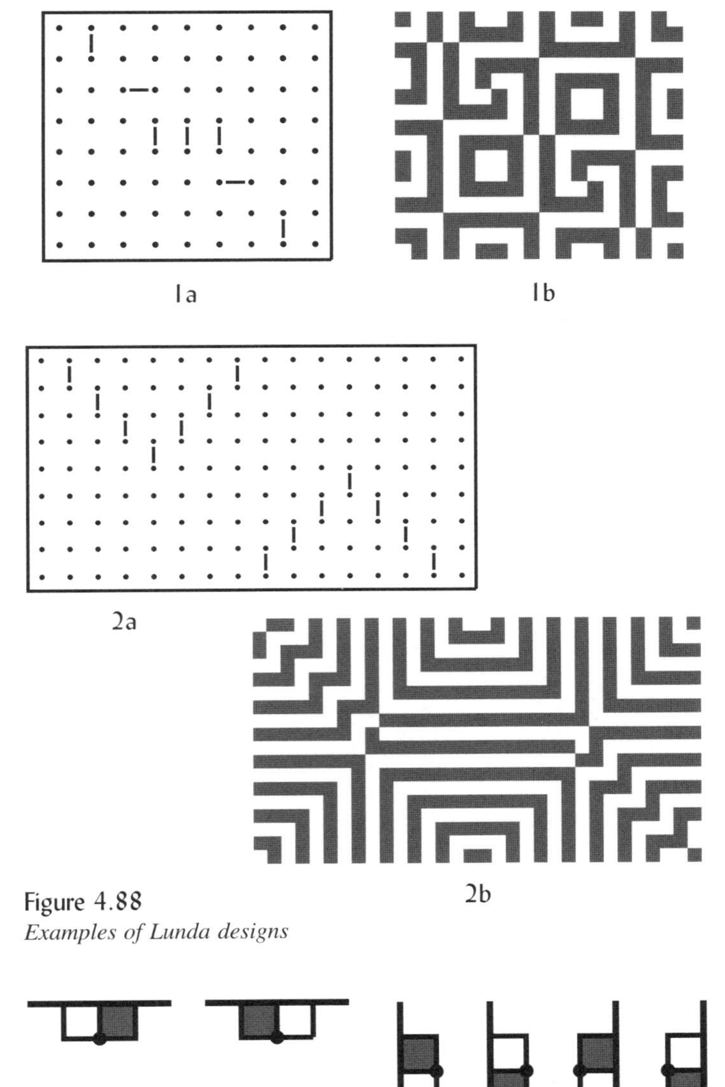

Figure 4.88
Examples of Lunda designs

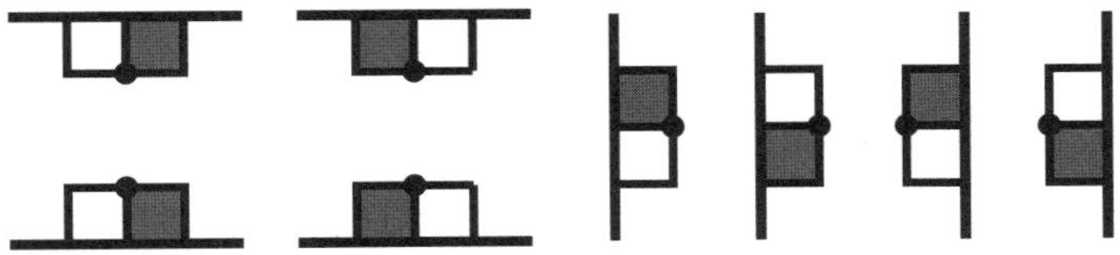

Figure 4.89
Possible border situations

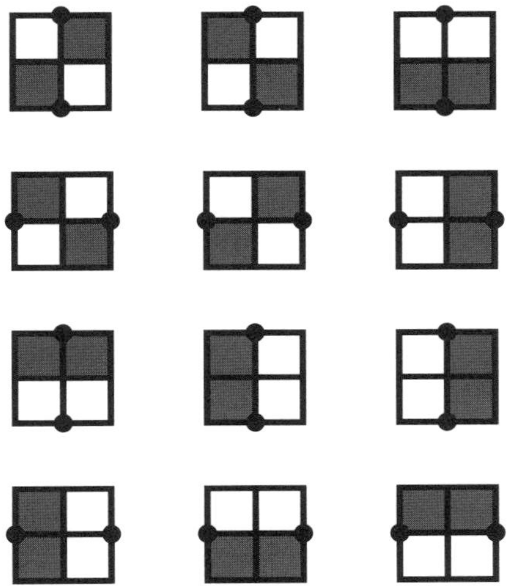

Figure 4.90
Possible situations between vertical and horizontal neighboring grid points

iv) Of the four unit squares between two arbitrary (vertical or horizontal) neighboring grid points, two are always black (see Figure 4.90).

Inversely, the following theorem can be proved (cf. Gerdes, 1996):

- any rectangular black-and-white design that satisfies the properties (i), (ii), (iii), and (iv) is a Lunda-design.

In other words, for any rectangular black-and-white design that satisfies the properties (i), (ii), (iii), and (iv) there exists a mirror curve that produces it in the discussed sense (cf. Figure 4.89). Moreover, in each case, such a mirror curve may be constructed.

The characteristics (i), (ii), (iii), and (iv) may be used to define Lunda-designs (of dimensions $m \times n$). In fact, it may be proven that the characteristics (iii) and (iv) are sufficient for this definition, as they imply (i) and (ii).

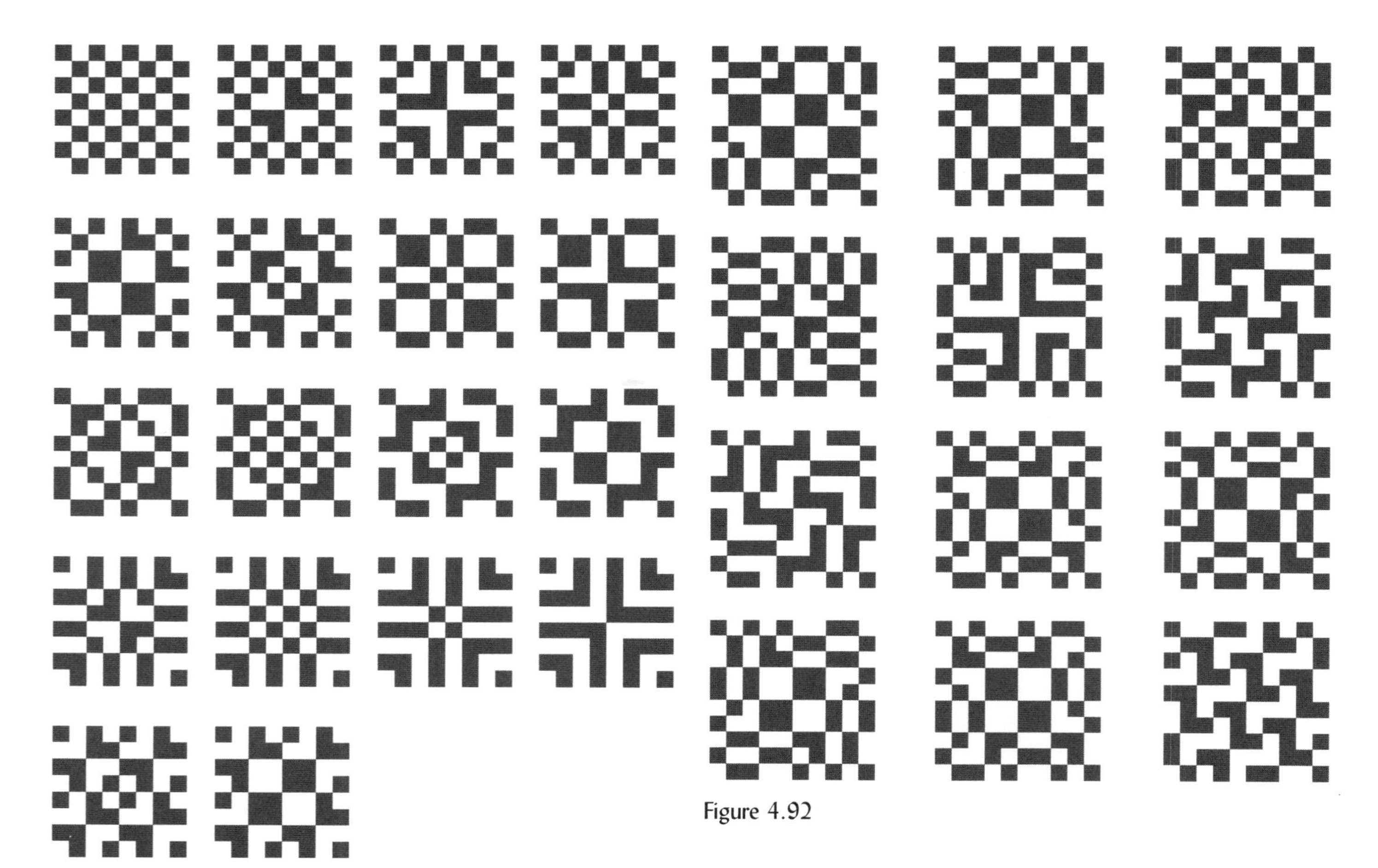

Figure 4.91

Figure 4.92

Especially attractive are Lunda-designs which display further symmetries. For instance, Figure 4.91 displays the eighteen 4 × 4 Lunda-designs which admit reflections in the diagonals (preserving the colors) and vertical and horizontal reflections (interchanging black and white). By consequence, a half-turn about the center preserves the colors and a quarter-turn reverses the colors.

Figure 4.92 shows some 5 × 5 Lunda-designs with a rotational 2-color symmetry: a quarter-turn about the center reverses the colors.

Square Lunda-designs may be used to build up fractals, that is geometrical figures with a built-in self-similarity. Figure 4.93 illustrates the first three phases of building up a fractal on the base of a symmetrical 2 × 2 Lunda-design.

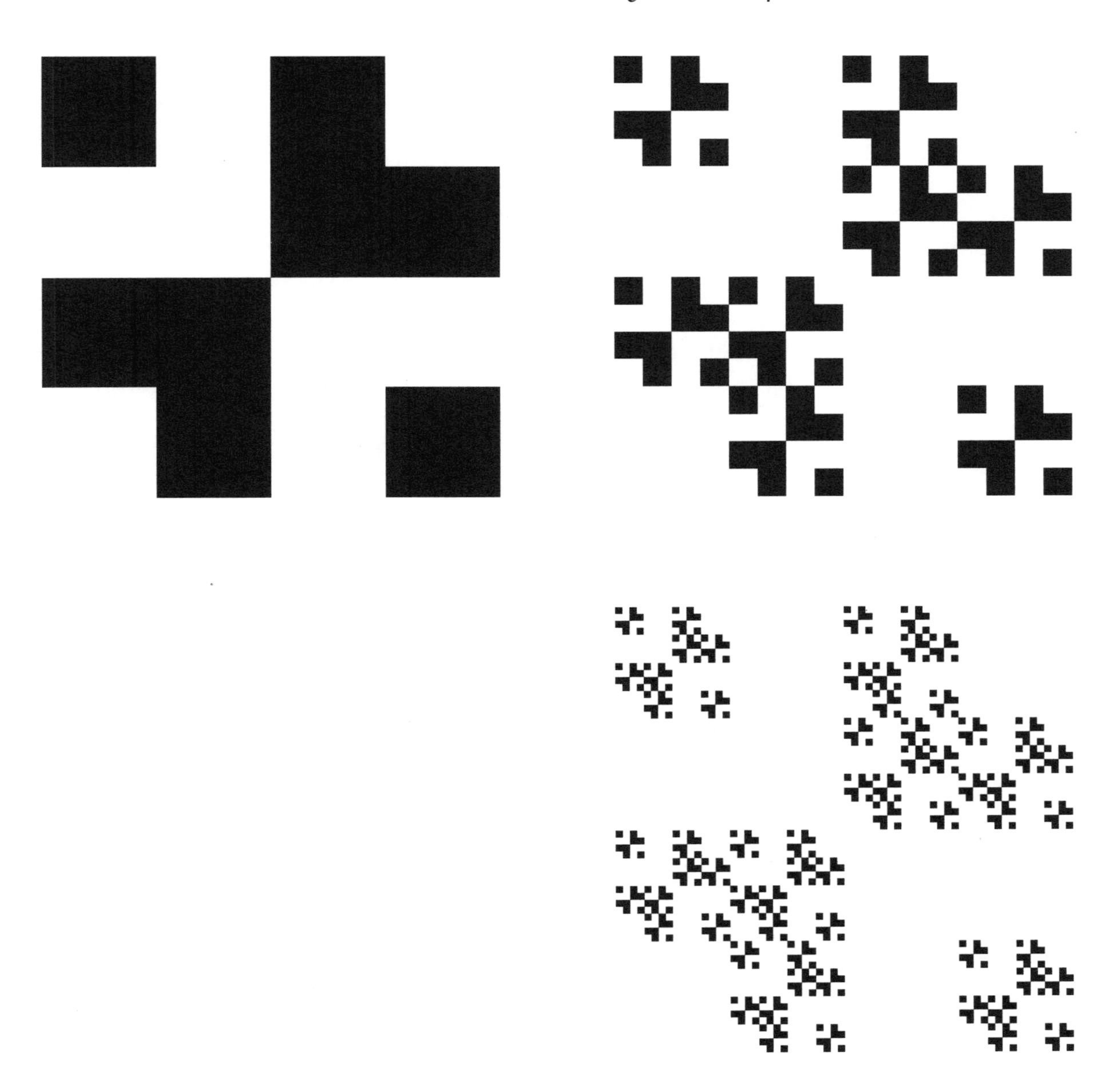

Figure 4.93
Bulding up a Lunda-fractal

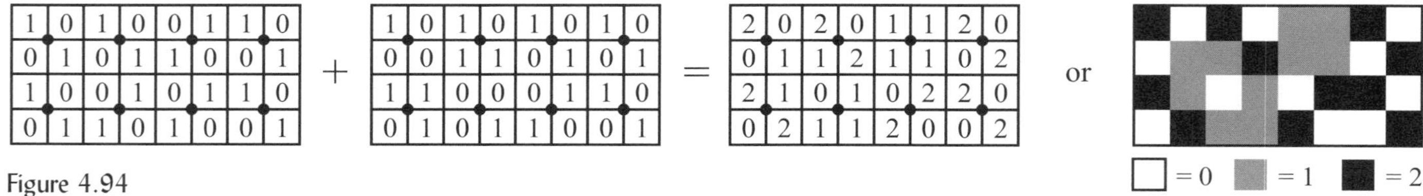

Figure 4.94
Example of the production of a Lunda-2-design of dimensions 2×4

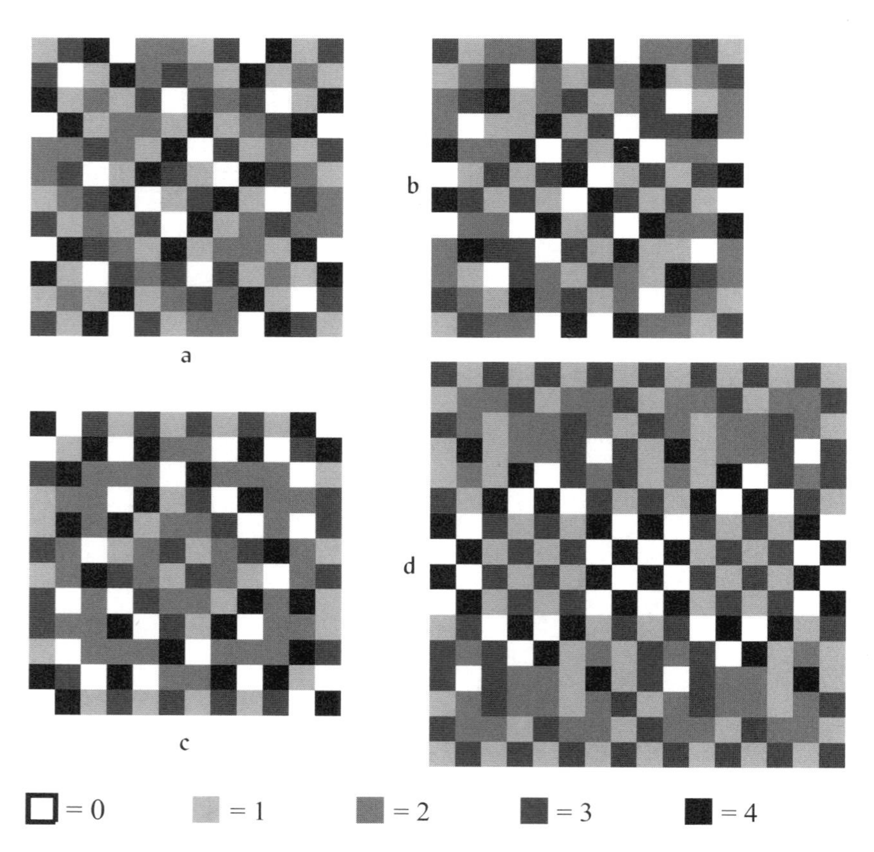

Figure 4.95
Examples of symmetrical square Lunda-4-designs

Generalizations

Lunda-designs may be generalized in several ways. Hexagonal and circular Lunda-designs, and Lunda-k-designs are some interesting possibilities.

Since Lunda-designs may be considered as matrices, it is quite natural to define addition of Lunda-designs in terms of matrix addition: the sum of two (or more) matrices (of the same dimensions) is the matrix in which the elements are obtained by adding corresponding elements (see the example in Figure 4.94). The sum of k Lunda-designs (of dimensions $m \times n$) may be called a *Lunda-k-design* (of dimensions $m \times n$). Figure 4.95 displays examples of symmetrical square Lunda-4-designs.

As is easy to see, the Lunda-k-designs inherit the following symmetry properties from the Lunda-designs:

ii) The sum of the elements in any column is equal to km;

i) The sum of the elements in any row is equal to kn;

iii) Along the border the sum of the elements in the two unit squares associated with a border grid point is equal to k;

iv) The sum of the integers in the four unit squares between two arbitrary (vertical

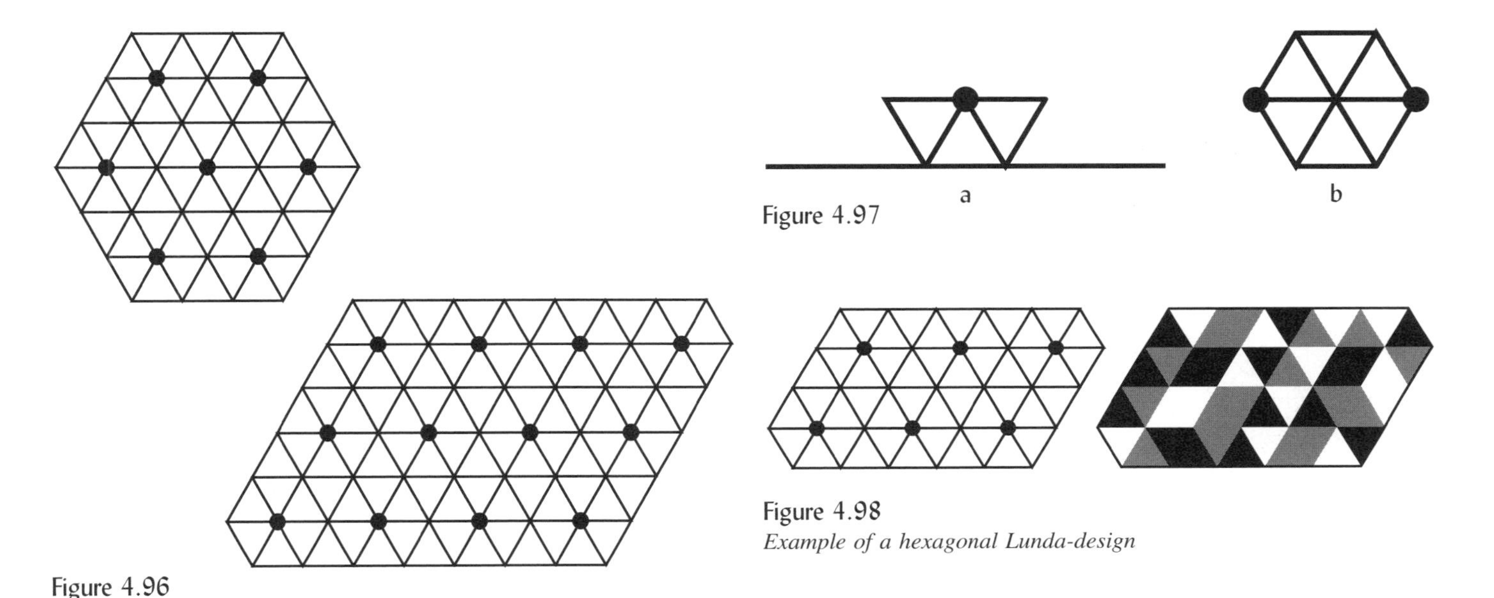

Figure 4.96

Figure 4.97

Figure 4.98
Example of a hexagonal Lunda-design

or horizontal) neighboring grid points is always $2k$.

Properties (i) and (ii) guarantee a global equilibrium for each row and column. Properties (iii) and (iv) guarantee more local equilibria.

The characteristics (i), (ii), (iii), and (iv) may be used to define Lunda-k-designs of dimensions $m \times n$. The characteristics (iii) and (iv) are sufficient for this definition, as they imply (i) and (ii).

Another way to expand the concept of Lunda-design is to start with hexagonal grids instead of rectangular ones. Figure 4.96 displays two examples. Each grid point is surrounded by six unit triangles. Each border grid point has three unit triangles that touch the border (see Figure 4.97a), and between two arbitrary neighboring grid points, there is always a hexagon composed of six unit triangles (see Figure 4.97b).

Suppose that to each unit triangle of a hexagonal grid we assign one of three colors (e.g., white, light brown, and brown). Then we obtain a three-colored design. If such a design satisfies the following two conditions:

i) Different colors are assigned to the three border unit triangles of any border grid point;
ii) Of the six unit triangles between two arbitrary neighboring grid points, there are two of each color,

it may be called a *hexagonal Lunda-design*. Figure 4.98 presents an example.

Figure 4.99 presents hexagonal Lunda-designs that have a three-color rotational symmetry: a clockwise rotation by 120° moves all the white to coincide with all the gray, moves gray to black, and black to white. In other words, the three colors occupy equivalent parts of the design.

Figure 4.100 displays an example of a circular grid of dimensions 4×12. It has 12 rays with 4 points on each ray.

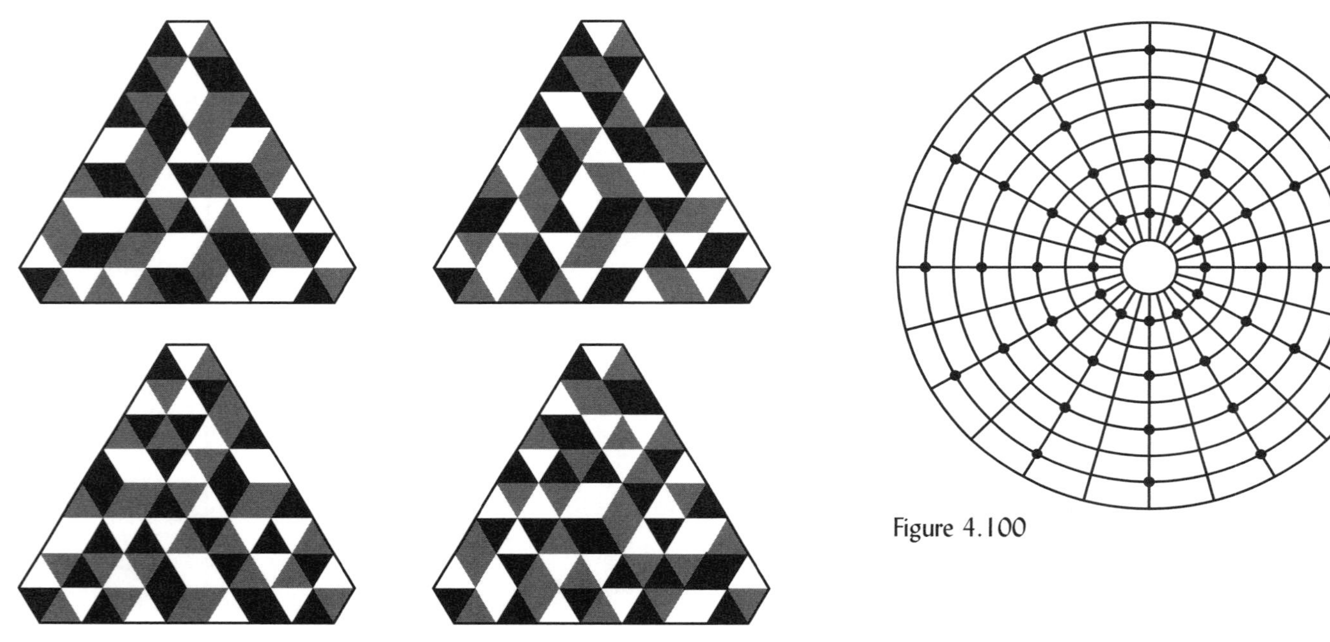

Figure 4.99
Hexagonal Lunda-designa with three-color symmetry

Figure 4.100

Analogously to the definition of rectangular Lunda-designs, circular Lunda-designs may now be defined as black-and-white designs of circular grids that satisfy the following two characteristics:

i) Of the two border unit disc segments of any border grid in the smallest or biggest circle, one is always white and the other black;
ii) Of the four unit disc segments between arbitrary neighboring grid points (belonging to the same circle or to the same ray), two are always black and two are white.

Figure 4.101 displays examples of symmetrical circular Lunda-designs (of dimensions 4×12). These may, in turn, inspire new decorative designs, for example, for coiled plates and baskets. Thus a cycle of inspiration and creation is completed: starting from attractive *sona* to the invention of Lunda-designs and returning to crafts and decorations....

A circular Lunda-design may also be considered a flattened cylindrical Lunda-design. Bending and closing a cylinder leads to a torus. How may Lunda-designs be defined on a torus? How may the concept of Lunda-design be extended to obtain black-and-white strip and plane patterns?

Lunda-polyhedral-designs

The concept of Lunda-design may also be extended to surfaces of regular polyhedra, considering the grid points

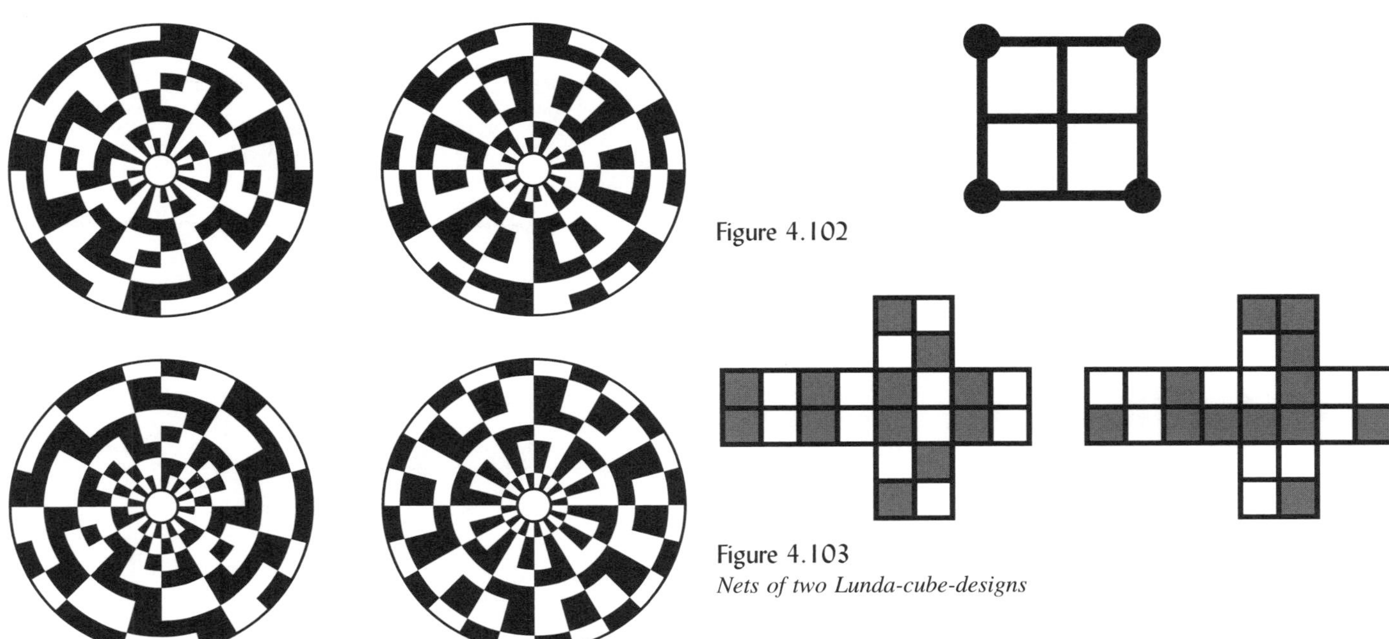

Figure 4.102

Figure 4.103
Nets of two Lunda-cube-designs

Figure 4.101
Examples of symmetrical circular Lunda-designs

placed at the vertices. As these surfaces do not have borders, only one characteristic is sufficient to define a Lunda-polyhedral-design:

i) Of the four unit quadrilaterals between two arbitrary neighboring grid points (= vertices), two are always black and two are white.

Two of the four unit squares belong to one face of the polyhedron, and the two other to another face.

A grid face of the cube looks like Figure 4.102, and two nets of Lunda-cube-designs are presented in Figure 4.103.

Figure 4.104 display the grid net of a regular tetrahedron (a) and the nets of two Lunda-tetrahedal-designs (b). A

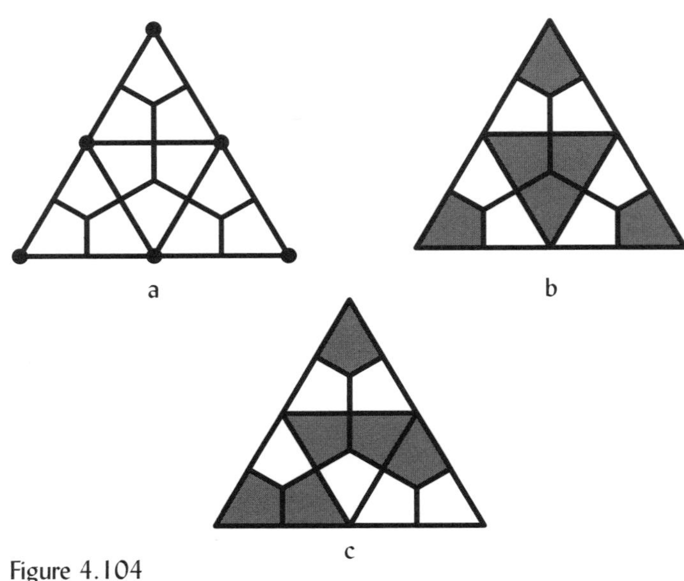

Figure 4.104

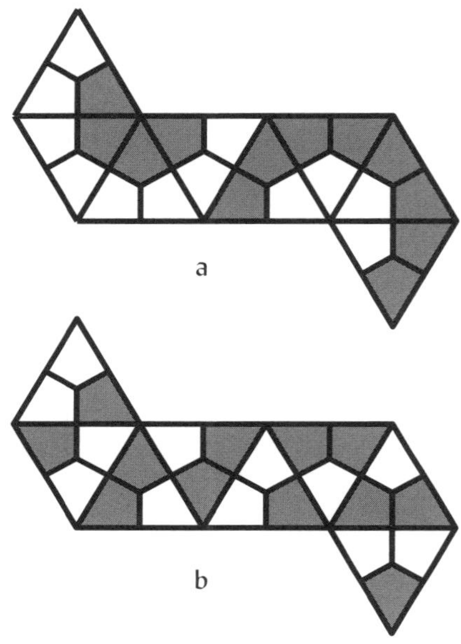

Figure 4.105
Nets of symmetrical Lunda-cotahedral-designs

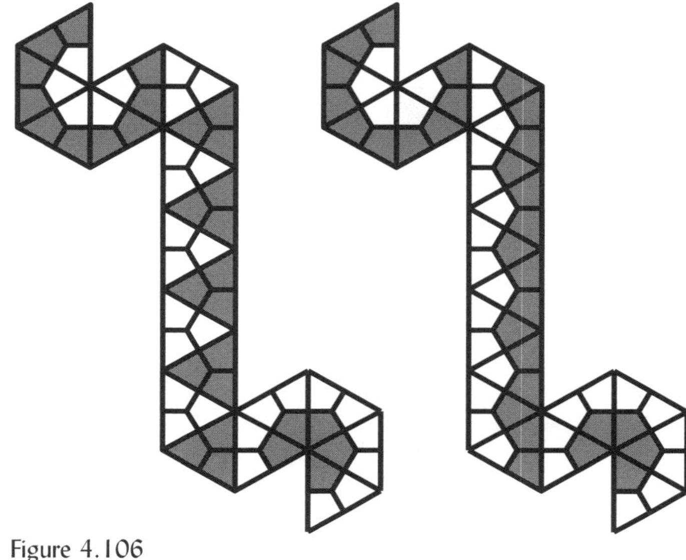

Figure 4.106
Nets of symmetrical Lunda-icosahedral-designs

natural question—in a mathematical context—is to ask how many of these designs exist and how many are symmetrical.

The nets of two symmetrical Lunda-octahedral-designs and of two symmetrical Lunda-icosahedral-designs are shown in Figures 4.105 and 4.106. How does the duality of the regular octahedron and the cube manifest itself in the Lunda-designs they generate?

Four nets of Lunda-dodecahedral-designs are presented in Figure 4.107. Which of the corresponding Lunda-dodecahedral-designs have symmetries? What symmetries? How many Lunda-dodecahedral-designs with these symmetries exist, etc. As the centers of the faces of a regular dodecahedron constitute the vertices of a regular icosahedron, it is possible to transform each Lunda-dodecahedral-design into a Lunda-icosahedral-design and vice-versa.

Spherical Lunda-designs may obviously be obtained through projection of regular polyhedral Lunda-designs. To which other polyhedra may the concept of Lunda-design be extended?

For (rectangular) Lunda-designs in the plane it is possible to construct monolinear *sona*-like drawings which will generate them. The question that emerges is: is it (always) possible to construct for any given polyhedral Lunda-design a generating monolinear *sona*-like drawing that generates the lunda-design?

For which polyhedra is it possible to increase the number of grid points beyond the number of vertices in such a way that Lunda-like designs may be defined. For instance, in the case of the cube, one may choose grid faces like the one in Figure 4.108. Figure 4.109 displays the net of a corresponding, symmetrical cubic Lunda-design.

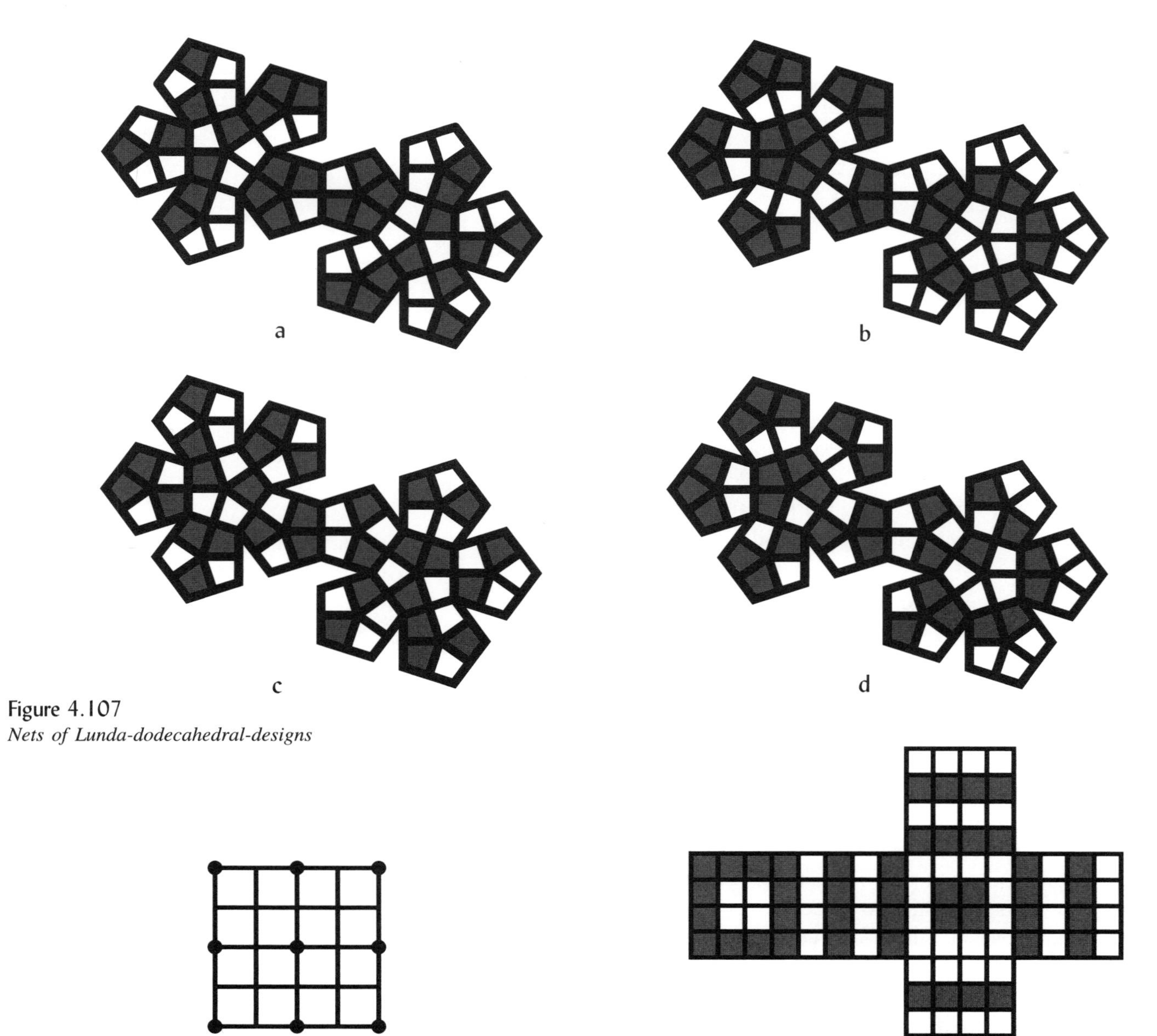

Figure 4.107
Nets of Lunda-dodecahedral-designs

Figure 4.108

Figure 4.109

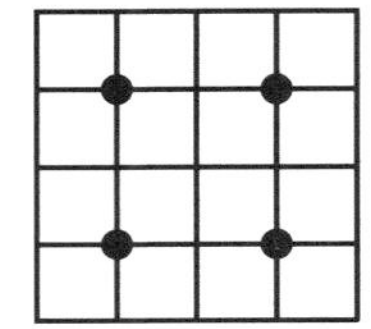

Figure 4.110

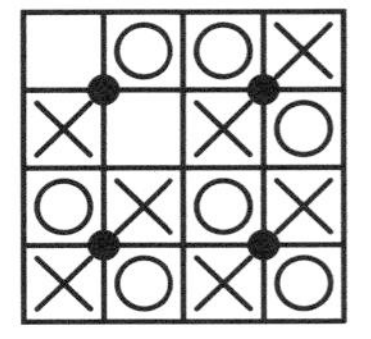

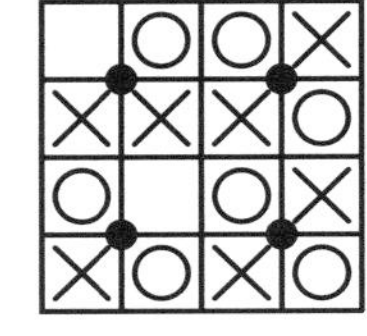

Figure 4.111

A Lunda-board game

Let us return to the traditional entertainment aspect of *sona*. The concept of Lunda-design discovered in the context of the analysis of the Chokwe *sona* sand drawings, may also be used to develop several games, as the following board game may illustrate.

Consider a square grid with sixteen unit squares and four points as shown in Figure 4.110.

It is a game for two players who mark, in turn, naughts and crosses, respectively, in the open unit squares, in such a way that:

i) there are never more than two naughts nor more than two crosses between two neighboring grid points;

ii) there are never two naughts nor two crosses in the two border unit squares of the same border grid point.

The game continues until one of the following situations occurs:

a) It is the turn of the second player, and (s)he does not have a free unit square in which to mark a cross, then the first player wins the game.

b) It is the turn of the first player, and (s)he does not have a free unit square in which to mark a naught, where as the second player has still a free unit square in which to mark a cross. The second player wins the game.

c) It is the turn of the first player, and (s)he does not have of a free unit square in which to mark a naught, neither the second player has a free unit square in which to mark a cross (see the examples in Figure 4.111). The game ends in a draw.

The game may be played on any other bigger rectangular grid, or any circular or polyhedral grid in the above discussed sense. As there are no border dots in the case of polyhedral grids, the second rule may be deleted.

The reader is invited to analyze if there exist winning strategies for the first or second player in the square 2×2 and other cases?

Is it possible to develop a hexagonal Lunda-board game? For how many players? What could be the rules? Would there be winning strategies?

Lunda-animals

Another way to 'play' with Lunda-designs is to analyze how many Lunda-polyominoes of certain dimensions or how many Lunda-paths exist (in the plane or on a certain surface). For instance, consider the Lunda-design in Figure 4.112. Some black and some white polyominoes that appear in it, are shown in Figure 4.113. Such polyominoes may be called Lunda-polyominoes. Will the question of how many Lunda-*n*-ominoes there are, be as difficult to answer as the general question of how many n-ominoes there are? Or, what about questions like 'How many symmetrical Lunda-*n*-ominoes exist?'

An interesting question for an investigation by school children may be to discover how many possible paths there

Figure 4.112

are for a Lunda-animal starting from a given position. Here a Lunda-animal may be defined as black pentomino (consisting of 5 cells) with one unit square at one of its ends marked as head (H). This animal walks through a(n infinite) grid, where (horizontal or vertical) neighboring points are marked at a distance of two units. All unit squares of the grid are white. The animal may only do one step in any direction that does not 'violate' the following Lunda-characteristic:

> Of the four unit squares between two arbitrary neighboring grid points, no more than two may be black.

When the animal starts from the position shown in Figure 4.114a, it may turn its head only to the unit squares marked 1 or 2, as going to the third one would violate the Lunda-characteristic. How many paths $p(m)$ of m steps are possible?

Figure 4.114 illustrates the first possible steps: $p(1) = 2$, $p(2) = 3$, $p(3) = 5, \ldots$ Successive numbers of the famous Fibonacci sequence are appearing. How do we prove this?

What would happen if Lunda-animals had been defined as 6-ominoes? Or, as 9-ominoes?

Bibliography

Ascher, Marcia (1988), Graphs in cultures (II): a study in ethnomathematics, in: *Archive for History of Exact Sciences*, Berlin, Vol.39, No.1, 75–95.

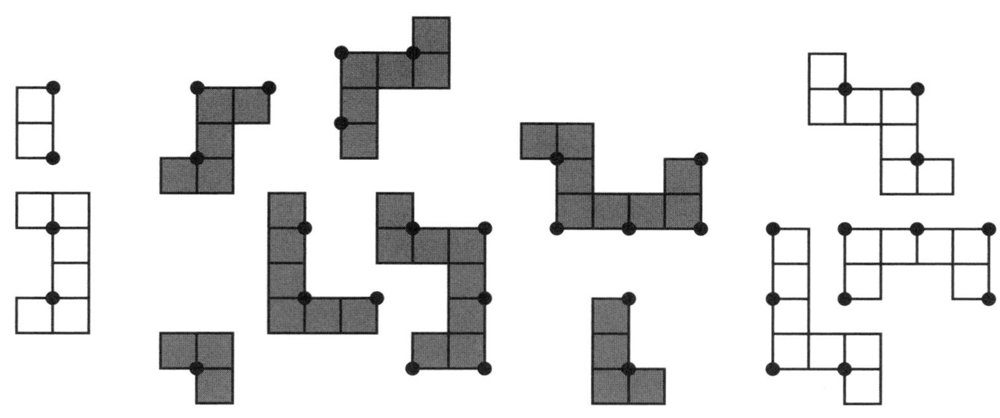

Figure 4.113
Lunda-animals

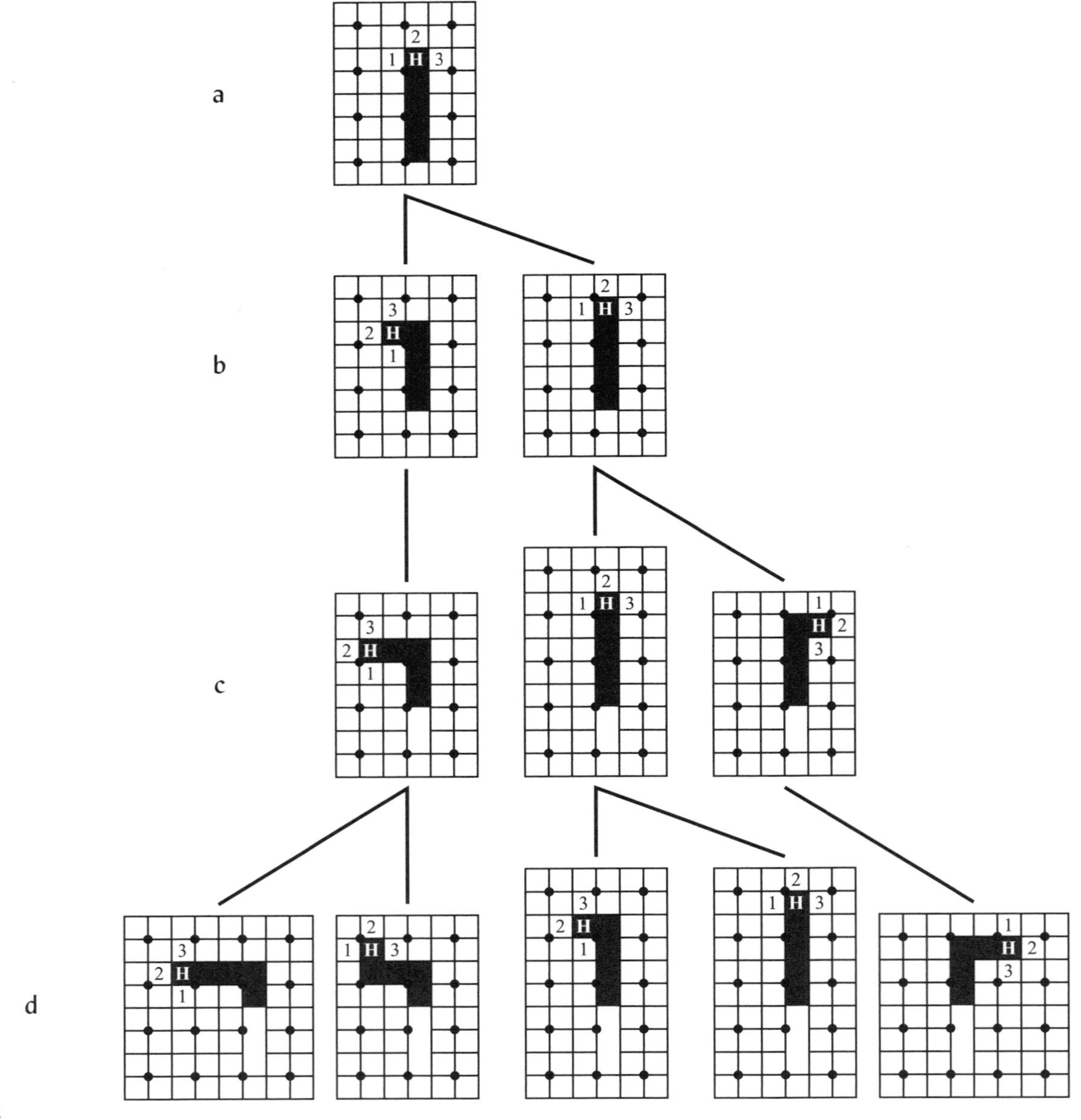

Figure 4.114

Fontinha, Mário (1983), *Desenhos na areia dos Quiocos do Nordeste de Angola*, Instituto de Investigação Científica Tropical, Lisbon.

Gerdes, Paulus (1988), On possible uses of traditional Angolan sand drawings in the mathematics classroom, *Educational Studies in Mathematics*, Dordrecht, Vol.19, No.1, 3–22.

—— (1990), On ethnomathematical research and symmetry, *Symmetry: Culture and Science*, Vol.1, No.2, 154–170.

—— (1990), *Vivendo a matemática: desenhos da África*, Editora Scipione, São Paulo.

—— (1991), *Lusona: Geometrical Recreations of Africa/ Recréations géométriques d'Afrique*, Universidade Pedagógica, Maputo (bilingual edition French/English).

—— (1993/4), *Geometria Sona: Reflexões sobre uma tradição de desenho em povos da África ao Sul do Equador*, Universidade Pedagógica, Maputo, 3 volumes. *Sona Geometry: Reflections on the sand drawing tradition of peoples of Africa south of the Equator*, Universidade Pedagógica, Maputo, Volume 1 [translation by A.B.Powell].

—— (1993a), Exploring Angolan sand drawings (sona): stimulating cultural awareness in mathematics teachers, *Radical Teacher*, Boston, Vol.43, 18–24.

—— (1995), *Une tradition géométrique en Afrique. — Les dessins sur le sable*, L'Harmattan, Paris / Montreal, 3 volumes.

—— (1996), *Lunda Geometry:Designs, Polyominoes, Patterns, Symmetries*, Universidade Pedagógica, Maputo.

—— (1997), On mirror curves and Lunda-designs, *Computers and Graphics, An international journal of systems & applications in computer graphics*, Oxford, Vol. 21, No. 3, 371–378.

—— (1997a), *Recréations géométriques d'Afrique – Lusona – Geometrical recreations of Africa*, L'Harmattan, Paris/ Montreal (bilingual edition French / English).

—— (1997b), *Ethnomathematik dargestellt am Beispiel der Sonageometrie*, Spektrum Verlag, Berlin/Heidelberg/ Oxford (3 volumes in one).

—— (1998a), On Lunda-designs and some of their symmetries, in: *Symmetry: Culture and Science* (in press).

—— (1998b), On Lunda-designs and the construction of associated magic squares of order 4p, *The College Mathematics Journal*, Washington DC (in press).

—— (1999a), On the geometry of Celtic knots and their Lunda-designs, *Mathematics in School,* May 1999, 29–33.

—— (1999b), Exploring powers of Lunda-designs and matrices, *Mathematics Magazine,* Washington, DC (in press).

Hamelberger, E. (1952), A escrita na areia, in: *Portugal em Africa*, Lisbon, 53, 323–330.

Heintze, Beatriz (1989), A cultura material dos Ambundu segundo as fontes dos séculos XVI e XVII , in: *Revista Internacional de Estudos Africanos*, Lisbon, No. 10 / 11, 15–63.

Jablan, Slavik (1995), Mirror generated curves, *Symmetry: Culture and Science*, Vol.6, No.2, 275–278.

Jaritz, Wolfgang (1983), Über Bahnen auf Billardtischen - oder: Eine mathematische Untersuchung von ideogrammen Angolanischer Herkunft, in: *Berichte der mathematisch-statistische Sektion im Forschungszentrum Graz*, Graz, Vol. 207, 1–22.

Kubik, Gerhard (1987), African space/time concepts and the tusona ideographs in Luchazi culture with a discussion of possible cross-parallels in music, in: *African Music*, Grahamstown, 6, 4, 53–89.

Kubik, Gerhard (1987), African graphical systems, in: *Muntu, Revue Scientifique et Culturelle du Centre International des Civilisations Bantu (CICIBA)*, Libreville, No.4, 71–135.

Kubik, Gerhard (1988), *Tusona-Luchazi ideographs, a graphic tradition as practised by a people of West-Central Africa*, Verlag Stiglmayr, Fohrenau.

Pearson, Emil (1977), *People of the Aurora*, Beta Books, San Diego.

Santos, Eduardo dos (1961), Contribuição para o estudo das pictografias e ideogramas dos Quiocos, in: *Estudos sobre a etnologia do ultramar português*, Lisbon, Vol.2, 17–131.

Silva, Elísio (1995), *Jogos de quadrícula do tipo mancala com especial incidência nos praticados em Angola*, Instituto de Investigação Científica Tropical, Lisbon.

Sources of illustrations

Unless otherwise indicated, the drawings were made by the author using the Adobe Illustrator program. "Drawn after" means drawn by the author using this program. "After" refers to a reproduction. If no source is given, then it is a new drawing by the author. For full bibliographic reference to each work, look in the bibliography for each chapter.

Chapter 1

1.1a: after Dowson, p. X, 150, 151
1.1b: after Willcox, p. 107
1.1c: drawn after Redinha; reproduced in Willcox, p. 82
1.2: after Shaw et al., p. 308; p.425; Stössel, p. 43
1.3: drawn after Stössel, p. 79
1.4: after Bolland, p. 27, 29
1.5a: drawn after a photograph in Bolland, p. 218
1.6a: drawn after a photograph in Bolland, p. 127
1.7a: drawn after a photograph in Bolland, p. 125
1.8a: drawn after a photograph in Bolland, p. 267
1.9a: drawn after a photograph in Bolland, p. 122
1.10a: drawn after a photograph in Bolland, p. 127
1.11a: drawn after a photograph in Bolland, p. 127
1.12a: drawn after a photograph in Bolland, p. 145
1.13a: drawn after a photograph in Bolland, p. 137
1.14a: drawn after a photograph in Bolland, p. 133
1.15a: drawn after a photograph in Bolland, p. 120
1.16a: drawn after a photograph in Bolland, p. 136
1.17: drawn after photographs in Bolland, p. 130, 116
1.18: drawn after a photograph in Bolland, p. 136
1.19: after Bolland, p. 235
1.20: after Bolland, p. 107, 179, 179, 201, 238, 172
1.21: after Bolland, p. 179, 282, 235
1.22: after Bolland, p. 179, 181, 282
1.23a: drawn after photograph in Bolland, p. 123
1.23b: after Bolland, p. 174
1.24: drawn after photographs in Washburn, p. 107, 72
1.25: drawn after Meurant, p. 150, no. 19, 21, 4, 18; p. 134
1.26: drawn after Torday & Joyce, p. 144-148
1.27a, b, c, and d: British Museum: Kuba II/36
1.27e, f, and g: drawn after Torday & Joyce, p. 214
1.28: British Museum: Kuba I/11
1.29a: drawn after Torday & Joyce, p. 162
1.29b: British Museum: Kuba I/23
1.30: drawn after Torday & Joyce, p. 198
1.31: drawn after Torday & Joyce, p. 231

1.32: drawn after Torday & Joyce, p. 90
1.33: drawn after a photograph in Stössel, p. 107
1.34: drawn after Oliver, p. 157
1.35: drawn after a photograph in Jefferson, p. 136
1.36: drawn after a photograph in Gardi, p. 76
1.37a: drawn after a photograph in Jefferson, p. 95
1.37b: British Museum: Ashanti II/20
1.38 a, b, and c drawn after a photograph in Lamb, 1975, p. 129; d, e, f, and g drawn after a photograph in Gilfoy, p. 78
1.39: drawn after a photograph in Lamb, 1975, p. 153
1.40: drawn after a photograph in Lamb, 1975, p. 189
1.41
a: drawn after photograph in Sieber, 1972, p. 225
b: drawn after photograph in Sieber, 1972, p. 208
c: drawn after photograph in Robbins & Nooter, p. 131
d: drawn after a photograph in Northern, p. 51
e: drawn after photograph in Northern , p.136
f: drawn after a photograph in Sieber, 1997, p. 174
g: drawn after a photograph in Gilfoy, p. 57
h: drawn after photograph in Robbins & Nooter, p. 299
i: drawn after Newman, p. 78
j: drawn after a photograph in Jefferson, p. 89
k: drawn after a photograph in Northern, p. 89
1.42: drawn after a photograph in Schildkrout & Keim, p.132
1.43: Drawn after the photograph in Schildkrout & Keim, p.139
1.44: after Schweinfurth, vol. 2, p. 116)
1.45: after Schildkrout & Keim, Fig. 12.21
1.46: after Schweinfurth, vol. 2, p. 74
1.47: drawn after the photograph in Schildkrout & Keim, p.104
1.48: reproduced from Gerdes, 1995, p. 165
1.49: after Guidoni, p. 167
1.50: Drawn after a photograph from 1911 in Schildkrout & Keim, p. 219
1.51: drawn after a photograph in Ghaidan, p. 47
1.52: drawn after Wenzel, p. 78
1.53: drawn after a photograph in Guidoni, p. 127
1.54: drawn after a photograph in Oliver, p.54
1.55: drawn after a photograph in Etienne-Nugue, p. 36
1.56: after Sieber, 1980, p. 30 (Wood, 1871)
1.57: after Guidoni, p. 134
1.58a: reproduced from Gerdes, 1995, p. 18
1.58b: Author's collection
1.60: drawn after Hauenstein, p. 69, 71
1.61: drawn after Newman, p. 145
1.62: drawn after a photograph in Schildkrout & Keim, p.108
1.63: drawn after photographs made in 1976, Nooter, p. 68, and photographs in Maurer & Roberts, p. 208, 209, 273, from baskets collected in 1953
1.64: drawn after a photograph in Newman, p. 169
1.65: drawn after a photograph in Newman, p. 173
1.66: drawn after photographs in Gardi, p. 147-152
1.67: drawn after a photograph in Newman, p. 191
1.68: after Gardi, p. 174
1.69: after Jefferson, p. 105
1.70: after Lestrange & Gessain, p. 250
1.71: drawn after a photograph in Newman, p. 162
1.72: drawn after a photograph in Northern, p. 59
1.73: after Chappel, p. 35, 127, 113, 79, 39
1.74
a: drawn after a photograph in Ben-Amos, p.14
b: drawn after photographs in Newman, p. 251
c: drawn after a photograph in Robbins & Nooter, p. 108
d: drawn after a phtograph in Guidoni, pl. XXIV
e: drawn after a photograph in Maurer, p.63
f: drawn after a photograph in Robbins & Nooter, p. 257
g: drawn after a photograph in Northern, p. 167

h: drawn after a photograph in Northern, p. 150
i: after Jefferson, p. 155
j: after Jefferson, p. 155
k: drawn after a photograph in Pruitt, p. 10
1.75a: drawn after photograph in Northern, p. 189
1.75b: drawn after a photograph in Newman, p. 243
1.75c: after Hambly, p. 27
1.76a: drawn after a photograph in Northern, p. 7
1.76b: reproduced from Gerdes, 1995, p. 75
1.77a: after Jefferson, p. 125
1.77b: reproduced from Gerdes, 1995, p. 49
1.78a: reproduced in Fischer, p. 68
1.78b: after Meurant, p. 178
1.78c: reproduced in Jefferson, p. 106 (Barth, 1849)

Chapter 2

All figures in sections 1, 2, 3, and 4 are reproduced from Paulus Gerdes, *African Pythagoras: A study in Culture and Mathematics Education,* Universidade Pedagógica, Maputo, [1992] 1994.

2.10:
a: drawn after Williams, p. 28
b: drawn after Williams, p. 6
c: drawn after Williams, p. 16
d: drawn after Cole & Aniakor, p. 46
e: drawn after Williams, p. 28
f: drawn after Baumann, p. 61
g: drawn after Baumann, p. 61
h: drawn after Mveng, p. 52
i: drawn after Williams, p. 15
j: drawn after Williams, p. 49
k: drawn after Williams, p. 24
l: drawn after Williams, p. 17
m: drawn after Williams, p. 65
n: drawn after Mveng, p. 51
o: drawn after Mveng, p. 34
p: drawn after Mveng, p. 79, 25
q: drawn after Mveng, p. 25
r: drawn after Denyer, p. 121
s: drawn after Mveng, p. 25
2.11: drawn after a photograph in Martins, p. 87
2.22: drawn after Meurant, p. 199
2.27: drawn after Meurant, p. 114, 126, 144, 176
2.28: drawn after Meurant, p. 176
2.30a: observed by the author in the Ethnographic Museum of Budapest, 1988
2.38: drawn after Wilson, p. 23
2.53a: drawn after a photograph in Bastin, p. 116
2.55: drawn after a basket in the Ethnographic Museum, Budapest

Chapter 3

All figures in section 1 are based on figures drawn in P. Gerdes, *Women and Geometry in Southern Africa,* Universidade Pedagógica, Maputo, 1995; new edition: *Women, Art and geometry in southern Africa*, Africa World Press, Trenton, NJ., 1998.

3.39: drawn after Weule, Table 19
3.64: photograph by the author
3.105: drawn after a photograph in Schildkrout & Keim, p. 168
3.106b: drawn after Trowell & Wachsmann, p. 146
3.107: after Coart, Pl. LIII, no. 383
3.108
a1: drawn after Coart, Pl.LIV, no. 386
b1: drawn after Coart, Pl.XLVI, no. 369
3.109a: drawn after Coart, Pl. LV, no. 387
b: drawn after Coart, Pl. LVI, no. 389

c: drawn after Coart, Pl. LIII, no. 383
d: drawn after Coart, Pl. LVI, no. 390
e: drawn after Coart, Pl. LV, no. 387
f: drawn after Coart, Pl. LV, no. 388
3.110: drawn after Coart, Pl. LIV, no. 385
3.111: drawn after Coart, Pl. LIV, no. 385
3.112: drawn after Coart, Pl. LVII, no. 392
3.116: after Coart, Pl. XLVI, no. 369
3.117a: drawn after Coart, Pl. XLVI, no. 369
3.117b: drawn after Coart, Pl. XLVII, no. 371
3.118: drawn after Coart, Pl. XLVI, no. 369
3.124: drawn after Coart, Pl. XLVIII, no. 373
3.125: drawn after Coart, Pl. XLVIII, no. 372 [slightly corrected]
3.126: drawn after Coart, Pl. L, no. 375
3.127: drawn after Coart, Pl. L, no. 375)
3.128b: drawn after Coart, Pl. XLVI, no. 370
3.129: reconstructed on the base of a schematic drawing in Coart, Pl. XLI, no. 350
3.130: drawn after Coart, Pl. XXXIV, no. 330 [slightly corrected]
3.133: Drawn after Lestrange & Gessain, p. 165
3.136: Drawn after Lestrange & Gessain, p. 165
3.139: Author's collection

Chapter 4

All figures in the sections 1 and 2 are reproductions of figures in Paulus Gerdes, *Sona Geometry: Reflections on a drawing tradition in Africa south of the Equator,* Universidade Pedagógica, Maputo, 1994, Vol. 1, 2, 3.

4.4: after Heintze, p. 35
4.5: drawn after Fontinha, p. 211, 151
4.6: drawn after Fontinha, p. 257, 225; Dos Santos, p. 95
4.7: drawn after Silva, p. 63, 64; Pearson, p. 74
4.8: drawn after Fontinha, p. 259; Dos Santos, p.110
4.9a: drawn after Fontinha, p. 277
4.10a: drawn after Dos Santos, p. 87
4.12: drawn after Fontinha, p. 281; Pearson, p. 160; Dos Santos, p. 115
4.13: drawn after Fontinha, p. 171, 193, 185, 187, 199
4.14: drawn after Hamelberger, p. 327; Dos Santos, p. 49; Pearson, p. 162
4.16 drawn after Fontinha, p. 109
4.18a: drawn after Fontinha, p. 189
4.19a: drawn after Fontinha, p. 287; Pearson, p. 145
4.21: drawn after Fontinha, p. 279
4.23b: drawn after Fontinha, p. 281
4.24: drawn after Fontinha, p. 225
4.26: drawn after Fontinha, p. 157, 239, 147, 169, 169, 159
4.29: drawn after Fontinha, p. 177
4.31: drawn after Fontinha, p. 203
4.32: drawn after Pearson, p. 120
4.33b: drawn after Fontinha, p. 271
4.36: drawn after Fontinha, p. 239; Dos Santos, p. 93; Fontinha, p. 199
4.38: drawn after Fontinha, p. 201
4.39: drawn after Fontinha, p. 143
4.42b: drawn after Fontinha, p. 179
4.43b: drawn after Fontinha, p. 191
4.45: drawn after Fontinha, p. 263
4.50: drawn after Fontinha, p. 215
4.75: drawn after Fontinha, p. 149

Most drawings in section 3 are reproductions from figures in Paulus Gerdes, *Lunda Geometry: Designs, Polyominoes, Patterns, Symmetries*, Universidade Pedagógica, Maputo, 1996.